Nano-catalysts for Energy Applications

Editor

Rohit Srivastava

Assistant Professor
School of Petroleum Technology
Pandit Deendayal Petroleum University
Gandhinagar, Gujarat, India

CRC Press
Taylor & Francis Group
Boca Raton London New York

CRC Press is an imprint of the
Taylor & Francis Group, an **informa** business
A SCIENCE PUBLISHERS BOOK

Cover credit: Image used on the cover has been made by the editor.

First edition published 2021
by CRC Press
6000 Broken Sound Parkway NW, Suite 300, Boca Raton, FL 33487-2742

and by CRC Press
2 Park Square, Milton Park, Abingdon, Oxon, OX14 4RN

Library of Congress Cataloging-in-Publication Data

Names: Srivastava, Rohit, editor.
Title: Nano-catalysts for energy applications / editor, Rohit Srivastava.
Description: First edition. | Boca Raton : CRC Press ; Taylor & Francis Group, 2021. | "CRC Press is an imprint of the Taylor & Francis Group, an Informa business." | Includes bibliographical references and index.
Identifiers: LCCN 2021004510 | ISBN 9780367536435 (hardcover)
Subjects: LCSH: Nanostructured materials. | Catalysts. | Renewable energy sources--Technological innovations. | Clean energy--Technological innovations.
Classification: LCC TA418.9.N35 N248555 2021 | DDC 621.042028/4--dc23
LC record available at https://lccn.loc.gov/2021004510

ISBN: 978-0-367-53643-5 (hbk)
ISBN: 978-0-367-53644-2 (pbk)
ISBN: 978-1-003-08272-9 (ebk)

Typeset in Times New Roman
by Radiant Productions

Preface

The worldwide consumption of continuous fossil fuels and petrochemical-based products are increasing. Incessant removal of fossil fuels from earth and rapidly growing industrialization, found on such fuel generated energy are expected to create a crucial problem in terms of environmental imbalance and scarcity of such fuels within decades. The scientific community now therefore must yield more in search of alternative renewable energy to comply with the pace of current civilization. It is expected that within the next 30 years the demand of global fossil would become double and the energy produced from crude oil, natural gas, coal and nuclear energy, will fall short of fulfilling the increasing demand of energy for sustaining human civilization. Moreover, continuous release of CO_2 in the environment due to combustion of fossil fuels and increasing population is now a major threat to mankind and the animal world on account of global warming and allied consequences. One is now left with no other option, except searching for clean and renewable alternatives of usable energy and manipulating the utilization of available energy resources such as, solar, wind, water and biomass Now a days among all the research areas, the worldwide scientific community is committed to bringing down the requirement of fossil fuels and compensate the same with alternative renewable energies from various pollutants and waste materials that are constantly threatening lives on earth. The research in the field of nanoscience and nanotechnology is interdisciplinary and now becomes very promising among the fields that have experienced impressive advances undergoing explosive growth. Among the current subjects in nanoscience and nanotechnology, the design and fabrication of advanced nanomaterials as catalysts are of great interest to the scientific community for energy applications. The current advancements in this field can be noticed through development of catalysts for water electrolysis, solar cells and efficiently enhance the overpotential of CO_2 reduction into other green fuels, researchers have demonstrated that a nanostructured catalyst can exhibit very different behavior as compared to its bulk counterpart. A nanostructured catalyst has recently shown emerging applications and have become promising potential candidates to enhance the technologies for making energy devices such as solar cells, fuel cells, supercapacitors, Li-ion batteries and CO_2 reduction. The proposed book entitled "**Nano-catalyst for Energy Applications**", aims at an investigation of recent findings published during the last 5 to 10 years on the most recent development in catalysis studies for energy applications. The proposed book comprises of 11 contributing chapters from authors who are actively involved in this field. Chapter 1 "***Strategies in Nanocatalyst Towards Water Splitting and Reduction of CO_2***" describes various experimental approaches about the synthesis of nanohybrid and nanocomposite as electro-catalysts that have a promising application for hydrogen production and CO_2 reduction into green fuels. The chapter also emphasize the challenges and issues that occur during electrolysis of water and CO_2 reduction. **Chapter 2** "***Hybrid Perovskite Photocatalysis for Energy Harvesting and Energy Saving***" describes the conversion of sunlight into electric energy by the fabrication of Perovskite photocatalysis. This chapter gives a brief overview of perovskite photocatalysis, materials synthesis, branches of perovskites, various device architectures, device engineering and challenges to improve the power conversion efficiency and potential of perovskites in other electronic applications. **Chapter 3** "***Porphyrins Based Nanostructured Material for the Conversion of CO_2 into Value Added Products***" demonstrates the utility of porphyrins as a catalyst for CO_2 reduction into value added products. In recent years, porphyrins-based materials have attracted attention in order to improve the high selectivity and reliability for CO_2 conversion application. **Chapter 4** "***Application of Metallic Foam in Solar Power System***" describes the preparation methods of different kinds of metallic foam and its application in solar power system. The fabrication of metal foams are lightweight and naturally inspired cellular materials. **Chapter 5** "***Valorization Chemistry: A Compendium on Photoreduction of CO_2 to Biofuels over Nano TiO_2***" describes the reduction of CO_2 by photochemical methods. This chapter deals with the photoreduction mechanism of CO_2 over pristine and modified TiO_2 that have excelled in the reduction efficacy with high yield of product. **Chapter 6** "***Concept of Nanocatalyst and its Application in the Energy Domain***". This chapter deals with various nanomaterials for hydrogen storage, catalytic behavior and plasmonic application and most important biomedical advantages of nano-catalysts are presented. The chapter also focuses on future aspects of nano-catalysts along with their energy and environmental concern. **Chapter 7** "***Recent Advancement of Electrocatalyst System in CO_2 Reduction: Insights of Fe, Co and Ni Metallo-ligand Clusters in Homogeneous Molecular Level***" describes the electrochemical reduction of CO_2 into useful products which could be converted into fuel. The multi-functionality and catalytic behavior of electrocatalyst strongly enhanced through multinuclear metal clusters interaction. In this chapter properties and behaviors of Fe, Co and Ni multinuclear metal clusters catalysts for CO_2 reduction are also discussed. Chapter 8 "***Design of Alloy Electrocatalysts for Oxygen Reduction Reaction from First-principles Viewpoint***" discusses the way to design the alloy structures and study the Oxygen Reduction Reaction (ORR) properties such as intermediates, reaction mechanisms, reaction rates and the stability of alloyed materials in the ORR environment by using Density Functional Theory (DFT). Chapter 9 "***Metal-Organic Framework Catalyst for Energy Applications***" gives an insight of a comprehensive understanding and knowledge on MOFs and their catalytic application in energy producing and storing devices. Chapter 10 "***Piezoelectric Energy Harvesting and Piezocatalysis***" gives an overview of the fundamental and recent developments of piezoelectric energy harvester. Finally, Chapter 11 "***Metal Hollow Sphere as Promising Electrocatalyst in Electrochemical Conversion of CO_2 to Fuels***" summarizes various synthesis protocols of hollow spheres along with electrocatalytic activity towards reduction of CO_2. It is believed that the proposed book would be of interest to graduate students, researchers, academicians and industrialists who are working in the areas of energy and environment, biomass conversion, sustainability, biofuels and chemical industries.

Last but not least, we would like to thank all contributors for their generous support, the publisher for accepting our book and the administrative heads of Pandit Deendayal Petroleum University (PDPU) Gandhinagar, Gujarat, India for their encouragement and cooperation, without which it would have been extremely difficult to complete this task on time.

Acknowledgment

I thank all the contributing authors for completing their chapters on time and without their incredible support it would not have been possible to edit this book. My grateful thanks to all my colleagues who worked tremendously to make this work successful. Their enormous effort and devotion motivated us to work harder. My special thanks to the publisher Taylor & Francis CRC press for providing such a great opportunity to share our knowledge with the entire scientific community. I was motivated and excited by the hard work, encouragement and passion of the people at Harvard University USA and PDPU India. I also thank my 2 years old child Aadvika Srivastava for being patient and a tremendous support. I would like to thank all my research group members Avni Goswami, Pragya Singh, Shaista Nouseen and Sneha Lavate for their patient and understanding with my overwhelming busy schedule and the consequent unavailability during the last couple of months. Thanks to CHRO, Director SPT and all the faculty members of SPT for their support and help. To the people who really matter my parents and elder brothers for their humble support. Love you all. Last but not least I would like to thank PDPU, India for providing the required infrastructure and start up research grant. This research work was funded by SERB-DST, New Delhi, under the scheme of "ASEAN-India Research & Development" (Grant No: IMRC/AISTDF/CRD/2018/000048).

Contents

Chapter 1

Strategies in Nanocatalyst towards Water Splitting and Reduction of CO_2

Sneha Lavate[1,2] and *Rohit Srivastava*[1,*]

Introduction

Nanoscience and nanotechnology research fields are an effective boost for reconstructing the nature on the atomic and molecular level related to the design, characterization, fabrication and applications of materials, devices and systems by controlling the shape and size at the 'nanoscale'. The fundamental properties of matter such as mass, weight, volume, density, etc., change at the nanoscale. The physical and chemical properties of nanomaterials can be quite different from those of larger particles of the same substance. These materials also have enhanced kinetics due to shortened diffusion pathways and large surface to volume ratios, favoring high power densities. For the growing concern on environmental pollution and its effects on the living system, nanotechnology can be a breakthrough. The use of smaller portions of potential nanomaterials and its highly precise manufacturing will break the tie between economic activity and resource use (Friends of Earth 2010, Subhra Jana 2015, Hao Ren et al. 2014).

Energy is considered as being of great value in this modern world due to the depletion of nonrenewable assets and extensive environmental pollution. This has enforced a look for an opportunity to supply and preserve the known asset of the word 'Energy' for future use in an ecofriendly manner. Utilization of fossil fuels by burning them for our energy needs, energy related industrialization, burning of wastes and deforestations are the primary concerns of growing attention of atmospheric CO_2 that leads to environmental pollution and severe greenhouse effect. The continuous

[1] Catalysis Research Lab, School of Petroleum Technology, Pandit Deendayal Petroleum University Gandhinagar-382421, Gujarat, India.
[2] School of Nano Science and Technology, Shivaji University, Kolhapur-416009, Maharashtra, India.
* Corresponding author: rohit.s@spt.pdpu.ac.in

increase of CO_2 in the atmosphere is found as a major cause for the process of global warming. Thus decarbonization of energy delivered with the aid of using the opportunity for alternative clean, sustainable and renewable energy needed for future energy sustainability and global security. Closing the carbon cycle by the way of means of making use of CO_2 as a feedstock for currently used commodities, in order to displace a fossil feedstock is an appropriate intermediate step towards a carbon-free future (Subramani et al. 2018, Murugananthan et al. 2015, Furat et al. 2020). Converting CO_2 into fuels through renewable power sources is a capacitive approach of mitigating future resource extraction and decreasing our global carbon footprint. Using electrolysis to drive the reduction of CO_2 is particularly appealing given the widespread use of electrolyzers for commodity chemical production (Yang et al. 2016, Jingfu et al. 2017a). Hydrogen will be the main source for energy use in future which can be considered as clean energy with high energy content as compared to hydrocarbon fuels (Ilgi and Fikret 2006). Steam reforming and water electrolysis are the methods in trend to produce H_2. Steam reforming has a major disadvantage over water electrolysis in that it produces CO_2 along with H_2 and hence is not ecofriendly. Water electrolysis is carried out by two half-cell reactions; reduction of H^+ ions, i.e., HER and oxidation of water, i.e., OER (Sengeni et al. 2016). On the other hand, CO_2 reduction has many challenges as it needs an efficient catalyst to mediate multiple electron and proton transfers without resorting to excessive reducing potentials; reducing CO_2 in the presence of H_2O; and selectively producing one main possible byproduct. The possibility of an electrochemical reduction process depends on thermodynamic value as well as kinetic properties (Yihong et al. 2012, Jinhui Hao and Weidong 2018). Reduction of CO_2 results in formyl, methylene and methyl groups coupled with a generation of new C-N, C-C and C-O bonds in the presence of N-, C- and O-nucleophiles which enlarges the range of compounds directly obtained from CO_2 (Xiao-Fang et al. 2018).

A number of compounds with a wide range of nanostuctures have been synthesized and studied for energy applications. In this chapter, some common synthesis strategies of nanostructured materials such as nanocomposites and nanohybrids electrocatalyst for H_2 production and CO_2 reduction have been described. Moreover, the electrochemistry for HER, OER and CDRR were overviewed. Nanostructured electrocatalysts are reviewed on the basis of the parameters such as overpotential, current density and the Tafel slope. Finally, the challenges and future prospects for H_2 production and CO_2 reduction are discussed.

Various methods of CO_2 reduction

Various types of nanomaterials-based catalysts have been developed for the reduction of CO_2 by the physical, chemical or biological route. The efficiency of conversion of CO_2 into various products like carbon monoxide (CO); primary, secondary or tertiary hydrocarbons by a suitable highly efficient nanocatalyst depends on various parameters such as size, shape, porosity and reactivity (Guodong et al. 2018, Shulin et al. 2019, Sheng et al. 2014). Utilization of these nanocatalyst materials in the reduction process of CO_2 can be carried out by using various pathways. Each pathway may result into a different product. A source of energy or reagent depends

Table 1.1: Types of approaches used for reduction of CO_2 into their respective products (Farihahusnah and Mohamed 2019).

Conversion method	Reagent or energy source	Products
Biochemical	Bacteria	CH_4
Bioelectrochemical	Enzyme + Methyl viologen/ CO_2 + Oxoglutaric acid	Isocitric acid
Biophotoelectrochemical	*hv*, light, enzyme + methyl viologen	HCOOH
Chemical	Mg^{2+}, Sn^{4-}, Na^+	C, CO, $Na_2C_2O_4$
Electrochemical	Electrons (electricity) + protons	Hydrocarbons
Photochemical	*hv*, light	CO, Hydrocarbons
Photoelectrochemical	*hv*, light + electrons (electricity)	CO
Radiochemical	Gamma radiations	HCOOH, HCHO

on the selection of the pathway and the requirement of the final product. The resultant products can be further used as feedstock or alternative fuel. Table 1.1 represents the ways of conversion of CO_2 using reagent or energy source.

Methods for H_2 production by water splitting

The generation of hydrogen from a source which is available in abundant amounts in the environment such as water is a promising approach in the field of catalysis. The water splitting process requires energy of four electrons to carry out the reaction. This energy can be provided in the form of electricity, thermal energy or photons. A catalyst is used to speed up the process. Nanostructured catalysts can improve the efficiency of water splitting. According to the source used, water splitting can be divided into three types; electrolysis, thermolysis and photolysis as shown in Fig. 1.1.

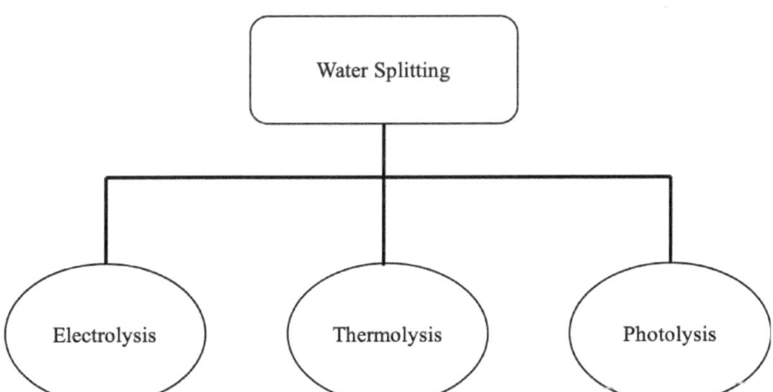

Figure 1.1: Flow chart for H_2 production methods by the water splitting process.

Electrolysis process

While being an established and well-known method for water splitting, electrolysis involves an endothermic reaction. Hence an external supply of electricity is required to overcome this drawback. A typical electrolysis unit consists of an anode, cathode suspended in the electrolyte and connected to an external electric supply. When electricity is provided, water splitting takes place producing hydrogen at the cathode and oxygen gets evolved at the anode. The electrolysis can be carried out in three possible ways; (1) in ordinary water, (2) solid oxide electrolysis cell and (3) alkaline/acidic electrolysis. High purity hydrogen can be produced by this method as well as high conversion efficiency can be achieved upto 75–85%.

Thermolysis process

The decomposition of water when heated to a high temperature is known as thermolysis or thermochemical water splitting. Various thermochemical water splitting cycles have been proposed to reach the required energy needed which is 2500°C in accordance with Gibbs energy to become zero. At such a high temperature, it is not possible to separate the H_2 and O_2. To overcome this drawback, researchers now use thermoelectrolysis where a combination of electricity and thermal energy is used to split the water. Currently, hydrogen is produced mainly by thermal energy through steam reforming of fossil fuel and only 4% through electrolysis using electrical energy. Alkaline electrolysis operates at nearly 200°C on a large scale in industries, whereas PEM electrolyzers operate at 100°C and is commercially available.

Photolysis process

Photolysis is carried out in the presence of light by the use of a photocatalyst which has the ability to decompose water into H_2 and O_2. In photolysis, a semiconductor material absorbs the photons of high or equal energy to generate an electron-hole pair. When the photon irradiation is greater than or equal to the bandgap of a semiconductor, these electrons and holes cause redox reactions. When this is applied in case of water molecules, splitting of water takes place by electrons reducing water to generate H_2, while holes oxidize water to generate O_2. Photolysis process is schematically represented in Fig. 1.2. The mechanism of photolysis is similar to electrolysis, only being different in the source of energy used to activate the water splitting process.

Ultra-Violet (UV) light has photonic energy greater than visible light which significantly implies better efficiency for water splitting. For a clearer understanding, a photocatalyst that absorbs visible light and shows less efficiency is always better than a photocatalyst absorbing only UV light and showing more efficiency.

Synthesis strategies for the preparation of nanocomposites and nanohybrids

On the basis of structure that one requires for an electrocatalyst, there are many approaches to synthesize nanostructures considering the parameters consisting of

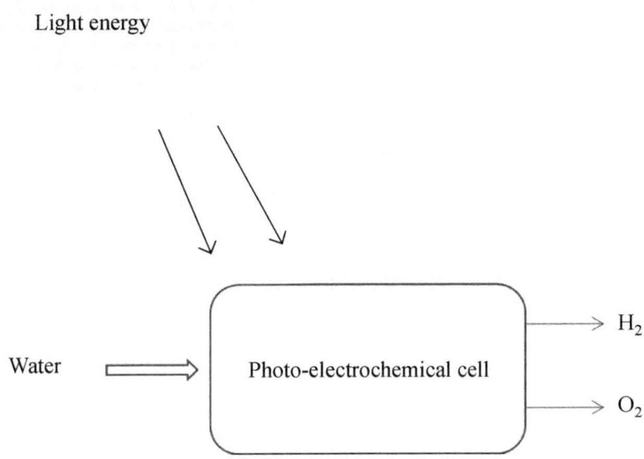

Figure 1.2: Schematic representation of photolysis process.

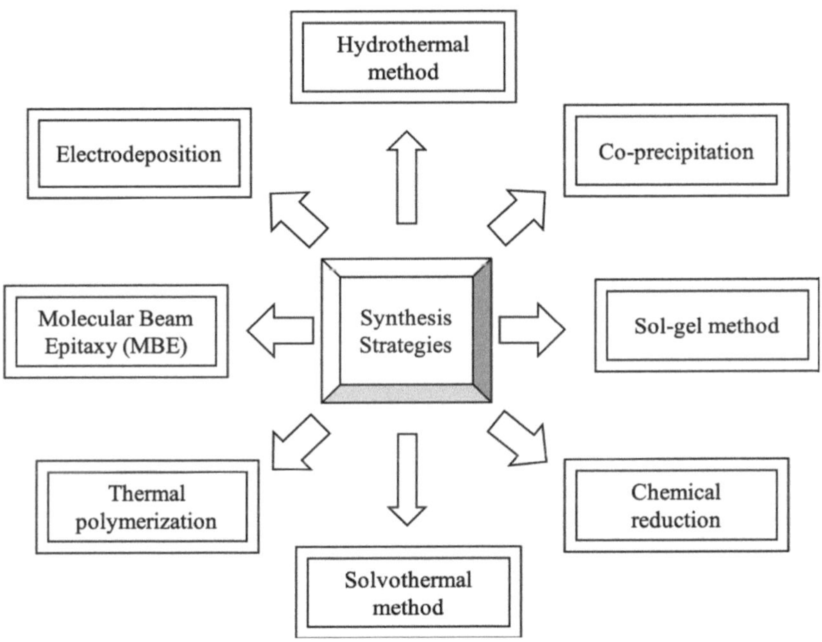

Figure 1.3: Flow chart of synthesis strategies for nanostructured electrocatalyst.

reaction time, temperature, reaction kinetics, pH, etc. A few common strategies to prepare nanostructured electrocatalyst are mentioned here. Figure 1.3 described the various synthesis strategies for nanostructured eletrocatalyst.

Hydrothermal method

The hydrothermal method is the most common for the synthesis of nanomaterials with a wide range of temperatures. The word 'hydrothermal' itself describes its meaning 'Hydro' = water and 'Thermal' = temperature. The method of synthesis carried out in the presence of water as a solvent and at various temperatures is known as the hydrothermal method. In hydrothermal synthesis, water is used as a solvent. The reaction is carried out in a Teflon lined stainless steel autoclave under high vapor pressure and at various ranges of temperatures. Steps followed for the synthesis by hydrothermal method are shown below in Fig. 1.4.

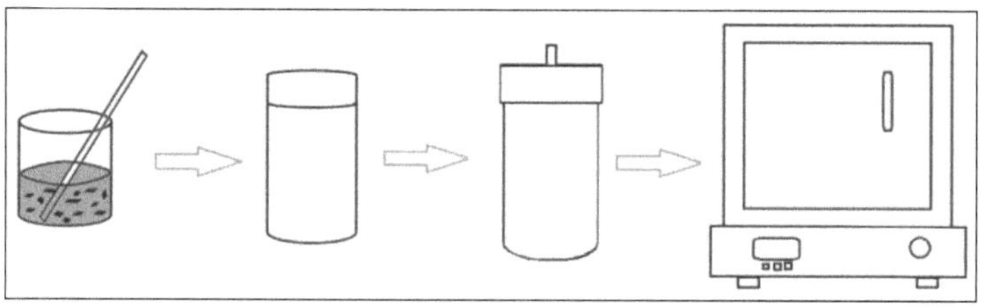

Figure 1.4: Schematic representation of hydrothermal method (a) reaction mixture (Precursor and water), (b) Teflon chamber, (c) Teflon lined stainless steel autoclave, (d) heating oven with autoclave.

Co-precipitation

The most frequently used method for preparation of heterogeneous catalyst is the co-precipitation method. One or more metals are precipitated collectively with help of its precursor. This method may be appealing for large scale synthesis. On the contrary, it is difficult and requires accurate control of conditions. One of the assets of this method is high metal dispersion. This method is also used for the synthesis of magnetic materials with catalytic properties. Figure 1.5 shows the schematic representation for co-precipitation method.

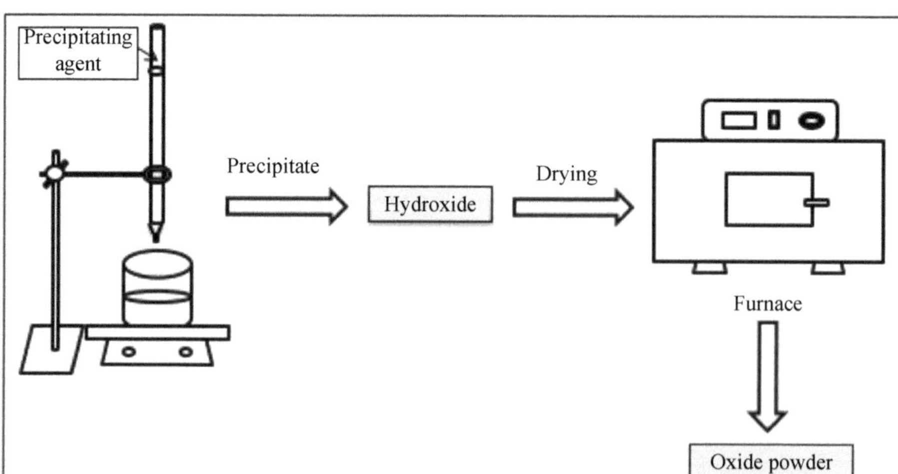

Figure 1.5: Schematic representation for co-precipitation method.

Sol-gel method

Over the years, a large number of beneficial strategies have been applied to synthesize nanoparticles and nanocomposites; the sol-gel method is one of them. In this method, the formation of two phases such as 'sol' and 'gel' takes place. Sol is the liquid form of colloidal solution whereas gel is a colloidal solution in its solid or semi-solid state. The sol-gel method is used for the synthesis of nanoparticles and nanocomposites. Metal alkoxides are mostly used as precursors which are hydrolyzed in an alkali or acidic environment accompanied by condensation which ends in the formation of polymeric network with solvents molecules embedded in it. The slower the rate of hydrolysis, the smaller is the particle size. Figure 1.6 is the schematic representation of the sol-gel method. Properties of products acquired are affected by the rate of reaction which relies on a range of qualities consisting of a nature of alkoxy group, electronegativity of the metal atom, structure of precursor and the solvent used.

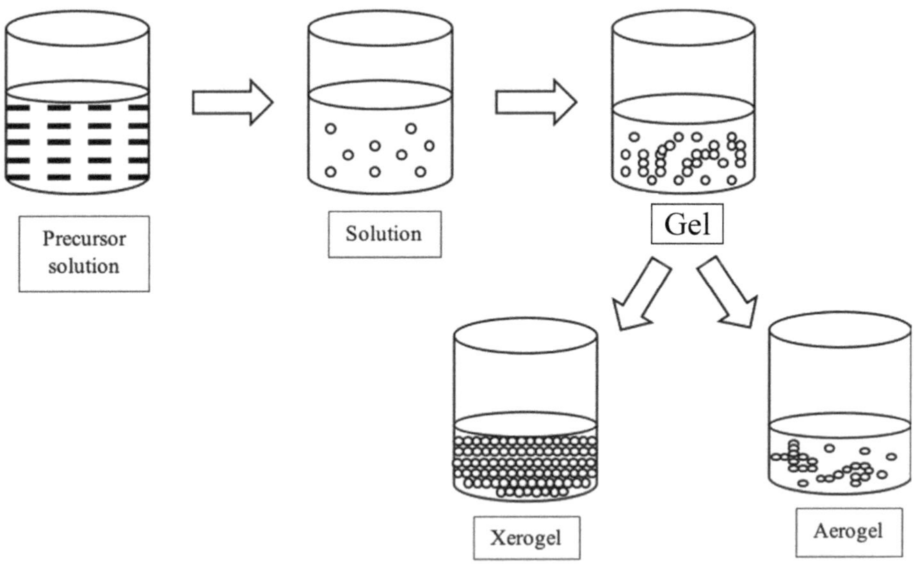

Figure 1.6: Schematic representation of sol-gel method.

Chemical reduction

Chemical reduction is generally the most used method for the synthesis of metal nanoparticles using its precursor inside the presence of a stabilizer. It is of low-cost and viable to expand on a large scale. Reducing agents such as sodium borohydride, hydrazine, formaldehyde and many others are frequently used. The advantage of this process is the scale of the nanoparticles that can be controlled. Synthesis of nanoparticles by chemical reduction method by using reducing agent and capping agent is represented schematically in Fig. 1.7.

Figure 1.7: Schematic representation of chemical reduction method.

Solvothermal method

Unlike the hydrothermal method, this method entails the use of a solvent in preference to water. The identical protocol is accomplished as the hydrothermal method and the reaction is carried out in Teflon lined stainless steel autoclave under high vapor pressure and different ranges of temperatures. Figure 1.8 shows steps involved in solvothermal method.

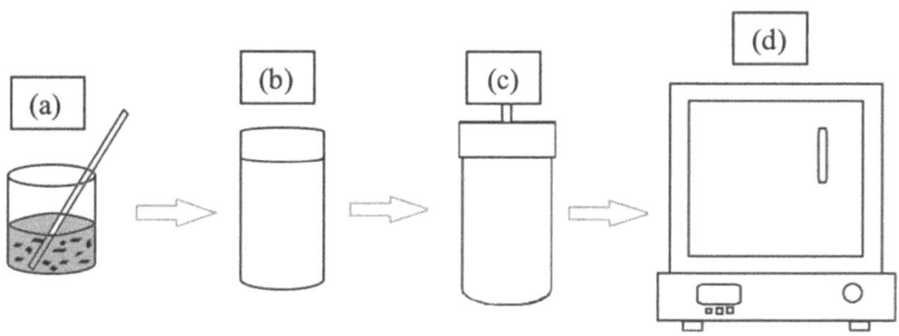

Figure 1.8: Schematic representation of the solvothermal method (a) reaction mixture (Precursor and solvent), (b) Teflon chamber, (c) Teflon lined stainless steel autoclave, (d) heating oven with autoclave.

Thermal polymerization

As the name indicates this method requires high thermal strength to perform the response. The precursor is heated at excessive temperatures in a muffle furnace at which, the precursor gets decomposed, breaking the low energy bonds left in the back of a polymer-based product. Different systems may be acquired through various temperatures. The drawback of this method is its high temperature because of which the reactant itself gets expanded if it reaches it sublimating point and additionally this response may also be time consuming. A purely thermal polymerization is one in which the monomer is converted to the polymer by thermal energy alone. Self-initiated or spontaneous homo- and co-polymerization has been reported for many monomers and monomer pairs. Homo-polymerization generally requires substantial thermal energy whereas copolymerization between certain electron-acceptor and electron-donor monomers can occur at ambient temperatures. As an example, thermal polymerization of PVP is shown in Fig. 1.9.

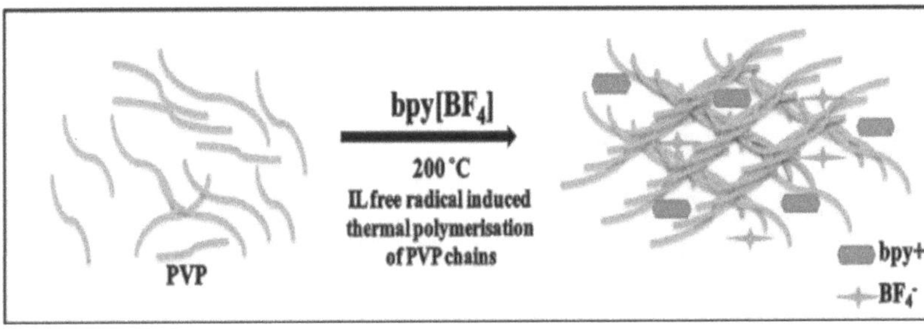

Figure 1.9: Schematic representation of bpy$^{+\cdot}$ cation radical initiated thermal polymerization in PVP chains at 200°C is shown above (Awasthy et al. 2017, Reproduced with permission, Copyright 2020, Royal Society of Chemistry).

There are five types of polymerizations: mass or bulk, emulsion, solution, suspension and gas phase polymerization. In mass or bulk polymerization, the reaction mixture consists mainly of monomers, and in the case of free-radical or ionic polymerization. Thus, the polymerization is carried out in undiluted monomers. This type of polymerization is frequently used for step-growth polymers.

Molecular Beam Epitaxy (MBE)

MBE is the atomic layer by atomic layer crystal growth technique which requires a ultra-high vacuum environment. It is the most reliable deposition method in thermal evaporation. The MBE system involves a growth chamber, analysis chamber and sample chamber. The evaporation rate of the source material is controlled by the computerized process control unit. The vapor source material can be any metalorganic compound. The main advantages of this system are its extremely great purity and highly crystalline thin film heterostructures. Figure 1.10 is the schematic of all principal elements of MBE system.

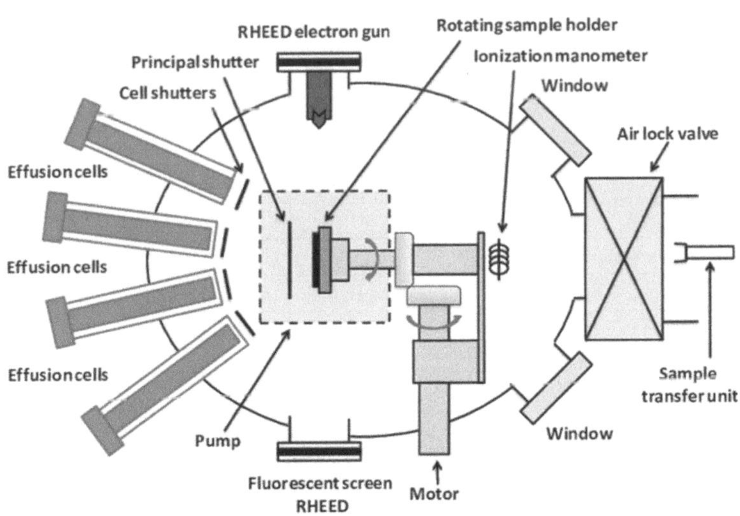

Figure 1.10: Schematic of principal elements of MBE (Jorge and Puebla 2012).

Electrodeposition

Electrodeposition is the deposition of materials on conducting substrates by exploiting interfacial electrochemical reactions. It is a popular method to supply *in situ* metal coatings. In addition, the method is extensively utilized for electrometallurgy. There are two types of electrodeposition cathodics and anodics. The electrolyte is the ionic conductor in this process. The thickness of the electrodeposited layer depends at the time length of plating. The experimental setup for the electrodeposition technique is represented in Fig. 1.11.

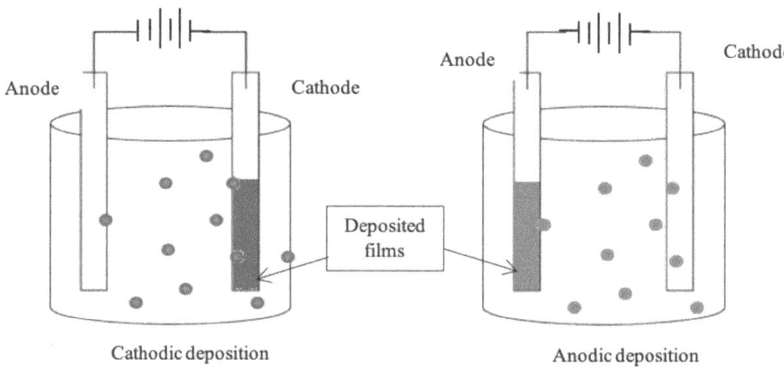

Figure 1.11: Schematic representation of electrodeposition technique.

Electrochemistry for HER, OER and CO_2 RR

Steps involved in Hydrogen Evolution Reaction

HER mechanism depends largely on pH value of the electrolyte and requires more energy than that depicted by the Nernstian equation. Overpotential is the most important parameter to compare and evaluate the working of electrodes and electrolysers. Using an electrocatalyst, the value of overpotential for HER can be significantly reduced. HER is based on either Volmer-Heyrosky or Volmer-Tafel mechanism. Reaction steps involved in HER for both acid and alkaline medium are shown in Table 1.2.

Table 1.2: Reactions steps in HER in acidic and alkaline medium.

	Acid	**Alkaline**
Volmer reaction	$H_3O^+ + e^- \rightarrow H_{ad} + H_2O$	$H_2O + e^- \rightarrow OH^- + H_{ad}$
Heyrovsky reaction	$H_{ad} + H^+ + e^- \rightarrow H_2$	$H_{ad} + H_2O + e^- \rightarrow OH^- + H_2$
Tafel reaction	$H_{ad} + H_{ad} \rightarrow H_2$	$H_{ad} + H_{ad} \rightarrow H_2$

Where H_{ad} represents adsorbed H species on the catalyst surface.

Steps involved in Oxygen Evolution Reaction

Like HER mechanism, the overpotential of OER is decided by the highest kinetic activation barrier of individual reaction steps, i.e., the overpotential is limited by the

Table 1.3: Reaction steps for OER in acidic and alkaline medium.

Acid	Alkaline
$H_2O(l) + * \rightarrow OH* + (H^+ + e^-)$	$OH^- + * \rightarrow OH* + e^-$
$OH* \rightarrow O* + (H^+ + e^-)$	$OH* + OH^- \rightarrow O* + H_2O\ (l) + e^-$
$O* + H_2O\ (l) \rightarrow OOH* + (H^+ + e^-)$	$O* + OH^- \rightarrow OOH* + e^-$
$OOH* \rightarrow * + O_2\ (g) + (H^+ + e^-)$	$OOH* + OH^- \rightarrow * + O_2\ (g) + H_2O\ (l) + e^-$

Here (*l*) and (*g*) refers to liquid and gas phases respectively. * indicates active sites on catalyst and O*, OH* and OOH* are adsorbed species.

step with maximum change in Gibbs free chemisorption energy of two subsequent adsorbed species. Table 1.3 shows the reaction steps involved in OER for both acidic and alkaline medium.

CO_2 reduction reaction

Depending on the different electron/proton transfer amounts, the reaction can be divided into two-, four-, six-, eight- and twelve-electron pathways in an aqueous medium. The table below shows the CDRR and their respective products in all pathways including the electrode potentials applied. Standard pathways for CDRR reaction are shown in Table 1.4.

Table 1.4: Standard CDRR reaction pathways.

Reactant	Product	Electrode potentials (V vs SHE)
$CO_2 + 2H^+ + 2e^-$	$CO + H_2O$	–0.52 V
$CO_2 + 2H^+ + 2e^-$	HCOOH	–0.61
$CO_2 + 4H^+ + 4e^-$	$HCHO + H_2O$	–0.51 V
$CO_2 + 6H^+ + 6e^-$	$CH_3OH + H_2O$	–0.38 V
$CO_2 + 8H^+ + 8e^-$	$CH_4 + 2H_2O$	–0.24 V
$2CO_2 + 12H^+ + 12e^-$	$C_2H_4 + 4H_2O$	–0.34 V

Electrocatalyst for OER and HER

Nanocomposites

The inorganic nanocomposites are combinations of either metal–metal or metal–non-metal. Electrochemical activity of the metals or non-metals can be enhanced by introducing a foreign element. In the case of nanocomposites, the foreign element for a metal is another metal or a non-metal and vice-versa.

Metal sulfide

Recent studies have shown that different structures of MoS_2 would need to be taken into account in the place of Pt for water electrolysis. Nanoboxes made up of Co_9S_8-MoS_2 by surfactant mediated method showed excellent HER performance at low overpotential of 106 mV, low tafel slope of 51.8 mV dec^{-1} and long term stability (Vinod et al. 2018). The ultra-small MoS_2-Au nanohybrid by the simple solvothermal method which exhibited low onset potential of 17 mV, a tafel slope of 40 mV dec^{-1} and a current density of 10 mA cm^{-2} at an overpotential of only 66 mV (Jinxuan et al. 2017). One pot synthesis of MoS_2-Ni_3S_2 heteronanorods on Ni foam as matrix after optimization showed low overpotential of 98 and 249 mV to obtain a current density of 10 mA cm^{-2} in 0.1M KOH (Yaqing et al. 2017). Ternary NiFeMoS anemone-like nanorods on nickel foam synthesized by electrodeposition and hydrothermal method as a bifunctional electrocatalyst exhibited current density of 150 mA cm^{-2} at overpotential of 280 mV for OER and an overpotential of 100 mV at 10 mA cm^{-2} for HER. Overall water splitting for the same has shown current density of 100 mA cm^{-2} at low cell voltage 1.52 V (Yan et al. 2018).

Hollow spheres of Co_9S_8 were obtained by the solvothermal method showed low overpotential of 285 mV reached 10 mA cm^{-2} and the Tafel slope of 58 mV dec^{-1} for OER. This superior catalytic activity for OER is due to the large surface area and abundant active sites (Xueting et al. 2018a). A bubble assisted solvothermal method is used to synthesize hierarchical $MoSe_2$ hollow structures. These hollow structures showed that the Tafel slope of 58.9 mV dec^{-1} for HER (Sha et al. 2018). $NiCo_2S_4$ hollow spheres obtained by one-pot synthesis without template and surfactant strategy (green route) had showed a low overpotential of 400 mV at a current density of 10 mA cm^{-2} with a small Tafel slope (Xueting et al. 2018b). A partial sulfurization/phosphorization strategy was used to synthesize Cobalt monophospho-sulfide material ($Co_{0.9}S_{0.58}P_{0.42}$) showed a low overpotential of 140 mV at the Tafel slope of 70 mV dec^{-1} for HER in both acidic and alkaline medium. For OER, $Co_{0.9}S_{0.58}P_{0.42}$ showed a low overpotential of 266 mV at 10 mA cm^{-2} with Tafel slope of 48 mV dec^{-1} (Zhengfei et al. 2017). Ball shaped $NiCo_2S_4$ hollow spheres synthesized by the solvothermal method showed onset overpotential of 27.9 mV and a Tafel slope of 60.4 mV dec^{-1} (Yiqing et al. 2018).

Metal oxides

Different structures with different morphologies of Cu_2O nanocrystals with a variety of range of nanocubes, sphere like truncated nanocubes and mesoporous Cu_2O showed electrocatalytic activity towards ORR (Oxygen Reduction Reaction) with a Tafel slope ranging between –120 to –59 mV dec^{-1} (Qing et al. 2013). Reduced titanium oxide ($Ti_{1.23}$) film fabricated by cathode reduction process showed low onset potential of 75 mV vs. RHE with the Tafel slope of 88 mV dec^{-1} and exhibited overpotential of 198 mV vs. RHE at 10 mA cm^{-2} for HER (Jayashree et al. 2016).

A novel structure of NiO/MnO_2@PANI showed a low overpotential of 345 mV at a current density of 10 mA cm^{-2} for OER and for ORR, the same material showed a potential of 820 mV at –3 mA cm^{-2}. The synthesis strategy for this material is different from the more commonly used methods. The NiO/MnO_2@PANI structure was synthesized by one step facile UCT (University of Connecticut) method (Junkai et al. 2017). Crystalline/amorphous Ni/NiO core/shell nanosheets fabricated by an optimized synthesis method which included the hydrothermal method, annealing and hydrogenation exhibited overpotential of 110 mV with current density of 5 mA cm^{-2} for HER (Xiaodong et al. 2015). A core-shell IrO_2/RuO_2 prepared by surface modification and precipitation in ethanol showed OER activity and the highest current density of 10.8 mA cm^{-2} at 1.5 V vs RHE (Thomas et al. 2016). NiFe-Prussian Blue analog (PBA) nanocubes were obtained by the modified precipitation method and the porous NiFe NCs were synthesized by annealing NiFe-PBA further. Porous structured NiFe NCs showed bifunctional activity with current density of 10 mA cm^{-2} at cell potential 1.67 V (Ashwani and Bhattacharyya 2017).

Metal phosphides

Transition Metal Phosphides (TMPs) as a new kind of anode catalyst for OER that have been recently reported (Xin et al. 2020). A platinum-free Ru catalyst based on phosphide material synthesized by the solvothermal method exhibited better catalytic efficiency towards HER and OER than Pt/C and Rh/C (Haohong et al. 2017). Metal phosphides such as CoP, FeP, MoP, Ni_2P and their composites have showed remarkable HER performance as shown in the table below.

Table 1.5: Metal phosphide based electrocatalysts.

Electrocatalyst	Method of synthesis	Overpotential mV	Current density mA cm^{-2}	Tafel slope mV dec^{-1}	Reference
Co-Pi/Co-P/Ti nanorod bundles	Low temperature phosphorization	310	10	58	Lunhong et al. 2017
Cobalt Phosphide deposited on copper substrate	Cathodic deposition	85	10	50	Fadl et al. 2014
CoP nanowire array on Ti mesh	CVD	72 (HER) 310 (OER)	10	85	Libin et al. 2016
Urchin-like CoP nanocrystals	Hydrothermal method and phosphidation	50	10	46	Hongchao et al. 2015
FeP nanoparticles	Optimized method	150	10.1	65	Lihong et al. 2016
Fe-tuned Ni_2P	Hydrothermal method and phospharization	158.9	10	55	Huawei et al. 2017
MoP	Temperature-programmed reduction method	125	10	54	Zhicai et al. 2014
Multishell MnCO oxyphosphide	Phosphidation	320	10	52	Bu et al. 2017
Multishelled Ni_2P hollow microspheres	Self-templating and phosphorization	270	10	40.4	Hongming et al. 2017
Ni_2P/Ni composite	Reflux	80	10	68	Yanmei et al. 2014
Ni_xP_y-325	Optimized method	157	10	46.1	Jiayuan et al. 2016
Amorphous NiFeOH/NiFeP/ Ni foam	Plasma treatment	258	300	39	Hanfeng et al. 2017

Metal selenide

$Mn_3O_4/CoSe_2$ hybrid nanocomposite derived by the polyol reduction method exhibited a small overpotential of ~ 0.45 V at a current density 10 mA cm^{-2} (Min-Rui et al. 2012). The $MoSe_2$-$NiSe_2$ nanohybrids showed HER activity with a low onset potential of ~ 150 mV with a current density of 10 mA cm^{-2} at a low overpotential of 210 mV and a Tafel slope of 56 mV dec^{-1} (Xiaoli et al. 2016). A film of nickel diselenide nanoparticles electrodeposited on conductive Ti plate derived 10 mA cm^{-2} at overpotential of 96 mV for HER and 20 mA cm^{-2} at overpotential of 295 mV for OER (Zonghua et al. 2016).

Metal nitride

Nickel nitride grown on nickel foam derived overpotential of 260 mV at a higher current density of 100 mA cm^{-2} for both HER and OER (Menny et al. 2015). To synthesize Co_3N electrocatalyst, an oxidation-nitridation method on a cobalt plate was followed. The catalyst showed efficient OER activity with a low overpotential of 330 mV at 10 mA cm^{-2} and a Tafel slope of 70 mV dec^{-1} (Zhe et al. 2018). Nanoparticles stacked porous cobalt-iron nitride nanowires fabricated by the hydrothermal method followed by annealing in the presence of NH_3 performed as bifunctional electrocatalyst for both HER and OER. Low overpotential of 23 mV at current density of 10 mA cm^{-2} for HER whereas 222 mV at 20 mA cm^{-2} for OER (Yanyong et al. 2016). The same methodology was followed to synthesize porous nickel-cobalt nanowires supported on carbon cloth derived low overpotential of ≈ 145 mV for HER and 360 mV for OER at current density of 10 mA cm^{-2} with Tafel slopes of 105.2 mV dec^{-1} for HER and 46.9 mV dec^{-1} for OER (Lei et al. 2017).

Metal carbide

N and O surface terminated 2D molybdenum carbide nanomeshes (Mo_2CT_xNMs) and microflowers (Mo_2CT_xMFs) were fabricated by an optimized method. An overpotential of 180 mV for current density of 10 mA cm^{-2} exhibited by Mo_2CT_x NMs for OER and overpotential of 1.7 V at 10 mA cm^{-2} for HER by Mo_2CT_xMFs (Zongkui et al. 2019). A bamboo-structured N-doped CNT co-encapsulated with metallic cobalt and Mo_2C nanoparticles synthesized by the pyrolysis method derived a 10 mA cm^{-2} for overpotential of ~ 186 and ~ 377 mV for HER and OER respectively (Lunhong et al. 2018). Highly crystalline Mo_2C nanoparticles supported on carbon sheets as an electrocatalyst prepared by a simple method showed onset potential of –60 mV for HER and an overpotential of 320 mV for OER at a current density of 10 mA cm^{-2} (Hao et al. 2107). Mo_2C embedded N-doped porous carbon nanosheets showed low onset potential of 0 mV and current density of 10 mA cm^{-2} with an overpotential of 45 mV and Tafel slope of 46 mV (Chenbao et al. 2017). N-doped tungsten carbide nanoarrays obtained by the hydrothermal method showed a current density of –200 mA cm^{-2} for the overpotential of –190 mV for HER activity (Nana et al. 2018).

Metal chalcogenides

$CoTe_2$@CdTe nanowire arrays fabricated using the hydrothermal method obtained an onset potential of 1.38 V and overpotential of 140 mV with a Tafel slope of 68 mV

dec^{-1} in an alkaline medium HER, whereas in an acidic medium it derived an onset potential of –0.078 V and overpotential of 110 mV with a Tafel slope of 98 mV dec^{-1} at a current density of 10 mA cm^{-2} (Kartick et al. 2019). The cobalt nickel iron oxides obtained from the precursors of sulfides and selenides exhibited low overpotential of 232 mV for current density of 10 mA cm^{-2} with Tafel slope of 35 mV dec^{-1} (Wei et al. 2016).

Metal borides

Amorphous Ni-B nanoparticles obtained by the chemical reduction method and deposited on glass carbon by electroless plating showed overpotential of 132 mV at current density of 20 mA cm^{-2} for HER (Min et al. 2016). An aqueous reaction method and further annealing was used for the synthesis of core-shell nickel boride nanoparticles-nickel borate exhibited overpotential of 302 mV at current density of 10 mA cm^{-2} and a Tafel slope of 52 mV dec^{-1} (Wen-Jie et al. 2017). Nickel boride nanosheets anchored on multi walled carbon nanotubes fabricated by using an optimized room temperature method derived a low overpotential of 286 and 116 mV for HER and OER resp. at a current density of 10 mA cm^{-2} (Xuncai et al. 2019). Amorphous Cobalt Boride synthesized using the chemical reduction method showed a current density of 10 mA cm^{-2} at 1.61 V for OER (Justus et al. 2016). Similarly, iron diboride nanoparticles fabricated by the chemical reduction method achieved overpotential of 61 mV for HER with a Tafel slope 87.5 mV dec^{-1} and for OER it showed overpotential of 296 mV with Tafel slope of 52.4 mV dec^{-1} at the current density of 10 mA cm^{-2} for both (Hui et al. 2017).

Metal hydroxides

Optimized iron hydroxide modified nickel hydroxylphosphate single walled nanotubes obtained by surfactant free solvothermal method derived low overpotential of 248 mV at a current density of 10 mA cm^{-2} and 323 mV at a large current density of 100 mA cm^{-2} with a Tafel slope of 45.4 mV dec^{-1} for OER (Wei et al. 2018). Highly nanostructured α-$Ni(OH)_2$ showed current density of 10 mA cm^{-2} at a small overpotential of 0.331 mV and a Tafel slope of ~ 42 mV dec^{-1} for OER (Min-rui et al. 2014a).

A series of layered double hydroxides materials were carried out for synthesis as an electrocatalyst for HER and OER. CoFe LDH obtained by using dry exfoliation of bulk CoFe LDH into nanosheets followed by Ar plasma etching showed low overpotential of 266 mV at current density of 10 mA cm^{-2} (Yanyong et al. 2017). Whereas CoFe LDH undergoes delamination followed by exfoliation in DMF-ethanol solvent and forms CoFe LDH nanosheets exhibiting overall water splitting at a current density ~ 10 mA cm^{-2} at the applied voltage of 1.63 V (Peng et al. 2016). NiFe LDH hollow microspheres fabricated by one step *in situ* growth method using SiO_2 as a sacrificial template showed OER activity at a current density of 10 mA cm^{-2} for a small onset overpotential of 239 mV and a Tafel slope of 53 mV dec^{-1} (Cong et al. 2016). A monolithic zeolitic imidazolate framework@LDH precursor on Ni foam derived overpotential of 318 mV at a current density of 10 mA cm^{-2} for OER, whereas for HER it showed –106 mV at –10 mA cm^{-2}. The overall water splitting was obtained at 1.59 V at 10 mA cm^{-2} using the catalyst as both the cathode and anode (Yanqun et al. 2017). The $NiCo_2S_4$ nanotubes, NiFe LDH nanosheets and $NiCo_2S_4$@NiFe LDH synthesized on Ni foam exhibited overpotentials of 306 mV, 260 mV and 201 mV at 60 cm^{-2} respectively, whereas for HER activity an overpotential of 200 mV at 10 mA cm^{-2} (Jia et al. 2017).

Metal and non-metal alloys

A carbon support free binary NiFe nanoparticles alloy synthesized by the hydrothermal method exhibited a much lower overpotential of 298 mV at current density of 10 mA cm^{-2} (Dongwook et al. 2020). A self-supported Au based alloy AuCuMn fabricated by the hydrothermal method tuning the electronic structure of Au showed better catalytic activity as compared to commercial Pt/C catalysts (Hongyu et al. 2020). CuZn alloy catalyst prepared by varying the amount of Zn using the electrodeposition method achieved 60% Faradaic Efficiency (FE) with enhancement in selectivity of HCOOH due to synergistic effect between Cu and Zn (Saira et al. 2019). Table 1.6 represents the synthesis processes and the various electrochemical measurement values of metal and non-metal alloys based catalysts.

Table 1.6: Metal and non-metal alloys based electrocatalysts.

Electrocatalyst	Method of synthesis	Overpotential mV	Current density mA cm^{-2}	Tafel slope $mVdec^{-1}$	Reference
IrCo nanodendrites	Colloidally synthetic method	17 (HER) 281 (OER)	10	67.3	Luhong et al. 2018
Manganese doped sponge-like nickel material	Salt melt synthesis	360	10	-	Marc et al. 2015
Necklace-like hollow NiRu nanoalloy	Galvanic replacement	41	–10	~ 31	Caihua et al. 2017
Nanoporous palladium (PdX) alloy	Electrochemical dealloying	200–500	> 20	-	Swanendu et al. 2019
Pd@PtCu nanocomposite	Seed mediated method	60	10	26.2	Mingjun et al. 2018
CuSn alloy	Electrodeposition	600	1.0	-	Saad et al. 2016
Brass and bronze	Photodeposition	200	3	-	Jingfu et al. 2017b

Perovskites

The precursor of A-deficient double perovskite, PrBaCaCoFeCoO (A-PBCCF) spun onto silk-like nanowires synthesized by an advanced electrospinning technique and simple heat-treatment process exhibited current density 10 mA cm^{-2} for voltage of 1.62 V (Bin et al. 2017). The nano structured NiOCa doped La_2NiO_4 layered perovskite hybrid fabricated by one-pot combustion process showed overpotential of 0.373 mV at current density 10 mA cm^{-2} Tafel slope 42 mV dec^{-1} (Ruochen et al. 2015).

Nanohybrids

Nanohybrids are compounds that include an organic substance like C-based materials and an inorganic material. There are many types of carbon materials based on their dimension, morphology and structures.

N-doped carbon compounds

To achieve the catalytic activity on HER, OER and ORR with the same catalyst, research on the catalytic mechanism and active sites of heteroatom doped carbon materials was carried out by a spontaneous gas-foaming strategy that resulted into a neoteric nitrogen doped carbon nanosheets (NCN-1000-5); a unique ultrathin nanosheet architecture and an ultrahigh specific surface area (Hao et al. 2019). Types of nanohybrids are shown in Fig. 1.12. As activated carbon led to some drawbacks, a number of efforts have been made to fabricate highly porous, advance structured N-doped carbon materials (Feng et al. 2015). Figure 1.13 explains the classification of carbon allotropes according to their dimensionality. Electrocatalysts based on N-doped carbon are described in the Table 1.7.

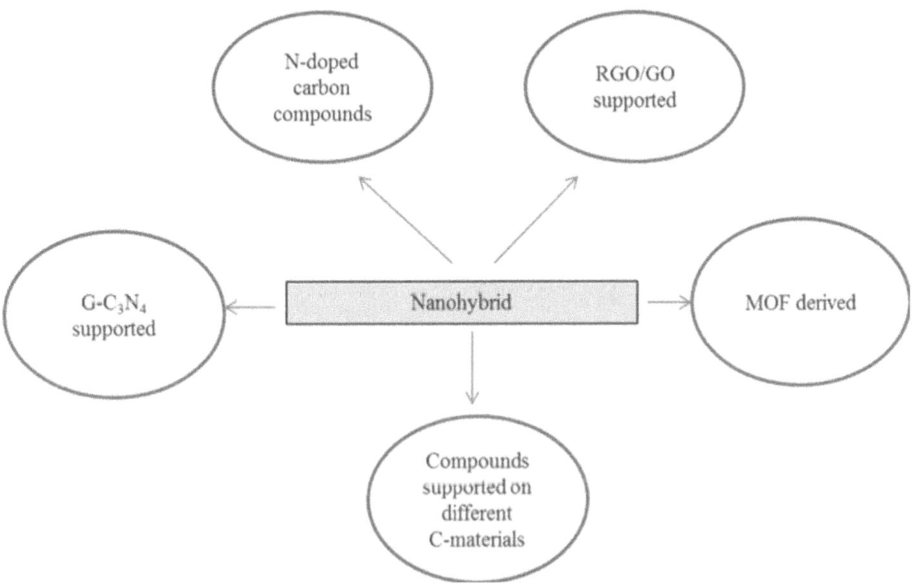

Figure 1.12: Types of nanohybrids.

g-C_3N_4 supported

A number of strategies have been developed to enhance the photocatalytic activity of g-C_3N_4. One pot hydrothermal synthesis is carried out to synthesize β-$Ni(OH)_2$/C_3N_4 nanohybrid to overcome the carrier separation inability of g-C_3N_4 (Junqing et al. 2016). Halide perovskite Quatum Dots (QDs) lack charge transportation efficiency and chemical stability. Amino assisted CsPbR3 QDs hybridized with g-C_3N_4 by self-assembly to overcome its drawbacks yielded 149 μmol h^{-1} g^{-1} under visible light irradiation for photoreduction of CO_2 to CO (Man et al. 2018). A 2D/2D

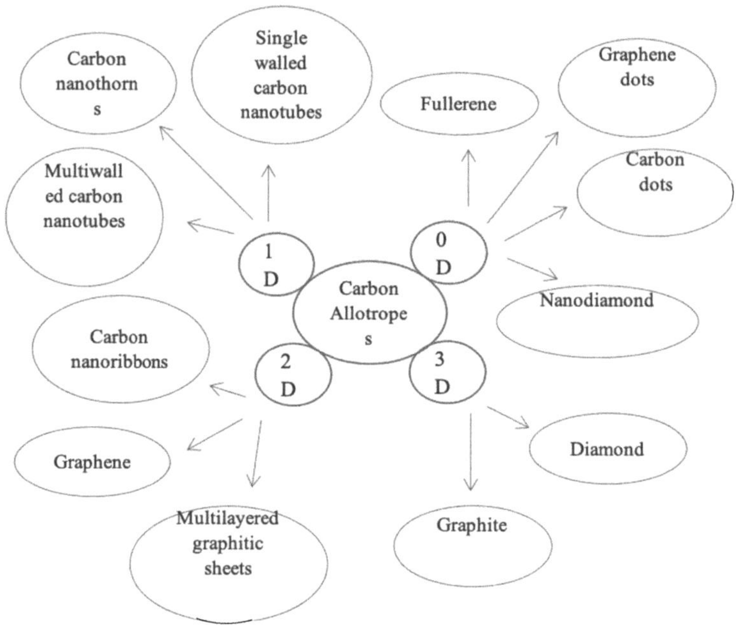

Figure 1.13: Classification of carbon allotropes according to their dimensionality.

Table 1.7: Electrocatalyst based on N-doped carbon.

Electrocatalyst	**Method of synthesis**	**Overpotential mV**	**Current density mA cm^{-2}**	**Tafel slope mVdec^{-1}**	**Reference**
Co on N-doped graphene nanosheets	Optimized method	30	10	82	Huilong et al. 2015
N-doped carbon coated Co-Mn oxide nanoparticle	Solvent evaporation-induced self-assembly method	263 (OER) 71 (HER)	20	97 (OER) 152 (HER)	Jun et al. 2015
N-doped Graphdyine-Co nanoparticles	Chemical reduction	284	10	207	Yurui et al. 2016
Iron oxide based nanoparticles embedded in N-doped carbon matrix	One pot method	330	10	114	Su et al. 2015
MoP@NPC-H	Induced phosphorization	198	10	94	Jing-Qi et al. 2018
N-doped carbon nanotube co-encapsulated with metallic cobalt and Mo_2C nanoparticles	Optimized method	186 (HER) 377 (OER)	10	134	Hong et al. 2017

heterojunction of phosphorene/g-C_3N_4 fabricated by a facile self-assembly method producing H_2 photocatalytically under visible light and yielding 571 μmol h^{-1} g^{-1} in 18 v% lactic acid aqueous solution (Jingrun et al. 2018). Crumpled cuprous oxide anchored on g-C_3N_4 nanosheets obtained by the chemical reduction method carried out water photosplitting under visible light liberating 842 μmol h^{-1} g^{-1} of hydrogen under visible light illumination in the presence of methanol as an agent (Sambandam et al. 2017).

RGO/GO supported

Graphene is an excellent two-dimensional support for dispersion of nanomaterials and maximization of their activity with different sizes, shapes, chemical compositions and heterojunctions resulting into its large specific surface area, excellent conductivity and good stability (Shan-Shan et al. 2014). Graphene sheets have proved to be an outstanding matrix to support foreign substances, leading to advanced materials for electrocatalysis and other energy-related applications because after growing foreign materials on graphene, it forms a strong chemical and electrical coupling, as well as can bring significant performance gains (Min-Rui et al. 2014b). RGO/GO supported electrocatalysts are represented in Table 1.8.

Table 1.8: Electrocatalyst supported on RGO/GO.

Electrocatalyst	**Method of synthesis**	**Overpotential mV**	**Current density mA cm^{-2}**	**Tafel slope mVdec^{-1}**	**Reference**
Cobalt-Iron layered double hydroxide/RGO	Co-precipitation and hydrothermal	325	10	43	Xiaotong et al. 2016
CoP/G	Low temperature phosphorization	120 (HER) 292 (OER)	10	57 (HER) 80 (OER)	Huawei et al. 2016
Cobalt porphyrin ring on electrochemically reduced graphene oxide hybrid film	Optimized method	316	1	96	Juanjuan et al. 2017
MoS_2 nanoflowers on RGO	Solvothermal method	126	-	90	Feng et al. 2015
Fe modulated CoOOH nanoparticles on graphene	Optimized conversion method	330	10	37	Xiaotong et al. 2017
Molybdenum carbide and reduced graphene coupled with polyoxometalate-polypyrrole/RGO	Optimized conversion method	34	1.09	33.6	Ji-Sen et al. 2016
MoS_2 nanosheets/GO	Hydrothermal method	107	-	86.3	Weijia et al. 2014
Ni_3FeN/RGO	Optimized method	94 (HER) 270 (OER)	10	90	Yu et al. 2018
NiO-$NiFe_2O_4$/RGO	Thermal annealing	296	10	43	Guoquan et al. 2016
NiFe LDH/RGO	Solvothermal method	250	10	91	Tianrong et al. 2016

MOF derived

The newly emerging MOFs made from metal ions and polyfunctional organic ligands have proved to be promising self-sacrificing templates and precursors for preparing various carbon-based nanomaterials, benefiting from their high BET surface areas, abundant metal/organic species, large pore volumes and extraordinary tunability of structures and compositions (Kui et al. 2016). Table 1.9 gives the details of MOF derived electrocatalysts.

Table 1.9: Electrocatalyst derived from MOF.

Electrocatalyst	**Method of synthesis**	**Overpotential mV**	**Current density mA cm^{-2}**	**Tafel slope mVdec^{-1}**	**Reference**
CoS_2@N/S-codoped porous carbon	Electrodeposition, sulfuration	95	10	68	Shi et al. 2018
$CuCo_2O_4$@CQDs	Hydrothermal method	331 (HER) 290 (OER)	10	64	Guijuan et al. 2018
Porous MoC_x nano-octahedrons	Carburization	142	10	53	Hao et al. 2015
3D CoCN framework	Annealing	181	10	96	Jie et al. 2016
CoP nanoparticles embedded N-doped CNT hollow polyhedron	Pyrolysis-oxidation-phosphidation strategy	164	10	53	Yuan et al. 2018

Compounds supported on different carbon materials

The most commonly used carbon materials such as GO, RGO, g-C_3N_4, etc., have been studied greatly. Other carbon structures which can be used as a matrix or support to synthesize hybrids are newly investigated viz., CNTs, carbon cloth, carbon fiber, carbon nanowires, carbon shell, carbon black and many more. Brief information about electrocatalysts supported on carbon compounds is given in Table 1.10.

Challenges and issues of CO_2 reduction and hydrogen production

To carry out the reduction of CO_2 and production of hydrogen, noble and expensive metals such as platinum, iridium or ruthenium were earlier used and are still being used. But one cannot afford such expensive materials for simple water electrolysis, during which there are chances of corrosion or getting consumed in the reaction because OER takes place at very high anodic potential as compared to HER. There are several thermodynamic and kinetic barriers and they cause a huge energy loss during the process which further increases overpotential that need to carry out the reaction which should be low. On the other hand, since CO_2 is linear and extremely stable, some problems should be overcome like large overpotential, low selectivity and inferior stability. Hence one requires an efficient catalyst to overcome these drawbacks. Nanomaterials due to their enhanced properties can be used as a catalyst since they have high potential to process the reaction. Also these processes are quite

Table 1.10: Carbon compounds supported electrocatalysts.

Electrocatalyst	Method of synthesis	Overpotential mV	Current density mA cm^{-2}	Tafel slope mVdec^{-1}	Reference
NiO/Ni-CNT	Low temperature hydrolysis, annealing	150	20	82	Ming et al. 2014
CoMn-LDH/MWCNT	Chemical bath deposition	300	10	73.6	Gan et al. 2016
NiMn-LDH/MWCNT	Chemical bath deposition	350	10	83.5	Gan et al. 2016
B-Mo_2 CNT	Carburization	112	10	55	Fei-Xiang et al. 2015
CNTs/S-N-C nanohybrid	Annealing	80	7.62	-	Qing et al. 2015
Co_2N nanocrystals embedded over carbon microfibers	Annealing	260	10	53.9	Tingting et al. 2018
CoF_2O_4 nanoparticles on 3D carbon fiber paper	Hydrothermal method	378	10	73	Alireza et al. 2015
MOF derived Co_3O_4 Carbon porous nanowire arrays	Carbonization	1520	10	123	Tian et al. 2014
Ternary $Fe_xCo_{1-x}P$ on carbon cloth	Optimized method	37	10	30	Chun et al. 2016
CuO/C hollow shell@3D copper dendrites	Direct heat treatment and surface modification	286	10	66.3	Bowei et al. 2018
Carbon shell coated FeP nanoparticles	Carbonization	71	10	52	Dong et al. 2017
Ni2P nanofilms grown on carbon nanosheets	Hydrothermal method	295	10	64	Yan et al. 2017

difficult to operate on a large scale; hence one needs to discover new ways to operate them on a large scale maintaining reaction conditions.

Conclusion

Here synthesis strategies of nanostructured electrocatalyst; chemical routes to reduce CO_2 and production of hydrogen by water splitting; electrochemistry of HER, OER and CDRR; and different types of catalysts with their catalytic parameters are summarized. It is hoped that this chapter provides an updated summary of the developments which rely on several different compounds as nanostructured catalysts and encompass all branches of catalysis.

Acknowledgment

This research work was funded by SERB-DST, New Delhi, under the scheme of "ASEAN-India Research & Development" (Grant No: IMRC/AISTDF/CRD/2018/000048).

References

Alireza Kargar, Serdar Yavuz, Tae Kyoung Kim, Chin-Hung Liu, Cihan Kuru, Cyrus S. Rustomji et al. 2015. Solution-processed $CoFe_2O_4$ nanoparticles on 3D carbon fiber papers for durable oxygen evolution reaction. ACS Applied Materials & Interfaces.

Ashwani Kumar and Sayan Bhattacharyya. 2017. Porous NiFe-oxide nanocubes as bifunctional electrocatalyst for efficient water splitting. ACS Applied Materials & Interfaces 9: 41906–41915.

Aswathy Joseph, Marilyn Mary Xavier, Gaweł ˙Zyła, P. Radhakrishnan Nair, A. S. Padmanabhanad and Suresh Mathew. 2017. Synthesis, characterization and theoretical studies on novel organic-inorganic hybrid ion-gel polymer thin films from a γ-Fe_2O_3 doped polyvinylpyrrolidone-N-butylpyridinium tetrafluoroborate composite via intramolecular thermal polymerization. RSC Advances 7: 16623–16636.

Bin Huaa, Meng Lia, Yi-Fei Suna, Ya-Qian Zhanga, Ning Yanb, Jian Chenc et al. 2017. A coupling for success: Controlled growth of Co/CoOx nanoshoots on perovskite mesoporous nanofibres as high-performance trifunctional electrocatalysts in alkaline condition. Nano Energy 32: 247–254.

Bowei Zhang, Chaojiang Li, Guang Yang, Kang Huang, Junsheng Wu, Zhong Li et al. 2018. Nanostructured CuO/C hollow shell@3D copper dendrites as a highly efficient electrocatalyst for oxygen evolution reaction. ACS Applied Materials & Interfaces 10: 23807–23812.

Bu Yuan Guan, Le Yu and Xiong Wen (David) Lou. 2017. General synthesis of multishell mixed-metal oxyphosphide particles with enhanced electrocatalytic activity in the oxygen evolution reaction. Angewandte Chemie International Edition 129: 2426–2429.

Caihua Zhang, Ying Liu, Yingxue Chang, Yanan Lu, Shulin Zhao, Dongdong Xu et al. 2017. Component-controlled synthesis of necklace-like hollow NiRu nanoalloys as electrocatalysts for hydrogen evolution reaction. ACS Applied Materials & Interfaces.

Chenbao Lu, Diana Tranca, Jian Zhang, Fermin Rodrig uez Hernan dez, Yuezeng Su, Xiaoding Zhuang et al. 2017. Molybdenum carbide-embedded nitrogen-doped porous carbon nanosheets as electrocatalysts for water splitting in alkaline media. ACS Nano. 11: 3933–3942.

Chun Tang, Linfeng Gan, Rong Zhang, Wenbo Lu, Xiue Jiang, Abdullah M. Asiri et al. 2016. Ternary $Fe_xCo_{1-x}P$ nanowire array as a robust hydrogen evolution reaction electrocatalyst with Pt-like activity: Experimental and theoretical insight. Nano Letters 16: 6617–6621.

Cong Zhang, Mingfei Shao, Lei Zhou, Zhenhua Li, Kaiming Xiao and Min Wei. 2016. Hierarchical NiFe layered double hydroxide hollow microspheres with highly-efficient behavior toward oxygen evolution reaction. ACS Applied Materials & Interfaces 8: 33697–33703.

Dong Young Chung, Samuel Woojoo Jun, Gabin Yoon, Hyunjoong Kim, Ji Mun Yoo, Kug-Seung Lee et al. 2017. Large-scale synthesis of carbon-shell-coated FeP nanoparticles for robust hydrogen evolution reaction electrocatalyst. Journal of American Chemical Society 139: 6669–6674.

Dongwook Lim, Euntaek Oh, Chaewon Lim, Sang Eun Shim and Sung-Hyeon Baeck. 2020. Bimetallic NiFe alloys as highly efficient electrocatalysts for the oxygen evolution reaction. Catalysis Today 352: 27–33.

Fadl H. Saadi, Azhar I. Carim, Erik Verlage, John C. Hemminger, Nathan S. Lewis and Manuel P. Soriaga. 2014. CoP as an acid-stable active electrocatalyst for the hydrogen-evolution reaction: electrochemical synthesis, interfacial characterization and performance evaluation. Journal of Physical Chemistry C 118: 29294–29300.

Farihahusnah Hussin and Mohamed Kheireddine Aroua. 2019. Recent development in the electrochemical conversion of carbon dioxide: Short review. AIP Conference Proceedings 2124. 030017: 1–9.

Fei-Xiang Ma, Hao Bin Wu, Bao Yu Xia, Cheng-Yan Xu and Xiong Wen (David) Lou. 2015. Hierarchical β-Mo_2C nanotubes organized by ultrathin nanosheets as a highly efficient electrocatalyst for hydrogen production. Angewandte Chemie 127: 15615–15619.

Feng Li, Le Zhang, Jing Li, Xiaoqing Lin, Xinzhe Li, Yiyun Fang et al. 2015. Synthesis of Cu-MoS_2/rGO hybrid as non-noble metal electrocatalysts for the hydrogen evolution reaction. Journal of Power Sources 292: 15–22.

Friends of the Earth. 2010. Nanotechnology, climate and energy: over-heated promises and hot air? Friends of the Earth, UK edition. 1–77.

Furat Dawood, Martin Anda and G. M. Shafiullah. 2020. Hydrogen production for energy: An overview. International Journal of Hydrogen Energy 45: 3847–3869.

Gan Jia, Yingfei Hu, Qinfeng Qian, Yingfang Yao, Shiying Zhang, Zhaosheng Li et al. 2016. Formation of hierarchical structure composed of (Co/Ni)Mn-LDH nanosheets on MWCNT backbones for efficient electrocatalytic water oxidation. ACS Applied Materials & Interfaces 8: 14527–14534.

Guijuan Wei, Jia He, Weiqing Zhang, Xixia Zhao, Shujun Qiu and Changhua An. 2018. Rational design of Co(II) dominant and oxygen vacancy defective $CuCo_2O_4$@CQDs hollow spheres for enhanced overall water splitting and supercapacitor performance. Inorganic Chemistry 57: 7380–7389.

Guodong Shi, Lin Yang, Zhuowen Liu, Xiao Chen, Jianqing Zhou and Ying Yu. 2018. Photocatalytic reduction CO_2 to CO over copper decorated g-C_3N_4 nanosheets with enhanced yield and selectivity. Applied Surface Science 427: 1165–1173.

Guoquan Zhang, Yanfang Li, Yufei Zhou and Fenglin Yang. 2016. NiFe layered double hydroxide-derived NiO-$NiFe_2O_4$/reduced graphene oxide architectures for enhanced electrocatalysis of alkaline water splitting. ChemElectroChem. 3: 1927–1936.

Hanfeng Liang, Appala N. Gandi, Chuan Xia, Mohamed N. Hedhili, Dalaver H. Anjum, Udo Schwingenschlogl et al. 2017. Amorphous NiFe-OH/NiFeP electrocatalyst fabricated at low temperature for water oxidation applications. ACS Energy Letters 2: 1035–1042.

Hao Bin Wu, Bao Yu Xia, Le Yu, Xin-Yao Yu and Xiong Wen (David) Lou. 2015. Porous molybdenum carbide nano-octahedrons synthesized via confined carburization in metal-organic frameworks for efficient hydrogen production. Nature Communications 6.

Hao Jiang, Jinxing Gu, Xusheng Zheng, Min Liu, Xiaoqing Qiu, Liangbing Wang et al. 2019. Defect-rich and ultrathin N doped carbon nanosheets as advanced trifunctional metal-free electrocatalysts for the ORR, OER and HER. Energy and Environmental Science 12: 322–333.

Hao Ren, Ranbo Yu, Jiangyan Wang, Quan Jin, Mei Yang, Dan Mao et al. 2014. Multishelled TiO_2 hollow microspheres as anodes with superior reversible capacity for lithium ion batteries. Nano Letters 14: 6679–6684.

Hao Wang, Yingjie Cao, Cheng Sun, Guifu, Jianwen Huang, Xiaoxiao Kuai et al. 2017. Strongly coupled molybdenum carbide@carbon sheets as a bifunctional electrocatalyst for overall water splitting. ChemSusChem. 10: 3540–3546.

Haohong Duan, Dongguo Li, Yan Tang, Yang He, Ji shu fang, Rongyue Wang et al. 2017. High performance Rh_2P electrocatalyst for efficient water splitting high performance Rh_2P electrocatalyst for efficient water splitting. Journal of American Chemical Society.

Hong Wang, Shixiong Min, Qiang Wang, Debao Li, Gilberto Casillas, Chun Ma et al. 2017. Nitrogen-doped nanoporous carbon membranes with Co/CoP janus-type nanocrystals as hydrogen evolution electrode in both acidic and alkaline environments. ACS Nano. 11: 4358–4364.

Hongchao Yang, Yejun Zhang, Feng Hu and Qiangbin Wang. 2015. Urchin-like CoP nanocrystals as HER and ORR dual-electrocatalyst with superior stability. Nano Letters 15: 7616–7620.

Hongming Sun, Xiaobin Xu, Zhenhua Yan, Xiang Chen, Fangyi Cheng, Paul S. Weiss et al. 2017. Porous multishelled Ni_2P hollow microspheres as an active electrocatalyst for hydrogen and oxygen evolution. Chemistry of Materials 29: 8539–8547.

Hongyu Gong, Ruizhi Yang, Bo Yang, Fan Li and Ling Li. 2020. Boosting the catalysis of AuCuMo for oxygen reduction: Important roles of an optimized electronic structure and surface electrochemical stability. Journal of Alloys and Compounds 837.

Huawei Huang, Chang Yu, Juan Yang, Changtai Zhao, Xiaotong Han, Zhibin Liu et al. 2016. Strongly coupled architectures of cobalt phosphide nanoparticles assembled on graphene as bifunctional electrocatalysts for water splitting. ChemElectroChem. 3: 719–725.

Huawei Huang, Chang Yu, Changtai Zhao, Xiaotong Han, Juan Yang, Zhibin Liu et al. 2017. Iron-tuned super nickel phosphide microstructures with high activity for electrochemical overall water splitting. Nano Energy 34: 472–480.

Hui Li, Peng Wen, Qi Li, Chaochao Dun, Junheng Xing, Chang Lu et al. 2017. Earth-abundant iron diboride (FeB_2) nanoparticles as highly active bifunctional electrocatalysts for overall water splitting. Advanced Energy Materials 7: 1–12.

Huilong Fei, Juncai Dong, M. Josefina Arellano-Jimenez, Gonglan Ye, Nam Dong Kim, Errol L. G. Samuel et al. 2015. Atomic cobalt on nitrogen-doped graphene for hydrogen generation. Nature Communications 6.

Ilgi Karapinar Kapdan and Fikret Kargi. 2006. Bio-hydrogen production from waste materials. Enzyme and Microbial Technology 38: 569–582.

Jayashree Swaminathan, Ravichandran Subbiah and S. Vengatesan. 2016. Defect-rich metallic titania (TiO)—An efficient hydrogen evolution catalyst for electrochemical water splitting. ACS Catalysis 6: 2222–2229.

Jia, C., K. Dastafkan, W. Ren, W. Yang and C. Zhao. 2017. Carbon-based catalysts for electrochemical CO_2 reduction. Sustainable Energy Fuels 7: 7855–7865.

Jia Liu, Jinsong Wang, Bao Zhang, Yunjun Ruan, Lin Lv, Xiao Ji et al. 2017. Hierarchical $NiCo_2S_4$@NiFe LDH heterostructures supported on nickel foam for enhanced overall-water-splitting activity. ACS Applied Materials & Interfaces.

Jiayuan Li, Jing Li, Xuemei Zhou, Zhaoming Xia, Wei Gao, Yuanyuan Ma et al. 2016. Highly efficient and robust nickel phosphides as bifunctional electrocatalysts for overall water-splitting. ACS Applied Materials & Interfaces 8: 10826–10834.

Jie Chen, Haifeng Zhou, Yongjun Huang, Huiwu Yu, Fuying Huang, Fengying Zheng et al. 2016. A 3D Co-CN framework as high performance electrocatalyst for hydrogen evolution reaction. RSC Advances 6: 42014–42018.

Jingfu He, Kevan E. Dettelbach, Danielle A. Salvatore, Tengfei Li and Curtis P. Berlinguette. 2017a. High-throughput synthesis of mixed-metal electrocatalysts for CO_2 reduction. Angewandte Chemie 56: 6068–6072.

Jingfu He, Kevan E. Dettelbach, Aoxue Huang and Curtis P. Berlinguette. 2017b. Brass and bronze as effective CO_2 reduction electrocatalysts. Angewandte Chemie International Edition 129: 16806–16809.

Jing-Qi Chi, Wen-Kun Gao, Li-Ming Zhang, Bin Dong, Kai-Li Yan, Jia-Hui Lin et al. 2018. Induced phosphorization-derived well-dispersed molybdenum phosphide nanoparticles encapsulated in hollow N-doped carbon nanospheres for efficient hydrogen evolution. ACS Sustainable Chemistry & Engineering 6: 7676–7686.

Jingrun Ran, Weiwei Guo, Hailong Wang, Bicheng Zhu, Jiaguo Yu and Shi-Zhang Qiao. 2018. Metal-free 2D/2D phosphorene/g-C_3N_4 Van der waals heterojunction for highly enhanced visible-light photocatalytic H_2 production. Advanced Materials 30.

Jinhui Hao and Weidong Shi. 2018. Transition metal (Mo, Fe, Co, and Ni)-based catalysts for electrochemical CO_2 reduction. Chinese Journal of Catalysis 39: 1157–1166.

Jinxuan Zhang, Tanyuan Wang, Lu Liu, Kuangzhou Du, Wanglian Liu, Zhiwei Zhu et al. 2017. Molybdenum disulfide and Au ultrasmall nanohybrids as highly active electrocatalysts for hydrogen evolution reaction. Journal of Materials Chemistry A 5: 4122–4128.

Ji-Sen Li, Yu Wang, Chun-Hui Liu, Shun-Li Li, Yu-Guang Wang, Long-Zhang Dong et al. 2016. Coupled molybdenum carbide and reduced graphene oxide electrocatalysts for efficient hydrogen evolution. Nature Communications 7.

Jorge Luis and Puebla Nunez. 2012. Spin phenomena in semiconductor quantum dots. Ph.D. Thesis, Department of Physics and Astronomy, University of Sheffield, UK.

Juanjuan Ma, Lin Liu, Qian Chen, Min Yang, Danping Wang, Zhiwei Tong et al. 2017. A facile approach to prepare crumpled CoTMPyP/electrochemically reduced graphene oxide nanohybrid as an efficient electrocatalyst for hydrogen evolution reaction. Applied Surface Science 399: 535–541.

Jun Li, Yongcheng Wang, Tong Zhou, Hui Zhang, Xuhui Sun, Jing Tang et al. 2015. Nanoparticle superlattices as efficient bifunctional electrocatalyst for water splitting. Journal of American Chemical Society 137: 14305–14312.

Junkai He, Mingchao Wang, Wenbo Wang, Ran Miao, Wei Zhong, Sheng-Yu Chen et al. 2017. Hierarchical mesoporous NiO/MnO_2@PANI core–shell microspheres, highly efficient and stable bifunctional electrocatalysts for oxygen evolution and reduction reactions. ACS Applied Materials & Interfaces 9: 42676–42687.

Junqing Yana, Huan Wua, Hong Chena, Liuqing Panga, Yunxia Zhanga, Ruibin Jianga et al. 2016. One-pot hydrothermal fabrication of layered β-$Ni(OH)_2$/g-C_3N_4 nanohybrids for enhanced photocatalytic water splitting. Applied Catalysis B: Environmental 194: 74–83.

Justus Masa, Philipp Weide, Daniel Peeters, Ilya Sinev, Wei Xia, Zhenyu Sun et al. 2016. Amorphous cobalt boride (Co_2B) as a highly efficient nonprecious catalyst for electrochemical water splitting: oxygen and hydrogen evolution. Advanced Energy Materials 6: 1–10.

Kartick Chandra Majhi, Paramita Karfa and Rashmi Madhuri. 2019. Bimetallic transition metal chalcogenide nanowire array: An effective catalyst for overall water splitting. Electrochimica Acta. 318: 901–912.

Kui Shen, Xiaodong Chen, Junying Chen and Yingwei Li. 2016. Development of MOF-derived carbon-based nanomaterials for efficient catalysis. ACS Catalysis 6: 5887–5903.

Lei Han, Kun Feng and Zhongwei Chen. 2017. Self-supported cobalt nickel nitride nanowires electrode for overall electrochemical water splitting. Energy Technology 5: 1908–1911.

Libin Yang, Honglan Qi, Chengxiao Zhang and Xuping Sun. 2016. An efficient bifunctional electrocatalyst for water splitting based on cobalt phosphide. IOP Nanotechnology 27: 1–7.

Lihong Tian, Xiaodong Yan and Xiaobo Chen. 2016. Electrochemical activity of iron phosphide nanoparticles in hydrogen evolution reaction. ACS Catalysis 6: 5441–5448.

Luhong Fu, Xiang Zeng, Gongzhen Cheng and Wei Luo. 2018. IrCo nanodendrite as an efficient bifunctional electrocatalyst for overall water splitting under acidic condition. ACS Applied Materials & Interfaces 10: 24993–24998.

Lunhong Ai, Zhiguo Niu and Jing Jiang. 2017. Mechanistic insight into oxygen evolution electrocatalysis of surface phosphate modified cobalt phosphide nanorod bundles and their superior performance for overall water splitting. Electrochimica Acta. 242: 355–363.

Lunhong Ai, Jinfeng Su, Mei Wang and Jing Jiang. 2018. Bamboo-structured nitrogen-doped carbon nanotube co-encapsulating cobalt and molybdenum carbide nanoparticles: an efficient bifunctional electrocatalyst for overall water splitting. ACS Sustainable Chemistry & Engineering 6: 9912–9920.

Man Ou, Wenguang Tu, Shengming Yin, Weinan Xing, Shuyang Wu, Haojing Wang et al. 2018. Amino-assisted anchoring of $CsPbBr_3$ perovskite quantum dots on porous g-C_3N_4 for enhanced photocatalytic CO_2 reduction. Angewandte Chemie 54: 13570–13574.

Marc Ledendecker, Guylhaine Clavel, Markus Antonietti and Menny Shalom. 2015. Highly porous materials as tunable electrocatalysts for the hydrogen and oxygen evolution reaction. Advanced Functional Materials 25: 393–399.

Menny Shalom, Debora Ressnig, Xiaofei Yang, Guylhaine Clavel, Tim Patrick Fellinger and Markus Antonietti. 2015. Nickel nitride as an efficient electrocatalyst for water splitting. Journal of Materials Chemistry A 3: 8171–8177.

Min Zeng, Hao Wang, Chong Zhao, Jiake Wei, Kuo Qi, Wenlong Wang et al. 2016. Nanostructured amorphous nickel boride for high-efficiency electrocatalytic hydrogen evolution over a broad pH range. ChemCatChem. 8: 708–712.

Ming Gong, Wu Zhou, Mon-Che Tsai, Jigang Zhou, Mingyun Guan, Meng-Chang Lin et al. 2014. Nanoscale nickel oxide/nickel heterostructures for active hydrogen evolution electrocatalysis. Nature Communications 5.

Mingjun Bao, Ibrahim Saana Amiinu, Tao Peng, Wenqiang Li, Shaojun Liu, Zhe Wang et al. 2018. Surface evolution of PtCu-alloy-shell over Pd-nanocrystals leads to superior hydrogen evolution and oxygen reduction reactions. ACS Energy Letters 3: 940–945.

Mingming Li, Fan Xu, Haoran Li and Yong Wang. 2016. Nitrogen-doped porous carbon materials: promising catalysts or catalyst supports for heterogeneous hydrogenation and oxidation. Catalysis Science & Technology.

Min-Rui Gao, Yun-Fei Xu, Jun Jiang, Ya-Rong Zheng and Shu-Hong Yu. 2012. Water oxidation electrocatalyzed by an efficient $Mn_3O_4/CoSe_2$ nanocomposite. Journal of American Chemical Society 134: 2930–2933.

Min-Rui Gao, Wenchao Sheng, Zhongbin Zhuang, Qianrong Fang, Shuang Gu, Jun Jiang et al. 2014a. Efficient water oxidation using nanostructured αnickel-hydroxide as an electrocatalyst. Journal of American Chemical Society 136: 7077–7084.

Min-Rui Gao, Xuan Cao, Qiang Gao, Yun-Fei Xu, Ya-Rong Zheng, Jun Jiang et al. 2014b. Nitrogen-doped graphene supported $CoSe_2$ nanobelt composite catalyst for efficient water oxidation. ACS NANO. 8: 3970–3978.

Murugananthan, M., M. Kumaravel, Hideyuki Katsumata, Tohru Suzuki and Satoshi Kaneco. 2015. Electrochemical reduction of CO_2 using Cu electrode in methanol/$LiClO_4$ electrolyte. International Journal of Hydrogen Energy 40: 6740–6744.

Nana Han, Ke R. Yang, Zhiyi Lu, Yingjie Li, Wenwen Xu, Tengfei Gao et al. 2018. Nitrogen-doped tungsten carbide nanoarray as an efficient bifunctional electrocatalyst for water splitting in acid. Nature Communications 9.

Peng Fei Liu, Shuang Yang, Bo Zhang and Hua Gui Yang. 2016. Defect-rich ultrathin cobalt-iron layered double hydroxide for electrochemical overall water splitting. ACS Applied Materials & Interfaces 8: 34474–34481.

Puebla Nunez and Jorge Luis. 2013. Spin phenomena in semiconductor quantum dots. PhD thesis, University of Sheffield.

Qing Li, Ping Xu, Bin Zhang, Hsinhan Tsai, Shijian Zheng, Gang Wu et al. 2013. Structure-dependent electrocatalytic properties of Cu_2O nanocrystals for oxygen reduction reaction. Journal of Physical Chemistry C 117: 13872–13878.

Qing Zhu, Ling Lin, Yi-Fan Jiang, Xiao Xie, Cheng-Zong Yuan and An-Wu Xu. 2015. Carbon nanotubes/S-N-C nanohybrids as a high performance bifunctional electrocatalyst for both oxygen reduction and evolution reactions. New Journal of Chemistry 39: 6289–6296.

Ruochen Liu, Fengli Liang, Wei Zhoun, Yisu Yang and Zhonghua Zhun. 2015. Calcium-doped lanthanum nickelate layered perovskite and nickel oxide nano-hybrid for highly efficient water oxidation. Nano Energy 12: 115–122.

Saad Sarfraz, AngelT Gardia-Esparza, Abdesslem Jedidi, Luigi Cavallo and Kazuhiro Takanabe. 2016. Cu-Sn bimetallic catalyst for selective aqueous electroreduction of CO_2 to CO. ACS Catalysis 6: 2842–2851.

Saira Ajmal, Yang Yang, Kejian Li, Muhammad Ali Tahir, Yangyang Liu and Tao Wang. 2019. Zinc-modified copper catalyst for efficient (photo-)electrochemical CO_2 reduction with high selectivity of HCOOH production. Journal of Physical Chemistry C 123: 11555–11563.

Sambandam Anandan, Jerry J. Wu, Detlef Bahnemann, Alexei Emeline and Muthupandian Ashokkumar. 2017. Crumpled Cu_2O-g-C_3N_4 nanosheets for hydrogen evolution catalysis. Colloids and Surfaces A: Physicochemical & Engineering Aspects 527: 34–41.

Sengeni Anantharaj, Sivasankara Rao Ede, Kuppan Sakthikumar, Kannimuthu Karthick, Soumyaranjan Mishra and Subrata Kundu. 2016. Recent trends and perspectives in electrochemical water splitting with an emphasis on sulfide, selenide, and phosphide catalysts of Fe, Co, and Ni: A review. ACS Catalysis 6: 8069–8097.

Sha Hu, Qingqing Jiang, Shuoping Ding, Ye Liu, Zuozuo Wu, Zhengxi Huang et al. 2018. Construction of hierarchical $MoSe_2$ hollow structures and its effect on electrochemical energy storage and conversion. ACS Applied Materials & Interfaces.

Shan-Shan Li, Jing-Jing Lv, Li-Na Teng, Ai-Jun Wang, Jian-Rong Chen and Jiu-Ju Feng. 2014. Facile synthesis of PdPt@Pt nanorings supported on reduced graphene oxide with enhanced electrocatalytic properties. Electrochimica Acta. 143: 36–43.

Sheng Zhang, Peng Kang and Thomas J. Meyer. 2014. Nanostructured tin catalysts for selective electrochemical reduction of carbon dioxide to formate. Journal of American Chemical Society 136: 1734–1737.

Shi Feng, Xingyue Li, Jia Huo, Qiling Li, Chao Xie, Tingting Liu et al. 2018. Controllable synthesis of CoS_2@N/S-codoped porous carbon derived from ZIF-67 for highly efficient hydrogen evolution reaction. ChemCatChem. 10: 796–803.

Shulin Zhao, Sheng Li, Tao Guo, Shuaishuai Zhang, Jing Wang, Yuping Wu et al. 2019. Advances in Sn-based catalysts for electrochemical CO_2 reduction. Nano-Micro Letters 11: 1–66.

Su, D., J. Wang, H. Jin, Y. Gong, M. li, Z. Pang et al. 2015. From "waste to gold": one-pot way to synthesize ultrafinely dispersed Fe_2O_3-based nanoparticles on N-doped carbon for synergistically and efficiently water splitting. Journal of Materials Chemistry A 3: 11756–11761.

Subhra Jana. 2015. Advances in nanoscale alloys and intermetallics: low temperature solution chemistry synthesis and application in catalysis. Dalton Transactions 44: 18692–18717.

Subramani Surendran, Sathyanarayanan Shanmugapriya, Sangaraju Shanmugam, Leonid Vasylechko and Ramakrishnan Kalai Selvan. 2018. Interweaved nickel phosphide sponge as an electrode for flexible supercapattery and water splitting applications. ACS Applied Energy Materials 1: 78–92.

Swanendu Chatterjee, Charles Griego, James L. Hart, Yawei Li, Mitra L. Taheri, John Keith et al. 2019. Free standing nanoporous palladium alloys as CO poisoning tolerant electrocatalysts for the electrochemical reduction of CO_2 to formate. ACS Catalysis 9: 5290–5301.

Thomas Audichon, Teko W. Napporn, Kouakou Boniface Kokoh, Christine Canaff, Claudia Morais and Clement Comminges. 2016. IrO_2 coated on RuO_2 as efficient and stable electroactive nanocatalysts for electrochemical water splitting. Journal of Physical Chemistry C 120: 2562–2573.

Tian Yi Ma, Sheng Dai, Mietek Jaroniec and Shi Zhang Qiao. 2014. Metal–organic framework derived hybrid Co_3O_4 carbon porous nanowire arrays as reversible oxygen evolution electrodes. Journal of American Chemical Society 136: 13925–13931.

Tianrong Zhan, Yumei Zhang, Xiaolin Liu, SiSi Lu and Wanguo Hou. 2016. NiFe layered double hydroxide/reduced graphene oxide nanohybrid as an efficient bifunctional electrocatalyst for oxygen evolution and reduction reactions. Journal of Power Sources 333: 53–60.

Tingting Liu, Mian Li, Zhongmin Su, Xiangjie Bo, Wei Guan and Ming Zhou. 2018. Monodisperse and tiny $Co_2N_{0.67}$ nanocrystals uniformly embedded over two curving surfaces of hollow carbon microfibers as efficient electrocatalyst for oxygen evolution reaction. Applied Nano Materials 1: 4461–4473.

Vinoth Ganesan, Sunghyun Lim and Jinkwon Kim. 2018. Hierarchical nanoboxes composed of Co_9S_8-MoS_2 nanosheets as efficient electrocatalysts for hydrogen evolution reaction. Chemistry—An Asian Journal 10: 413–420.

Wei Bian, Yichao Huang, Xiaobin Xu, Muhammad Aizaz Ud Din, Gang Xie and Xun Wang. 2018. Iron Hydroxide-modified nickel hydroxylphosphate single wall nanotubes as efficient electrocatalysts for oxygen evolution reactions. ACS Applied Materials & Interfaces 10: 9407–9414.

Wei Chen, Yayuan Liu, Yuzhang Li, Jie Sun, Yongcai Qiu, Chong Liu et al. 2016. *In situ* electrochemically derived nanoporous oxides from transition matal dichalcogenides for active oxygen evolution catalysts. Nano Letters 16: 7588–7596.

Weijia Zhou, Kai Zhou, Dongman Hou, Xiaojun Liu, Guoqiang Li, Yuanhua Sang et al. 2014. Three-dimensional hierarchical frameworks based on MoS_2 nanosheets self-assembled on graphene oxide for efficient electrocatalytic hydrogen evolution. ACS Applied Materials & Interfaces.

Wen-Jie Jiang, Shuai Niu, Tang Tang, Qing-Hua Zhang, Xiao-Zhi Liu, Yun Zhang et al. 2017. Crystallinity-modulated electrocatalytic activity of a nickel(II) borate Thon layer on Ni_3B for efficient water oxidation. Angewwandte Chemie 56: 6572–6577.

Xiaodong Yan, Lihong Tian and Xiaobo Chen. 2015. Crystalline/amorphous Ni/NiO core/shell nanosheets as highly active electrocatalysts for hydrogen evolution reaction. Journal of Power Sources 300: 336–343.

Xiao-Fang Liu, Xiao-Xia Li and Liang-Nian He. 2018. Transition metal-catalyzed reductive functionalization of CO_2. European Journal of Organic Chemistry.

Xiaoli Zhou, Yun Liu, Huanxin Ju, Bicai Pan, Junfa Zhu, Tao Ding et al. 2016. Design and epitaxial growth of $MoSe_2$–NiSe vertical heteronanostructures with electronic modulation for enhanced hydrogen evolution reaction. Chemistry of Materials 28: 1838–1846.

Xiaotong Han, Chang Yu, Juan Yang, Changtai Zhao, Huawei Huang, Zhibin Liu et al. 2016. Mass and charge transfer coenhanced oxygen evolution behaviors in CoFe-layered double hydroxide assembled on graphene. Advanced Materials & Interfaces 3: 1–8.

Xiaotong Han, Chang Yu, Si Zhou, Changtai Zhao, Huawei Huang, Juan Yang et al. 2017. Ultrasensitive iron-triggered nanosized Fe-CoOOH integrated with graphene for highly efficient oxygen evolution. Advanced Energy Materials 7: 1–9.

Xin Ding, Waqar Uddin, Hongting Sheng, Peng Li, Yuanxin Du and Manzhou Zhu. 2020. Porous transition metal phosphides derived from Fe-based Prussian blue analogue for oxygen evolution reaction. Journal of Alloys and Compounds 814.

Xueting Feng, Qingze Jiao, Huiru Cui, Mengmeng Yin, Qun Li, Yun Zhao et al. 2018a. One-pot synthesis of $NiCo_2S_4$ hollow spheres via sequential ion-exchange as an enhanced oxygen bifunctional electrocatalyst in alkaline solution. ACS Applied Materials & Interfaces 10: 29521–29531.

Xueting Feng, Qingze Jiao, Tong Liu, Qun Li, Mengmeng Yin, Yun Zhao et al. 2018b. Facile synthesis of Co_9S_8 hollow spheres as a high performance electrocatalyst for the oxygen evolution reaction. ACS Sustainable Chemistry & Engineering 6: 1863–1871.

Xuncai Chen, Zixun Yu, Li Wei, Zheng Zhou, Shengli Zhai, Junsheng Chen et al. 2019. Ultrathin nickel boride nanosheets anchored on functionalized carbon nanotubes as bifunctional electrocatalysts for overall water splitting. Journal of Materials Chemistry A 7: 764–774.

Yan, K. -L., J. -F. Qin, Z. -Z. Liu, B. Dong, J. -Q. Chi, W. -K. Gao et al. 2018. Organic-inorganic hybrids-directed ternary NiFeMoS anemone-like nanorods with scaly surface supported on nickel foam for efficient overall water splitting. Chemical Engineering Journal 334: 922–931.

Yan Li, Pingwei Cai, Suqin Ci and Zhenhai Wen. 2017. Strongly coupled 3D nanohybrids with Ni_2P/carbon-nanosheets as pH-universal hydrogen evolution reaction electrocatalysts. ChemElectroChem. 4: 340–344.

Yang Song, Rui Peng, Dale K. Hensley, Peter V. Bonnesen, Liangbo Liang, Zili Wu et al. 2016. High-selectivity electrochemical conversion of CO_2 to ethanol using a copper nanoparticle/N-doped graphene electrode. Chemistry Select 1: 6055–6061.

Yanmei Shi, You Xu, Sifei Zhuo, Jingfang Zhang and Bin Zhang. 2014. Ni_2P nanosheets/Ni foam composite electrode for long-lived and pH-tolerable electrochemical hydrogen generation. ACS Applied Materials 118: 29294–29300.

Yanqun Tang, Xiaoyu Fang, Xin Zhang, Gina Fernandes, Yong Yan, Dongpeng Yan et al. 2017. Space-confined earth-abundant bifunctional electrocatalyst for high-efficiency water splitting. ACS Applied Materials & Interfaces 9: 36762–36771.

Yanyong Wang, Dongdong Liu, Zhijuan Liu, Chao Xie, Jia Huo and Shuangyin Wang. 2016. Porous cobalt-iron nitride nanowire as excellent bifunctional electrocatalyst for overall water splitting. RSC Chemical Communications 52: 12614–12617.

Yanyong Wang, Yiqiong Zhang, Zhijuan Liu, Chao Xie, Shi Feng, Dongdong Liu et al. 2017. Layered double hydroxide nanosheets with multiple vacancies obtained by dry exfoliation as highly efficient oxygen evolution electrocatalysts. Angewwandte Chemie International Edition 129: 5961–5965.

Yaqing Yang, Kai Zhang, Huanlei Lin, Xiang Li, Hang Cheong Chan, Lichun Yang et al. 2017. Heteronanorods of MoS_2-Ni_3S_2 as efficient and stable Bi-functional electrocatalysts for overall water splitting. ACS Catalysis 7: 2357–2366.

Yihong Chen, Christina W. Li and Matthew W. Kanan. 2012. Aqueous CO_2 reduction at very low overpotential on oxide-derived Au nanoparticles. Journal of American Chemical Society 134: 19969–19972.

Yiqing Jiang, Xing Qian, Changli Zhu, Hongyu Liu and Linxi Hou. 2018. Nickel cobalt sulfide double-shelled hollow nanospheres as superior bifunctional electrocatalysts for photovoltaics and alkaline hydrogen evolution. ACS Applied Materials & Interfaces 10: 9379–9389.

Yu Gu, Shuai Chen, Jun Ren, Yi Alec Jia, Cheng-Meng Chen, Sridhar Komarneni et al. 2018. Electronic structure tuning in Ni_3FeN/r-GO aerogel towards bifunctional electrocatalyst for overall water splitting. ACS Nano. 12: 245–253.

Yuan Pan, Kaian Sun, Shoujie Liu, Xing Cao, Konglin Wu, Weng-Chon Cheong et al. 2018. Core-shell ZIF-8@ZIF-67 derived CoP nanoparticles-embedded N-doped carbon nanotube hollow polyhedron for efficient over-all water splitting. Journal of American Chemical Society 140: 2610–2618.

Yurui Xue, Jiaofu Li, Zheng Xue, Yongjun Li, Huibiao Liu, Dan Li et al. 2016. Extraordinarily durable graphdiyne-supported electrocatalyst with high activity for hydrogen production at all values of pH. ACS Applied Materials & Interfaces 8: 31083–31091.

Zhe Xu, Wenchao Li, Yadong Yan, Hongxu Wang, Heng Zhu, Meiming Zhao et al. 2018. *In-situ* formed hydroxide accelerating water dissociation kinetics on Co_3N for hydrogen production in alkaline solution. ACS Applied Materials & Interfaces 10: 22102–22109.

Zhengfei Dai, Hongbo Geng, Jiong Wang, Yubo Luo, Bing Li, Yun Zong et al. 2017. Hexagonal-phase cobalt monophosphosulfide for highly efficient overall water splitting. ACS Nano. 11: 11031–11040.

Zhicai Xing, Qian Liu, Abdullah M. Asiri and Xuping Sun. 2104. Closely interconnected network of molybdenum phosphide nanoparticles: a highly efficient electrocatalyst for generating hydrogen from water. Advanced Materials 26: 5702–5707.

Zonghua Pu, Yonglan Luo, Abdullah M. Asiri and Xuping Sun. 2016. Efficient electrochemical water splitting catalyzed by electrodeposited nickel diselenide nanoparticles based film. ACS Applied Materials & Interfaces 8: 4718–4723.

Zongkui Kou, Lei Zhang, Yuanyuan Ma, Ximeng Liu, Wenjie Zang, Jian Zhang et al. 2019. 2D carbide nanomeshes and their assembling into 3D microflowers for efficient water splitting. Applied Catalysis B: Environmental 243: 678–685.

Chapter 2

Hybrid Perovskite Photocatalysis for Energy Harvesting and Energy Saving

Ankur Solanki[1,*] and *Silver-Hamill Turren-Cruz*[2]

Introduction

Photocatalysis has emerged as an effective artificial technique of solar-to-chemical energy conversion, inspired from natural and efficient energy conversion photosynthesis phenomenon (Kim et al. 2015). Since the very first report (Fujishima and Honda 1972) of hydrogen production on titanium-dioxide (TiO_2) using ultra-violet sensitized photocatalytics, research area has been explored for numerous applications such as water splitting, CO_2 reduction and chemical transformation, for various organic pollutants treatment. The fundamental photo-mechanims in photocatalysis materials is completed in three basic processes: (i) solar energy absorption and generation of excitons/electron-hole pairs (Ankur 2017), (ii) exciton/electron-hole pair dissociation into free charge carriers and diffusion towards respective reaction sites under an internal potential gradient, and (iii) occurrence of chemical reaction (i.e., oxidation and reduction) by free charge carriers at their respective reaction sites (Hisatomi et al. 2014).

For an efficient accomplishment of the mentioned mechanism, the photocatalytic materials should have lower energy band-gap for broader and stronger light absorption, lower exciton/electron-hole pair binding energy, efficient charge transport and strong redox ability for chemical reactions. Pristine TiO_2, graphitic carbon nitride

[1] Department of Physics, School of Technology Pandit Deendayal Petroleum University, Gandhinagar 38007 Gujarat, India.

[2] Helmholtz-Zentrum Berlin für Materialienund Energie GmbH, 14109 Berlin, Germany.
Email: silver.turren_cruz@helmholtz-berlin.de

* Corresponding author: oksolanki@gmail.com

(g-C_3N_4), and $BiVO_4$ are the most common materials in the photocatalysis family, however large energy band-gap (Ge et al. 2017), higher charge carrier recombination rates, poor charge carrier mobility and lower photo-reduction potential limit their wider potential (He et al. 2014). Therefore, the research community is continuously striving to develop new semiconducting photocatalysis with efficient charge carrier generation, dissociation and transport properties.

Owing to the outstanding optical/electronic/mechanical properties, structural tunability and easier processability, Hybrid Inorganic/Organic Perovskite (HIOP) has emerged as a strong candidate of photocatalytic materials and captured the attention of the scientific community. The structure of the ideal perovskite has AMX_3 as its general formula. Perovskite is also the name of a group of crystals that take the same structure; therefore perovskites are ceramic (solids that combine metallic with non-metallic elements). An organic–inorganic perovskite material has an AMX_3 structure as shown in Fig. 2.1. In the general structure A = (methylammonium (MA) $CH_3NH_3^+$; formamidinium (FA) $CH_3(NH_2)_2^+$) (Koh et al. 2014, Lee et al. 2012, Kim et al. 2012) is a bulky and electropositive cation, and M = (Pb^{2+}; Sn^{2+}; Ge^{2+}) (Lee et al. 2012, Hao et al. 2014) is a smaller cation, belonging to the transition metals or an element of the p block; X = (Cl^-; Br^-; I^-) (Hendon et al. 2015, Heo et al. 2014, Jiang et al. 2015, Nagane et al. 2014) is a halide, cation A is surrounded by 12 anions forming a cube-octahedron while cation M is located in the center of an octahedron; X anions are coordinated into two M and four A cations. The layout of the MX_6 octahedral unit plays a crucial role in determining the band-gap as well as the transport properties in this type of materials.

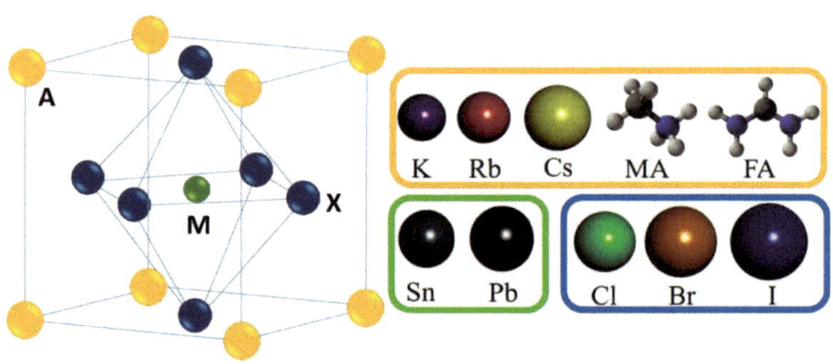

Figure 2.1: Perovskite ideal structure, AMX_3. Reported high-performance cations (in the blue box), metals (black box), and anions (green box) in an AMX_3 perovskite structure (right side) (Saliba 2019).

Symmetry and phase

The perovskites have a cubic geometry that is its identity. The distortion in geometry from an ideal structure plays a crucial role to tune the physical properties of the hybrid perovskites. As mentioned earlier, the M-X unit in AMX_3 type perovskite material plays a very important role in determining the electronic properties of materials. However, the A ion is almost neutral to such aspects, whereas the size of A can distort the M-X bond thus distorting the symmetry as well. In an ideal case, perovskite can be found in the highest symmetrized cubic phase associated with a pm3m space group. The crystal structure is connected with a coordination number of 12 and 6 for the cations A and M, respectively, with MX_6 octahedral residing at the corner.

The cubic symmetry offers the optimum electronic properties because of the high degree of ionic bonding, and thus it means that deviation from the cubic symmetry can alter the properties of the material as expected of which the first notable influence is the size effect. The geometrical relationship for the ideal cubic geometry of perovskites is as follows:

$$a = (r_A + r_X) = \sqrt{2(r_M + r_X)}$$

where 'a' is called the cell axis, r_A, r_M, r_X are the radii of A cation, M cation and X anion. The distortion and stability of perovskites can explained based on the ratio between two cell lengths, called the goldsmith tolerance (t) factor with formula (Correa-Baena et al. 2017).

$$t = \frac{R_A + R_X}{\sqrt{2(R_M + R_X)}}$$

where R_S are the ionic radii and the suffix denotes the corresponding ion (i.e., A, M, X, etc.). It is seen that cubic perovskite can exist in range of $0.89 < t < 1$. If the A ion is smaller than $t < 1$, resulting in a tilting of the octahedral geometry to fill the space giving orthorhombic and rhombohedra variants. Conditions, when there is mismatching between the ionic radii crystal symmetry, can be changed that result in a drastic change in the symmetry. For $t < 1$, the M-X and A-X bonds, respectively, undergo compression and tension, leading to rotation of octahedral in order to accommodate this induced stress. This will cause distortion and tilting of MX_6 octahedral breaking the symmetry. On the other hand, for $t > 1$ (that may occur as a result of large R_A or small R_M), more stabilized hexagonal structure is the outcome.

Also depending upon the size of the cation, A can influence the dimensionality, i.e., the shape of the perovskites. For instance, monovalent A cations result in a three-dimensional framework, whereas replacement of this with a large A cations leads in the destruction of 3D network giving a reduced dimensionality of two or three correspondings to $t << 1$, Fig. 2.2.

Materials synthesis and thin film deposition techniques

Materials processed in solution (one of the principal means for developing perovskite structures), both inorganic lead to halide matrix and mobile (organic) cations, such as cesium, methylammonium (MA) or formamidinium (FA), they have inherently soft structures. In a hybrid perovskite, the inorganic and organic components are interdependent; thus, a change or deformation in the organic cations immediately results in subsequent effects or modifications in its inorganic counterpart (subnet).

Since the first development of the HOIP, there has also been a lot of interest in the possibility of processing perovskite materials using a large number of techniques ranging from sping-coating (Lee et al. 2012), sol-gel (Burschka et al. 2013), 2-step interdiffusion (Xiao et al. 2014), and vacuum assisted evaporation (Bi et al. 2016).

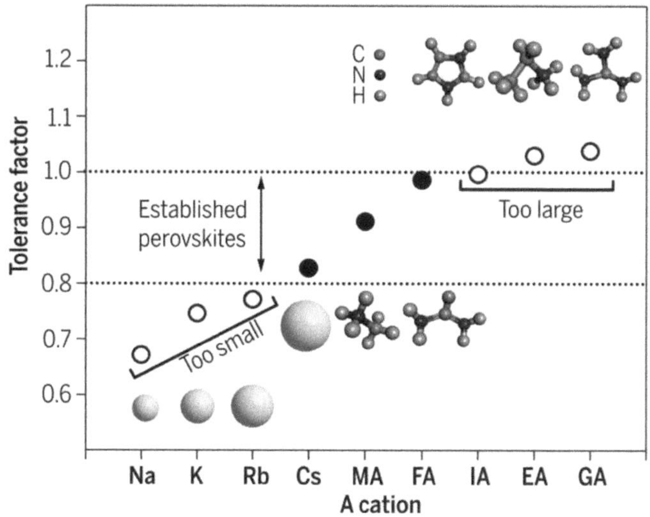

Figure 2.2: The tolerance factor of $APbI_3$ perovskite with A cations that are too small (Na, K, Rb), establishes (Cs, MA, FA) or too large [imidazolium (IA), ethylamine (EA), guanidinium (GA)]. The inset images depict the cation structures. Empirically, perovskites with a tolerance factor between 0.8 and 1.0 (dotted lines) show a photoactive black phase (solid circles) as opposed to non-photoactive phases (open circles). Rb is very close to this limit, making it a candidate for modification of the perovskite lattice by a multiplication approach (Saliba et al. 2016). Reprinted with permission from AAAS.

These techniques are compatible at low temperature (< 100°C) and can be solution-based, producing high quality semiconductors with a sharp band edge, shallow trap states, long diffusion length of electron-hole pairs, and an E_g that can be continuously tuned at 1–3 eV by exchange of cations, metals or halides (Saliba et al. 2016). The unusual behavior of perovskites, show a dramatically large diffusion length and therefore low recombination (Turren-Cruz et al. 2018a), which are believed to be the main reason for the success of this class of materials due to the implication of such smooth dynamic behavior and various hypotheses have been formulated to explain them. Actually, a hybrid perovskite currently has efficiencies greater than 20% efficiency (Jiang et al. 2019), and long-term stability greater than 1000 hours in constant light (Turren-Cruz et al. 2018b, Xu et al. 2020).

The commonly used techniques for hybrid perovskite are a solution-processed method that has proved to be one of the most effective in synthesizing crystalline organometal halide perovskites, where the precipitation of both metal halide, as well as organic halides, take place by evaporation of corresponding solution through heating as well as spin coating (Lee et al. 2012). This technique implements different types of solvents. N, N DMF and dimethylsulfoxide (DMSO) that are most frequently used solvents to dissolve the perovskite precursors. There are some reports of producing inorganic/organic perovskite at room temperature by dissolving the organic component as well as the inorganic part that include appropriate solvents or integrating inorganic cations like Rb (Turren-Cruz et al. 2018a, Dualeh et al. 2014). The drop-casting method using different antisolvents like toluene, anisole, ethyl acetate and chlorobenzene is also a widely reported strategy to develop the high efficiency perovskite solar cells (Yavari et al. 2018, Al-Ashouri et al. 2019).

Besides a solution process, another effective way of synthesizing hybrid halide perovskite is the vapor-evaporation method. It has the advantage over the other, in the sense that it always produces a sample with higher crystallinity compared to the solution process. However, the method has limitations in terms of the yield and thus is not very successfully used in the industry scale. It was found that the precursor flux has remarkable effects on the composition of the sample as procured in this manner, one should mention that unlike the tetragonal phase shown by $CH_3NH_3PbI_3$ at room temperature the same material synthesized by the vapor-evaporation method shows a cubic phase. In addition, *in situ* XRD confirms that there are gaps between different phases. Following the other methods, many have also used single and sequential steps in evaporation to crystallinity to the as-synthesized perovskite.

The thermal evaporation both single source and co-evaporation have been used since the early 1990s for synthesizing hybrid perovskite with good crystallinity. Both approaches have their respective challenges. For the single-source evaporation one needs a reliable instrument, for the other precise control on the flow rate, i.e., flux of either precursor is needed, which is itself a challenge as the vapor pressure of the organic component is much higher compared to the inorganic one. There are separate reports of synthesizing mixed halide perovskite with a smooth surface, small grain size and large surface coverage at low temperature vapor-assisted solution process reported by (Chen et al. 2014, Liu et al. 2013). Furthermore it was notable that Liu et al. were able to synthesize $CH_3NH_3PbI_{3-x}Cl_x$ with good crystallinity by employing dual source vapor-deposition method.

Solar cells based on perovskite photocatalysis and their operation mechanism

The operational mechanisms of perovskite solar cells are very similar to the organic photovoltaics and dye-sensitized solar cells. Figure 2.3 represents the operation mechanism of the PSC, where the perovskite layer as the main photoactive layer, is sandwiched between charge-transporting layers and electrodes. The function of the PSC can be divided into three categories: photon absorption, charge transport and charge extraction. The perovskite layer in the PSC absorbs the sunlight and excitons are generated and further dissociated into free charge carriers (electron and holes) at the ultrafast time scale. Owing to the potential gradient across the solar cell, free electrons and holes are injected into Electron and Hole Transport Layers (ETL and HTL), respectively. Subsequently, the preferred energy level matching between transport layers and electrodes allows holes to migrate towards the anode, while the electron towards the cathode. These charges are further extracted into the external circuit to generate electricity.

Device Configurations: In solar cells, device configuration is one of the crucial factors to determine the cell performance and stability. Depending on the type of transporting layer configured to transmit the incident radiations, PSCs are mainly divided into two configurations: as regular (n-i-p) and inverted (p-i-n). These

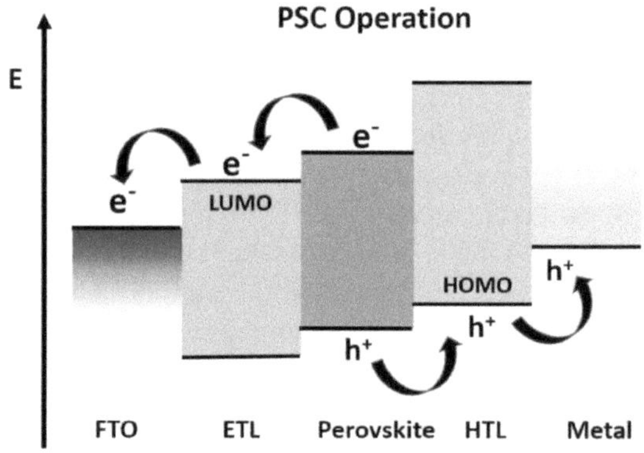

Figure 2.3: Operation principle of perovskite solar cells.

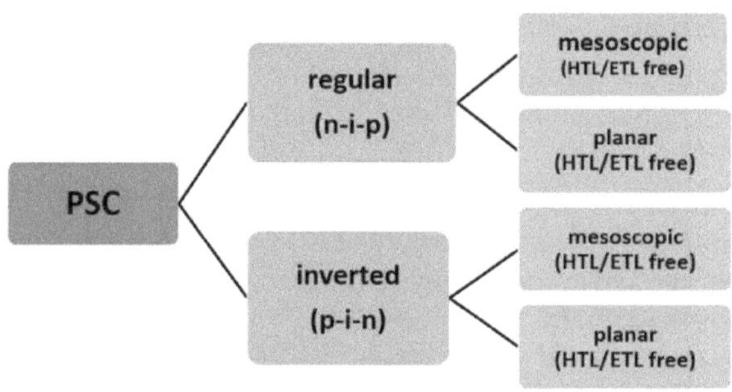

Figure 2.4: Different photovoltaic device structures based on the perovskite as an active layer.

structures are further divided into different categories to deliver HTL or ETL free configurations as summarized in Fig. 2.4.

Regular n-i-p structure: The conventional or regular n-i-p mesoscopic configuration for PSCs was adopted from traditional Dye-Sensitized Solar Cells (DSSC) by replacing the light-harvesting dye with a perovskite layer. This initial advancement of PSCs consist of the configuration as: glass/FTO (fluorine-doped tin oxide–transparent conducting oxide)/ETL (planer and mesoscopic titanium dioxide)/perovskite/HTL/Au as shown in Fig. 2.5. The planar configuration is an evolution of the mesoscopic device structure where the mesoscopic layer is filtered, and the perovskite layer is directly deposited on the planar TiO_2 followed by HTL. The high performance of PSCs in this configuration is still obtained by careful control of the perovskite composition, different layer's growth and interfaces across the device (Solanki et al. 2019a). Surprisingly, the planer PSC configuration delivers higher open-circuit voltage (V_{OC}), short-circuit current density (J_{SC}) compared to the mesoscopic configuration, at the cost of the current-voltage (*J-V*) hysteresis (Kim et al. 2015). However, in-depth investigation confirmed that *J-V* hysteresis not only depends on the various layers but also the voltage scan direction, rate and voltage range are other important parameters to control it (Chen et al. 2016). Additionally, the low-temperature processing of the planar structure is the main, unlike the mesoscopic configuration.

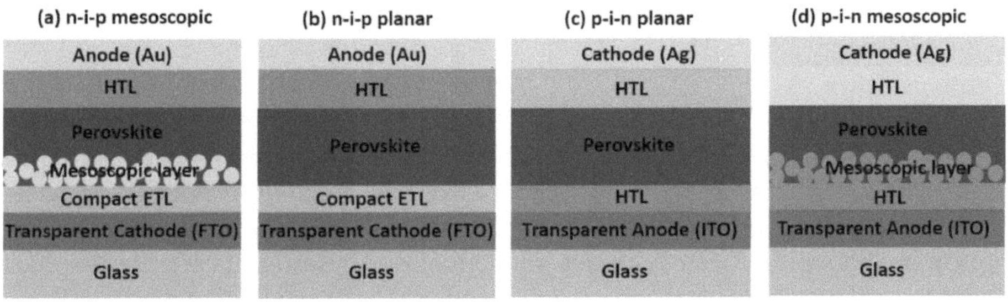

Figure 2.5: Device configuration of different perovskite solar cells (a) n-i-p mesoscopic, (b) n-i-p planar, (c) p-i-n planar, and (d) p-i-n mesoscopic.

Inverted p-i-n configuration: Like a regular structure, the p-i-n configuration in PSC is adopted from the organic photovoltaics (Jeng et al. 2013, Solanki et al. 2015, Solanki et al. 2016, Solanki et al. 2014), where HTL is in direct contact with the transparent electrode followed by perovskite layer and ETL. Based on previous literature, glass/ITO (transparent electrode)/PEDOT:PSS (HTL)/Perovskite/PCBM(ETL)/Ag is the widely used p-i-n structure. The p-i-n configuration was further expended to mesoscopic p-i-n device architecture by the selective organic layer replacement by oxide HTL. The planar p-i-n configuration always offers easy and low temperature processing along with negligible *J-V* hysteresis with competitive efficiency.

Besides the great success of the above-mentioned configurations, there are plenty of reports where a surface treatment or any salt/solvent additive in perovskite solution is used to replace either HTL or ETL to develop transport layer-free PSC. The absence of any charge transport layer reduces series resistance, the number of interfaces and therefore suppresses the recombination losses to deliver comparable or higher device performance. It is worth noting that these configurations consist of the other counterpart charge transport layer to balance the charge extraction.

Photoactive layer in perovskite solar cells: Since the first report in 2009, methylammonium lead iodide $MAPbI_3$ (band-gap ~ 1.50–1.61 eV) is a widely used light harvester in PSCs and crossed the benchmark of 20% Power Conversion Efficiency (PCE) (Kojima et al. 2009). It consists of a monovalent cation (methylammonium or $CH_3NH_3^+$ or MA^+) a divalent cation (lead or Pb^{2+}), and a halide ion (iodide or I^-). However, exposure to humidity causes this perovskite to decompose into their constituents and thus leads to the degradation of the device with poor performance (Solanki et al. 2019b). Many other monovalent cation-based perovskites such as $FAPbI_3$ (formamidinium lead iodide), $CsPbI_3$ (cesium lead iodide), etc., are explored for PSC, however their poor stability and inferior performance makes them inappropriate for further development.

Seok et al. developed a double or mixed cation-based halide perovskite to overcome the purity of the single cation and to deliver high efficiency with stronger environmental stability (Jeon et al. 2015). Double cation MA^+/FA^+ perovskites demonstrate that a small fraction of MA^+ with FA^+ in perovskites offers better crystallization of the photoactive black phases and consequently, superior thermal and structural stability than pure MA or FA perovskites. Later Saliba et al. introduced additional monovalent cations to develop triple (CsMAFA) (Saliba et al. 2016a) and quaternary (RbCsMAFA) (Solanki et al. 2019a, Saliba et al. 2016b) formulations and identified 5% cesium (Cs^+) and rubidium (Rb^+) to achieve efficiency more than 21% with > 500 hours operational stability. Beyond the monovalent cations, researchers also replaced lead partially or fully to reduce the toxicity at the cost of performance. Furthermore, mixed halide ($I^-/Cl^-/Br^-$) was also found to be a better composition for higher stability and efficiency than a single halide. These results opened the door of composition engineering to the use of a wide range of organic/inorganic cations, metal cations and halide anions to develop the efficient and stable solar cells.

Tandem solar cell

Regardless of the state-of-the art, none of the current single-junction photovoltaics can cross the Schockley-Queisser limit of 33%. The alternative is multi-junction or tandem (two junctions) solar cells: that consist of two or more solar cells connected in series or parallel to deliver the higher PCE. For efficient absorption of the sun spectrum, the top cell of the tandem structure is composed of a higher band-gap photoactive layer while a low band-gap photoactive layer is inserted in the bottom cell. In such a cell architecture, higher-energy photons from the incident sun spectrum are efficiently absorbed by the top cells, while lower-energy photons are transmitted and absorbed by bottom cells to maximize the energy harvesting from the solar spectrum. However, the materials availability, process applicability, cost and other factors limit the range of tandem solar cells. Due to the easier processability, hybrid perovskites have been used with various single junction solar cells such as: Perovskite/Perovskite, Perovskite/c-silicon, Perovskite/CIGS, Perovskite/DSSC, and Perovskite/polymer tandem solar cells. Perovskite as an active layer in tandem configuration (Perovskite/c-silicon) was reported for the first time in 2015 (Albrecht et al. 2016). Within the last five years, all these technologies underwent tremendous improvement from perovskite composition/processing to the device engineering and therefore PCEs have been reported recently with higher than 22% for 1 cm^2 active area in these tandem configurations (Jošt et al. 2020). However, moderate V_{OC}s, FFs and low values of current densities are the main challenges to approach the empirical limit of PCE. The optimization of the perovskite layer in the top cell to ensure the low recombination at the interface and energy level alignment are pathways to enhance the Voc and FF as well. While current density has to be improved from the low band-gap bottom cell by optical management to ensure the higher absorption and minimum photon loss. It is believed that it is possible to achieve the predicted PCEs of 34% in the near future by light coupling enhancement and suppression of the recombination loss.

Energy saving

Light-harvesting to generate electricity is extremely important, but energy saving by low power consumption and efficient energy storage are also equally important. Due to efficient electronic conductivity and unbelievable ion migration properties, perovskites have demonstrated their potential in other electronic devices such as light emission, high-density charge storage, photodetectors (Dou et al. 2014), supercapacitors (Molinari et al. 2017), teraherz photonic devices (Kumar et al. 2020), etc. The details of some of them are as follows:

Light Emission Diodes (LEDs): Light emission is the other best application of hybrid perovskites beyond light-harvesting where free charge carriers are injected in and radiatively recombine to emit the light. The inherent low defects density, higher radiative recombination probability and low power consumption have made these hybrid perovskites applicable for light emission as well. Within a span of a few years, these hybrid perovskite LEDs (Tan et al. 2014) have shown a record performance over 20% (Bao et al. 2020) External Quantum Efficiency (EQE). In LEDs, EQE, i.e., the ratio of the emitted photons to the number of the injected electrons into the device is the crucial parameter to determine the performance of LED (Van Le et al. 2018). The solution processability of perovskite and their outstanding photoluminescence quantum yield has made perovskite one of the best materials for light emission. Furthermore, low power consumption in addition to the flexibility of this class of materials has led perovskites to high color purity (Si et al. 2017) (intrinsic quantum well structured in mixed dimensional perovskites), band-gap tunability (Vashishtha et al. 2019) to cover the entire visible spectrum and moderate ionization potential for efficient and stable interfaces. A striking breakthrough was made when using solution-processed perovskite single-crystal nanowires that demonstrated a near 100% PLQE with a low threshold, high-quality factor and wide spectral tunability (Zhu et al. 2015). This highlighted the importance of controlling the crystallinity and structural defects of the perovskite materials in achieving high-performance light emitting devices. Keeping these considerations in mind, it seems that a divergence between PV and light-emitting research will be one of the scales of the perovskite crystal domains, where isolated nanocrystals in a host may prove to be more advantageous for LEDs or lasers, in contrast to large crystalline domains in homogeneous thin films for PV research. It is believed that the indepth understanding of the origin of non-radiative recombination losses and exploring the ways to suppress such pathways is the main bottleneck for further enhancing the performance of light emitting devices (Wolff et al. 2019).

Photodetectors: Photodetectors, a device to transform the optical signal into an electrical one, are one of the key semiconductor components of many fields such as remote sensing, electrocommunication, optical image sensor, environmental surveillance, etc. (Baeg et al. 2013, Konstantatos and Sargent 2010, García de Arquer et al. 2017, Konstantatos et al. 2006). Perovskite based photodetectors (PPDs) combine the advantages of perovskite semiconductor materials with superior optical and electronic properties and easier solution-processed manufacturing. Based on the device configuration and working mechanism, photodetectors are categorized as photodiodes (PDs), photoconductors (PCs) and phototransistors (PTs). Due to the capability to balance high gain and low dark current, PTs are the most popular devices compared to the other two. A material with high photo-electric conversion efficiency is the prime requirement for an active layer of photodetectors. The performance of any photodetector is evaluated by measuring the responsivity R, detectivity D, Linear Dynamic Range (LDR) and response time (rise/decay time). Till date, various materials such as organic molecules (Zhang et al. 2018), quantum dots (Bi et al. 2019), graphene (Xia et al. 2009), carbon nanotubes (Li et al. 2017), black phosphorous (Na et al. 2014), etc., have been used as the active layer in photodetectors. Xie and co-workers proposed the first organic–inorganic hybrid perovskite photoconductor, the active layer was processed on a flexible polyethylene terephthalate (PET) substrate patterned with Indium Tin Oxide (ITO) electrodes (Hu et al. 2014). Besides the great improvement across the different layers, more research efforts are required to develop the indepth understanding and resolving the aforementioned bottlenecks to promote the performance of future devices.

Alongwith the above-mentioned applications, hybrid perovskites have also demonstrated their potential by saving energy by low power consumption in different areas such as lasers, transistors, scintillators, etc., and energy storage devices such as supercapacitors and batteries (Zhang et al. 2019, Chen et al. 2015, Kostopoulou et al. 2018).

Challenges in perovskite and solutions

However two main challenges remain to overcome the commercilizations of these technologies, for the outstanding success of the perovskites in light harvesting, light emission and other electronic application. These are stability and toxicity of hybrid perovskites.

Stability: While PCE at the laboratory scale has achieved the landmark of 25% (Park 2020), PSCs must be insusceptible to outdoors under operating conditions such as: real sun radiation, rising temperature, environmental moisture and oxygen for more than 20 years to compete the traditional Si technology. Long-term stability of perovskite includes both extrinsic (owing to temperature and humidity) and intrinsic (owing to lattice distortion) caused by different environmental operating conditions and materials deterioration needs to be urgently resolved for commercialization. Goldschmidt tolerance factor (τ), used to determine the various crystal structure formation of perovskite compounds ABX_3 is also used to judge their structural stability (Bartel et al. 2019). A value of τ in the rage of $0.8 < \tau < 0.9$ represents the ideal cubic, while the value in the range $0.9 < \tau < 1$ represents a distorted cubic perovskite structure. However, the value of $\tau < 0.8$ and > 1 represents the non-perovskite structure. Therefore, it is recommended to maintain τ value between 0.8 and 1 to form the stable perovskite structure. For $FAPbI_3$, τ is close to 1, which is near the upper boundary for perovskite structure, and therefore, $FAPbI_3$ is prone to the formation of a hexagonal δ-phase which is photoinactive (Fan et al. 2019). $CsPbI_3$ with $\tau \approx 0.8$, is at the edge of lower boundary for perovskite structures, and normally crystallizes into a δ-phase. $MAPbI_3$ with $\tau \approx 0.9$ is close to the middle of the perovskite zone and forms a black photoactive perovskite phase.

Additionally, the stability of perovskite and consecutive layers against outdoor conditions such as real sun radiation, heat, moisture, oxygen, etc., are equally important. The vulnerability of perovskite against moisture is very well known where water molecules easily interact with the structure and break into its constituents as follows (Solanki et al. 2019b, Huang et al. 2017, Chen et al. 2019):

$$4CH_3NH_3PbI_3 + 4H_2O \leftrightarrow 4\{CH_3NH_3PbI_3.H_2O\}$$
$$\leftrightarrow (CH_3NH_3)_4PbI_6.2H_2O + 3PbI_2 + 2H_2O$$
$$(CH_3NH_3)_4PbI_6.2H_2O \rightarrow 4CH_3NH_3I + PbI_2 + 2H_2O$$

Therefore, degradation of perovskite highlights the importance of the sealing to protect them from the external environment or development of a new class of stable hybrid perovskites.

Till date, numerous approaches such as the compositional engineering, solvent additive (Solanki et al. 2020), variation in processability, etc., are reported to enhance the photovoltaic stability. In the last few years, Ruddlesden Popper (RP) perovskite has captured the attention in the field of photovoltaics (Tsai et al. 2016), light emission, memristor due to the structurally stable compound (Gao et al. 2019, Zhang et al. 2019). These perovskites have a general formula $(RNH_3)_2A_{\bar{n}-1}M_{\bar{n}}X_{3\bar{n}+1}$ ($\bar{n}$ = 1, 2, 3, 4……), where RNH_3 is a large organic ammonium insulating spacer cation, and A represents a monovalent organic cation (MA, FA, etc.). M is a divalent metal cation (Pb, Sn, etc.), while X represents a halide anion (I, Br, Cl), and $\bar{n}$ is the number of $[MX_6]^{4+}$ octahedral layers within each organic spacer. These perovskites are also known as layered or mixed dimensional or low dimension perovskites. RP perovskites have shown strong resistance to environmental degradation, owing to the unique quantum well structured and hydrophobicity by the insertion of bulky organic ammonium spacers. Furthermore, tunable exciton binding energy, out-of-plane oriented growth and phase distribution of these perovskites make it attractive. The continuous efforts and materials engineering has improved the PCEs > 18% (Ren et al. 2020) in RP perovskite-based PSC, comparable to bulk perovskite while LED quantum efficiency has also crossed the barrier of 20% (Bao et al. 2020) with long-term environmental stability.

Toxicity: Other than the stability of PSCs, lead toxicity is another major concern for commercialization. Although 'lead' based perovskite direct the PCE, its toxicity may cause irreversible health effects: interfering with several body functions, primarily affecting the central nervous, hematopoietic, hepatic and renal system producing serious disorders. Therefore, motivation to suppress or eliminate the lead toxicity has paved a research path towards less-lead and lead-free perovskites. Based on the Goldschmidt tolerance factor, a wide range of the metal cations are available to replace lead in perovskites. Specifically, group 14 elements: Sn, Ge, Bi, alkaline earth metals: Be, Mg, Ca, Ba, Sr, transition metals: V, Mn, Fe, Co, Ni Cu, Zn, lanthanides: Eu, Tm, Yb and many more are the strong alternative for lead-free perovskites. However, only selected elements, i.e., Sn^{2+}, Ge^{2+}, Mg^{2+}, Mn^{2+}, Ni^{2+}, and Co^{2+} are considered promising candidates based on solar cell properties and their stability.

Due to the similar ionic radii of lead (1.49 Å), tin (Sn) (1.35 Å) is the most popular element to explore lead-free perovskites. Importantly, higher charge carrier mobility (10^2–10^3 $cm^2/V{\cdot}s$) and energy band-gap (~ 1.6 eV) in Sn-based perovskites are analogous to their counterpart lead (Konstantakou and Stergiopoulos 2017). Sn-based perovskites exhibit binding energies (2–50 meV) similar to that of Pb perovskites, owing to the exceptionally low effective masses. Additionally, their excellent optoelectronic properties such as narrow optical band-gap for a wider absorption range, higher conductivity (~ 10^{-2} S cm^{-1}), mobility (~ 10^{3} cm^{2} V^{-1} s^{-1}) and extended charge carrier diffusion lengths, superior to traditional inorganic semiconductors, e.g., Si, CdTe, etc., make them attractive. However, environment stability is the biggest challenge due to the easier oxidation of Sn^{2+} to Sn^{4+} on exposure to the outdoor conditions (Konstantakou and Stergiopoulos 2017). Furthermore, the rapid crystallization nature of Sn perovskites leads to non uniform coverage, that results in large number of pinholes and thus reduce the PCE significantly.

Germanium is another popular element from group 14. However, its oxidation process, smaller ionic radii, wide energy gap, poor solubility, horrendous morphology and very low PCE have led to their rejection for photovoltaic applications. Till date, multiple perovskites were synthesized for PSC though $CH_3NH_3GeI_3$ and $FAGeI_3$ are the best-reported compositions for photovoltaics. Bismuth is another contender from group 15 elements for lead-free perovskites. Bi^{3+} possesses analogous properties such as electronic configuration, electronegativity and ionic radii of divalent lead ions. Furthermore, strong environmental stability, less/non-toxicity and easier processing have captured the attention of bismuth-based lead-perovskites. Methyl-ammonium Bismuth Iodide (MBI) is the most popular perovskite in this category which yielded 1.64% owing to the indirect band-gap (Zhang et al. 2017).

It is very challenging to replace lead and concurrently maintain the high PCE and stability of PSCs. Mixing of lead with another non-toxic element has captured the attention of researchers where the partial fraction of lead is replaced by group 14 elements. Till date, PCE close to 20% has been reported using Sn-Pb based perovskite (Wei et al. 2020), and further developments are required to compete for Pb-based perovskite solar cells.

Summary and future perspectives

Owing to the excellent light absorbing properties, simplicity of processing and wide tunability, hybrid perovskites have attracted great interest in light harvesting photovoltaics. Hybrid perovskites have emerged as strong competitors of conventional in-organic photovoltaics with impressive PCE. Nevertheless, perovskite solar cells may well be the best-performing low-cost photovoltaic technology. Even though very high efficiencies have been reported, the stability of these devices is still a matter of concern; hence the commercialization of PSCs will be only possible when instability under ambient conditions and toxicity of the perovskite solar cells are solved. These shortcomings have challenged researchers to suppress the intrinsic degradation and toxicity by developing new compositions, and also explore the encapsulation strategy to protect them from oxygen and moisture exposure under outdoor operation conditions.

The main prospects for the future cover strategies on perovskite stabilization and the suppression of toxicity, chemistry work on chemical composition has great value. In addition, exposing the charge transport properties and improving interfacial engineering methods in device fabrication are important to enhance the entire device stability. In this perspective, researchers should use iodine and lead free perovskite for portable photovoltaics with desirable optoelectronic properties. Energy saving is the other area of energy harvesting, where hybrid perovskites are also a playing pivotal role. Therefore, the above mentioned approaches can also be adopted in other optical and electrical devices such as LED, laser, memory, detector, etc.

References

Albrecht, S., M. Saliba, J. P. Correa Baena, F. Lang, L. Kegelmann, M. Mews et al. 2016. Monolithic perovskite/silicon-heterojunction tandem solar cells processed at low temperature. Energy & Environmental Science 9(1): 81–88.

Al-Ashouri, A., A. Magomedov, M. Ross, M. Jost, M. Talaikis, G. Chistiakova et al. 2019. Conformal monolayer contacts with lossless interfaces for perovskite single junction and monolithic tandem solar cells. Energy & Environmental Science 12(11): 3356–3369.

Ankur, S. 2017. Morphology dependent photophysics in bulk heterojunction organic solar cells. Nanyang Technological University, Singapore.

Baeg, K. -J., M. Binda, D. Natali, M. Caironi and Y. -Y. Noh. 2013. Organic light detectors: photodiodes and phototransistors. Advanced Materials 25(31): 4267–4295.

Baena, J. P., W. R. Tress, A. Abate, A. Hagfeldt and M. Gratzel. 2016. Incorporation of rubidium cations into perovskite solar cells improves photovoltaic performance. Science 354(6309): 206–209.

Bao, C., W. Xu, J. Yang, S. Bai, P. Teng, Y. Yang et al. 2020. Bidirectional optical signal transmission between two identical devices using perovskite diodes. Nature Electronics 3(3): 156–164.

Bartel, C. J., C. Sutton, B. R. Goldsmith, R. Ouyang, C. B. Musgrave, L. M. Ghiringhelli et al. 2019. New tolerance factor to predict the stability of perovskite oxides and halides. Science Advances 5(2): eaav0693.

Bi, D. Q., C. Y. Yi, J. S. Luo, J. D. Decoppet, F. Zhang, S. M. Zakeeruddin et al. 2016. Polymer-templated nucleation and crystal growth of perovskite films for solar cells with efficiency greater than 21%. Nature Energy 1(10): 16142.

Burschka, J., N. Pellet, S. J. Moon, R. Humphry-Baker, P. Gao, M. K. Nazeeruddin et al. 2013. Sequential deposition as a route to high-performance perovskite-sensitized solar cells. Nature 499(7458): 316–9.

Chen, B., M. Yang, S. Priya and K. Zhu. 2016. Origin of J–V hysteresis in perovskite solar cells. The Journal of Physical Chemistry Letters 7(5): 905–917.

Chen, Q., H. Zhou, Z. Hong, S. Luo, H. S. Duan, H. H. Wang et al. 2014. Planar heterojunction perovskite solar cells via vapor-assisted solution process. Journal of American Chemical Society 136(2): 622–5.

Chen, Q., N. De Marco, Y. Yang, T. -B. Song, C. -C. Chen, H. Zhao et al. 2015. Under the spotlight: The organic–inorganic hybrid halide perovskite for optoelectronic applications. Nano Today 10(3): 355–396.

Chen, S., A. Solanki, J. Pan and T. C. Sum. 2019. Compositional and morphological changes in water-induced early-stage degradation in lead halide perovskites. Coatings 9(9): 535.

Correa-Baena, J. P., M. Saliba, T. Buonassisi, M. Gratzel, A. Abate, W. Tress et al. 2017. Promises and challenges of perovskite solar cells. Science 358(6364): 739–744.

Correa-Baena, J. P. 2018a. Enhanced charge carrier mobility and lifetime suppress hysteresis and improve efficiency in planar perovskite solar cells. Energy & Environmental Science 11(1): 78–86.

Dou, L., Y. Yang, J. You, Z. Hong, W. -H. Chang, G. Li et al. 2014. Solution-processed hybrid perovskite photodetectors with high detectivity. Nature Communications 5(1): 5404.

Dualeh, A., N. Tetreault, T. Moehl, P. Gao, M. K. Nazeeruddin and M. Gratzel. 2014. Effect of annealing temperature on film morphology of organic-inorganic hybrid pervoskite solid-state solar cells. Advanced Functional Materials 24(21): 3250–3258.

Fan, Y., H. Meng, L. Wang and S. Pang. 2019. Review of stability enhancement for formamidinium-based perovskites. Solar RRL 3(9): 1900215.

Fujishima, A. and K. Honda. 1972. Electrochemical photolysis of water at a semiconductor electrode. Nature 238(5358): 37–38.

García de Arquer, F. P., A. Armin, P. Meredith and E. H. Sargent. 2017. Solution-processed semiconductors for next-generation photodetectors. Nature Reviews Materials 2(3): 16100.

Gao, X., X. Zhang, W. Yin, H. Wang, Y. Hu, Q. Zhang et al. 2019. Ruddlesden–Popper perovskites: synthesis and optical properties for optoelectronic applications. Advanced Science 6(22): 1900941.

Ge, M., Q. Li, C. Cao, J. Huang, S. Li, S. Zhang et al. 2017. One-dimensional TiO_2 nanotube photocatalysts for solar water splitting. Advanced Science 4(1): 1600152.

Hao, F., C. C. Stoumpos, D. H. Cao, R. P. H. Chang and M. G. Kanatzidis. 2014. Lead-free solid-state organic-inorganic halide perovskite solar cells. Nature Photonics 8(6): 489–494.

He, R. a., S. Cao, P. Zhou and J. Yu. 2014. Recent advances in visible light Bi-based photocatalysts. Chinese Journal of Catalysis 35(7): 989–1007.

Hendon, C. H., R. X. Yang, L. A. Burton and A. Walsh. 2015. Assessment of polyanion (BF4- and PF6-) substitutions in hybrid halide perovskites. Journal of Materials Chemistry A 3(17): 9067–9070.

Heo, J. H., D. H. Song and S. H. Im. 2014. Planar $CH_3NH_3PbBr_3$ hybrid solar cells with 10.4% power conversion efficiency, fabricated by controlled crystallization in the spin-coating process. Advanced Materials 26(48): 8179–83.

Hisatomi, T., J. Kubota and K. Domen. 2014. Recent advances in semiconductors for photocatalytic and photoelectrochemical water splitting. Chemical Society Reviews 43(22): 7520–7535.

Hu, X., X. Zhang, L. Liang, J. Bao, S. Li, W. Yang et al. 2014. High-performance flexible broadband photodetector based on organolead halide perovskite. Advanced Functional Materials 24(46): 7373–7380.

Huang, J., S. Tan, P. D. Lund and H. Zhou. 2017. Impact of H_2O on organic–inorganic hybrid perovskite solar cells. Energy & Environmental Science 10(11): 2284–2311.

Jeon, N. J., J. H. Noh, W. S. Yang, Y. C. Kim, S. Ryu, J. Seo et al. 2015. Compositional engineering of perovskite materials for high-performance solar cells. Nature 517(7535): 476–480.

Jiang, Q., D. Rebollar, J. Gong, E. L. Piacentino, C. Zheng and T. Xu. 2015. Corrigendum: pseudohalide-induced moisture-tolerance in perovskite $CH_3NH_3Pb(SCN)_2$ I thin films. Angewandte Chemie International Edition 54(38): 11006.

Jiang, Q., Y. Zhao, X. Zhang, X. Yang, Y. Chen, Z. Chu et al. 2019. Surface passivation of perovskite film for efficient solar cells. Nature Photonics 13(7): 460–466.

Jošt, M., L. Kegelmann, L. Korte and S. Albrecht. 2020. Monolithic perovskite tandem solar cells: a review of the present status and advanced characterization methods toward 30% efficiency. Advanced Energy Materials n/a(n/a): 1904102.

Kim, H. S., C. R. Lee, J. H. Im, K. B. Lee, T. Moehl, A. Marchioro et al. 2012. Lead iodide perovskite sensitized all-solid-state submicron thin film mesoscopic solar cell with efficiency exceeding 9%. Science Report 2: 591.

Kim, D., K. K. Sakimoto, D. Hong and P. Yang. 2015. Artificial photosynthesis for sustainable fuel and chemical production. Angewandte Chemie International Edition 54(11): 3259–3266.

Koh, T. M., K. W. Fu, Y. N. Fang, S. Chen, T. C. Sum, N. Mathews et al. 2014. Formamidinium-containing metal-halide: an alternative material for near-IR absorption perovskite solar cells. Journal of Physical Chemistry C 118(30): 16458–16462.

Kojima, A., K. Teshima, Y. Shirai and T. Miyasaka. 2009. Organometal halide perovskites as visible-light sensitizers for photovoltaic cells. Journal of the American Chemical Society 131(17): 6050–6051.

Konstantatos, G., I. Howard, A. Fischer, S. Hoogland, J. Clifford, E. Klem et al. 2006. Ultrasensitive solution-cast quantum dot photodetectors. Nature 442(7099): 180–183.

Konstantatos, G. and E. H. Sargent. 2010. Nanostructured materials for photon detection. Nature Nanotechnology 5(6): 391–400.

Kostopoulou, A., E. Kymakis and E. Stratakis. 2018. Perovskite nanostructures for photovoltaic and energy storage devices. Journal of Materials Chemistry A 6(21): 9765–9798.

Kumar, A., A. Solanki, M. Manjappa, S. Ramesh, Y. K. Srivastava, P. Agarwal et al. 2020. Excitons in 2D perovskites for ultrafast terahertz photonic devices. Science Advances 6(8): eaax8821.

Lee, M. M., J. Teuscher, T. Miyasaka, T. N. Murakami and H. J. Snaith. 2012. Efficient hybrid solar cells based on meso-superstructured organometal halide perovskites. Science 338(6107): 643–7.

Li, X., D. Yu, J. Chen, Y. Wang, F. Cao, Y. Wei et al. 2017. Constructing fast carrier tracks into flexible perovskite photodetectors to greatly improve responsivity. ACS Nano 11(2): 2015–2023.

Liu, M., M. B. Johnston and H. J. Snaith. 2013. Efficient planar heterojunction perovskite solar cells by vapour deposition. Nature 501(7467): 395–8.

Luther, J. M., Z. C. Holman and M. D. McGehee. 2020. Triple-halide wide-band gap perovskites with suppressed phase segregation for efficient tandems. Science 367(6482): 1097–1104.
Molinari, A., P. M. Leufke, C. Reitz, S. Dasgupta, R. Witte, R. Kruk et al. 2017. Hybrid supercapacitors for reversible control of magnetism. Nature Communications 8(1): 15339.
Na, J., Y. T. Lee, J. A. Lim, D. K. Hwang, G. -T. Kim, W. K. Choi et al. 2014. Few-layer black phosphorus field-effect transistors with reduced current fluctuation. ACS Nano 8(11): 11753–11762.
Nagane, S., U. Bansode, O. Game, S. Chhatre and S. Ogale. 2014. $CH_{(3)}NH_{(3)}PbI_{(3-x)}(BF_{(4)})x$: molecular ion substituted hybrid perovskite. Chemical Communication (Camb.) 50(68): 9741–4.
Park, N. -G. 2020. Research direction toward scalable, stable, and high efficiency perovskite solar cells. Advanced Energy Materials 10(13): 1903106.
Ren, H., S. Yu, L. Chao, Y. Xia, Y. Sun, S. Zuo et al. 2020. Efficient and stable Ruddlesden–Popper perovskite solar cell with tailored interlayer molecular interaction. Nature Photonics 14(3): 154–163.
Saliba, M., T. Matsui, K. Domanski, J. Y. Seo, A. Ummadisingu, S. M. Zakeeruddin et al. 2016. Incorporation of rubidium cations into peroskite solar cells improves photovoltaic performance. Science 354(6309): 206–209.
Saliba, M., T. Matsui, J. -Y. Seo, K. Domanski, J. -P. Correa-Baena, M. K. Nazeeruddin et al. 2016a. Cesium-containing triple cation perovskite solar cells: improved stability, reproducibility and high efficiency. Energy & Environmental Science 9(6): 1989–1997.
Saliba, M., T. Matsui, K. Domanski, J. -Y. Seo, A. Ummadisingu, S. M. Zakeeruddin et al. 2016b. Incorporation of rubidium cations into perovskite solar cells improves photovoltaic performance. Science 354(6309): 206–209.
Saliba, M. 2019. Polyelemental, multicomponent perovskite semiconductor libraries through combinatorial screening. Advanced Energy Materials 9(25).
Si, J., Y. Liu, Z. He, H. Du, K. Du, D. Chen et al. 2017. Efficient and high-color-purity light-emitting diodes based on *in situ* grown films of CsPbX3 (X = Br, I) nanoplates with controlled thicknesses. ACS Nano 11(11): 11100–11107.
Solanki, A., B. Wu, T. Salim, E. K. L. Yeow, Y. M. Lam and T. C. Sum. 2014. Performance improvements in polymer nanofiber/fullerene solar cells with external electric field treatment. The Journal of Physical Chemistry C 118(21): 11285–11291.
Solanki, A., B. Wu, T. Salim, Y. M. Lam and T. C. Sum. 2015. Correlation between blend morphology and recombination dynamics in additive-added P3HT:PCBM solar cells. Physical Chemistry Chemical Physics 17(39): 26111–26120.
Solanki, A., A. Bagui, G. Long, B. Wu, T. Salim, Y. Chen et al. 2016. Effectiveness of external electric field treatment of conjugated polymers in bulk-heterojunction solar cells. ACS Applied Materials & Interfaces 8(47): 32282–32291.
Solanki, A., P. Yadav, S. -H. Turren-Cruz, S. S. Lim, M. Saliba and T. C. Sum. 2019a. Cation influence on carrier dynamics in perovskite solar cells. Nano Energy 58: 604–611.
Solanki, A., S. S. Lim, S. Mhaisalkar and T. C. Sum. 2019b. Role of water in suppressing recombination pathways in $CH_3NH_3PbI_3$ perovskite solar cells. ACS Applied Materials & Interfaces 11(28): 25474–25482.
Solanki, A., M. M. Tavakoli, Q. Xu, S. S. H. Dintakurti, S. S. Lim, A. Bagui et al. 2020. Heavy water additive in formamidinium: a novel approach to enhance perovskite solar cell efficiency. Advanced Materials 32(23): 1907864.
Tan, Z. K., R. S. Moghaddam, M. L. Lai, P. Docampo, R. Higler, F. Deschler et al. 2014. Bright light-emitting diodes based on organometal halide perovskite. Nature Nanotechnology 9(9): 687–92.
Tsai, H., W. Nie, J. -C. Blancon, C. C. Stoumpos, R. Asadpour, B. Harutyunyan et al. 2016. High-efficiency two-dimensional Ruddlesden–Popper perovskite solar cells. Nature 536(7616): 312–316.
Turren-Cruz, S. H., M. Saliba, M. T. Mayer, H. Juarez-Santiesteban, X. Mathew, L. Nienhaus et al. 2018a. Enhanced charge carrier mobility and lifetime suppress hysteresis and improve efficiency in planar perovskite solar cells. Energy & Environmental Science 11(1): 78–86.
Turren-Cruz, S. H., A. Hagfeldt and M. Saliba. 2018b. Methylammonium-free, high-performance, and stable perovskite solar cells on a planar architecture. Science 362(6413): 449–453.
Van Le, Q., H. W. Jang and S. Y. Kim. 2018. Recent advances toward high-efficiency halide perovskite light-emitting diodes: review and perspective. Small Methods 2(10): 1700419.

Vashishtha, P., M. Ng, S. B. Shivarudraiah and J. E. Halpert. 2019. High efficiency blue and green light-emitting diodes using ruddlesden–popper inorganic mixed halide perovskites with butylammonium interlayers. Chemistry of Materials 31(1): 83–89.

Wei, M., K. Xiao, G. Walters, R. Lin, Y. Zhao, M. I. Saidaminov et al. 2020. Combining efficiency and stability in mixed tin–lead perovskite solar cells by capping grains with an ultrathin 2D layer. Advanced Materials 32(12): 1907058.

Wolff, C. M., P. Caprioglio, M. Stolterfoht and D. Neher. 2019. Nonradiative recombination in perovskite solar cells: the role of interfaces. Advanced Materials 31(52): 1902762.

Xia, F., T. Mueller, Y. M. Lin, A. Valdes-Garcia and P. Avouris. 2009. Ultrafast graphene photodetector. Nature Nanotechnology 4(12): 839–43.

Xu, J., C. C. Boyd, Z. J. Yu, A. F. Palmstrom, D. J. Witter, B. W. Larson et al. 2020. Triple-halide wide-band gap perovskites with suppressed phase segregation for efficient tandems. Science 367(6482): 1097–1104.

Yavari, M., M. Mazloum-Ardakani, S. Gholipour, M. M. Tavakoli, S. H. Turren-Cruz, N. Taghavinia et al. 2018. Greener, nonhalogenated solvent systems for highly efficient perovskite solar cells. Advanced Energy Materials 8(21).

Zhang, Z., X. Li, X. Xia, Z. Wang, Z. Huang, B. Lei et al. 2017. High-quality $(CH_3NH_3)_3Bi_2I_9$ film-based solar cells: pushing efficiency up to 1.64%. The Journal of Physical Chemistry Letters 8(17): 4300–4307.

Zhang, J., J. Jin, H. Xu, Q. Zhang and W. Huang. 2018. Recent progress on organic donor–acceptor complexes as active elements in organic field-effect transistors. Journal of Materials Chemistry C 6(14): 3485–3498.

Zhang, N., Y. Fan, K. Wang, Z. Gu, Y. Wang, L. Ge et al. 2019. All-optical control of lead halide perovskite microlasers. Nature Communications 10(1): 1770.

Zhu, H., Y. Fu, F. Meng, X. Wu, Z. Gong, Q. Ding et al. 2015. Lead halide perovskite nanowire lasers with low lasing thresholds and high quality factors. Nature Materials 14(6): 636–642.

CHAPTER 3

Porphyrins Based Nanostructured Material for the Conversion of CO_2 into Value Added Products

Santosh Kumar,[1,2,*] *Joonseok Koh,*[2] *Rohit Srivastava*[3] and *Dinesh Kumar Gupta*[4]

Introduction

Anthropogenic carbon dioxide is a primary greenhouse gas that is responsible for global warming, rising sea levels and increasing acidity of the oceans. The conversion of atmospheric carbon dioxide (CO_2) is still an open challenge to society. CO_2 storage has been extensively studied using a wide range of porous materials and it is now important to convert the captured CO_2 to value-added chemicals (Haszeldine 2009). CO_2 cannot compensate for emission-based CO_2, but this strategy provides access to high value-added products at potentially non-toxic, and at a low cost. In this chapter, the importance of studying and investigating porphyrins based nanostructured materials as catalysts are highlighted. It can act as a catalyst to convert CO_2 into useful and value added products with ease and selectivity.

The porphyrin nucleus consists of four pyrrole rings connected by four methine bridges to create an almost planar macrocycle based on an 18π-electron conjugated network. Porphyrins and their derivatives, decorated with self-assembly motifs, have been widely used as molecular scaffolds for the construction of well-defined organic nanostructures. Porphyrins in the skeleton can bind to various metal ions to form a heterogeneous metal-porphyrin catalyst (Kumar et al. 2015, Younus Wani

[1] Department of Chemistry, University of Coimbra, 3004-535 Coimbra, Portugal.
[2] Division of Chemical Engineering, Konkuk University, Seoul, 05029, South Korea.
[3] Pandit Deendayal Petroleum University, Gandhinagar, India.
[4] Department of Chemistry, U.P. Rajarshi Tandon Open University, Prayagraj, India.
* Corresponding author: santoshics@gmail.com

et al. 2015). These can be easily incorporated into the organic backbone by forming interconnects with rigid units and forming porous organic polymers due to the unique macrocyclic structure of porphyrin. A review on porphyrins as nanoreactor in CO_2 capture and conversion in 2015 (Kumar et al. 2015) has also been reported.

The CO_2 molecules can interact simultaneously with multiple pore surfaces in porphyrins. Several recent reviews have been reported on CO_2 capture or sequestration in Covalent-Organic Frameworks (COFs), Metal-Organic Frameworks (MOFs) and Porous Organic Polymers (POPs). This chapter gives an overview of fundamental aspects and recent research advances of porphyrins based nanostructured catalyst for CO_2 conversion systems into value added products.

Porphyrins as nanostructured materials

Nanostructured materials, as promising upcoming materials that have been used to solve numerous challenges faced by researchers worldwide. A nanostructured material is nanosized, which provides great advantages compared to the macroporous materials. The porphyrin based COF nanostructured materials are particularly exciting due to their various intrinsic unique characteristics such as predesignable periodical structures with atomic precision, tailored functionalities, good thermal stability, permanent porosity and extremely low density. These materials containing imine (C = N) bonds have attracted considerable attention due to their ability to capture of CO_2 through $N^{\delta-} - C^{\delta+} O_2$ interactions and high chemical stability and low skeleton density (Xu et al. 2003).

Porphyrins in conversion of carbon dioxide

Generally, a larger surface area of a photocatalyst makes more active sites for the surface CO_2 adsorption, and a high CO_2 adsorption level would lead to CO_2 reduction in both chemical kinetics and thermodynamics. The conversion of carbon dioxide to high value chemicals is one of the most important issues in the environmental and energy areas but remains a great challenge in chemistry. A lot of effort has been devoted to researching technologies for CO_2 conversion. Various types of chemicals, polymers or fuels can be synthesized from CO_2, for example, some industrialized products like urea, formic acid, methanol, organic carbonate and also carbamate (Mikkelsen et al. 2010, Krebs 2010, Wu et al. 2012, Morris et al. 2009, Darensbourg 2007, Sakakura and Kohno 2009, Sakakura et al. 2007). The electrochemical reduction of CO_2 in various types of catalysts has been intensively investigated as a way to convert waste CO_2 into useful products and contribute to efforts to control global warming while producing new materials. Among these catalysts, porphyrins and metalloporphyrins act as nanostructured materials with modified sizes and geometries, large core sizes, a greater number of pyrrolic subunits, connectivity and highly functionalized reactive sites to better analog and therefore is suitable as an advanced catalyst for CO_2 converted to value added products (Fig. 3.1). The first porphyrin-based homogeneous single site catalyst with an aluminum metal center was found in 1986 (Fig. 3.2) (Aida et al. 1986). Hu et al. showed controlled

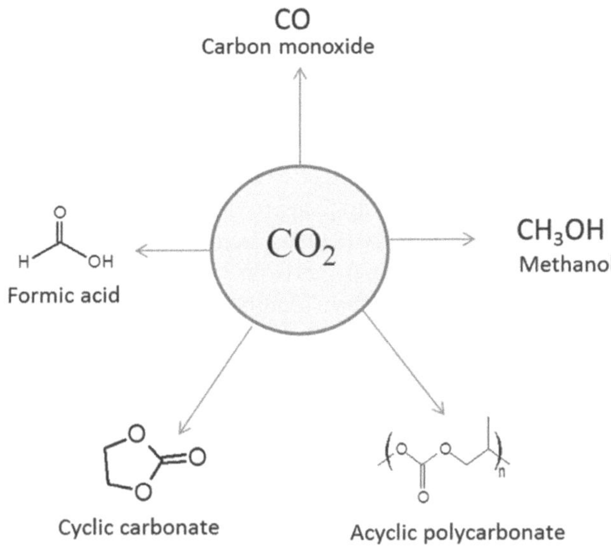

Figure 3.1: Porphyrine based nanostructured materials in conversion of CO_2 into value added products.

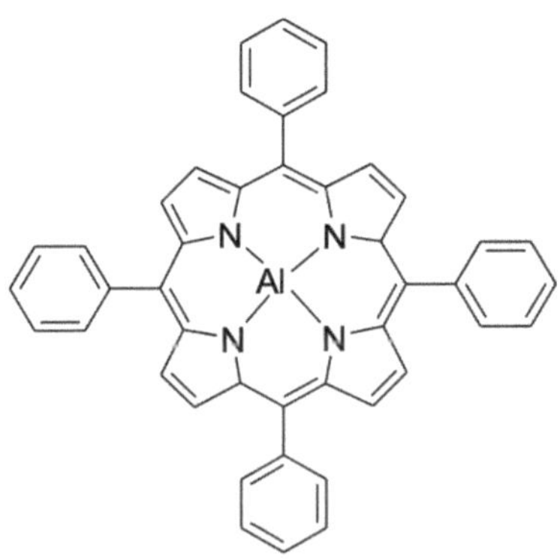

Figure 3.2: Porphyrin based homogeneous single-site catalysts for epoxide/CO_2 copolymerization.

electropolymerization of a carbazole-functionalized iron porphyrin electrocatalyst for CO_2 reduction (Hu et al. 2016).

Carbon monoxide and methane

Carbon monoxide (CO) is emerging as one of the most practical objectives due to its large current density and high selectivity compared to other CO_2 reduction pathways. Porphyrin-based catalysts received great scientific interest in combination with nitrogen donor co-catalysts (Klaus et al. 2011). It is also believed that porphyrin-based catalysts can have a dramatic effect on both electrochemical (Bhugun et al. 1994, Costentin et al. 2012, Ramírez et al. 2009, Ogura and Yoshida 1988, Bhugun et al. 1996, Tripkovic et al. 2013, Nielsen and Leung 2010, Leung et al. 2010, Costentin et al. 2012) and photochemical reduction of CO_2 (Dhanasekaran et al. 1999, Grodkowski and Neta 2000). Metallized porphyrin functionalized with graphene for electrochemical reduction of CO_2 to CO has been reported (Tripkovic et al. 2013). Cobalt (Co) porphyrin and iron (Fe) porphyrin are homogeneous catalysts and have been well studied for photocatalytic and electrocatalytic activity in CO_2 reduction (Grodkowski et al. 1997, Behar et al. 1998). Porphyrin transition metal complexes have been reported to be effective catalysts for CO_2 electroreduction in the form of gas diffusion electrodes, up to approximately ca. 70% (Magdesieva et al. 2002). Windle et al. (Windle et al. 2012) have demonstrated CO_2 reduction activity of two porphyrin based zinc-rhenium dyads (Fig. 3.3) [Dyad 1 and Dyad 2] and have compared their reactivity with that of two separate components. It was observed that there was no production of CO successfully.

Porphyrin cobalt complexes play an important role as an active catalyst in photoreduction of CO_2 to CO (Savéant 2008). Alenezi et al. demonstrated photocatalytic conversion of CO_2 to CO with efficiency of 90% or more on boron doped hydrogen terminated p-type silicon electrodes using mesotetraphenyl porphyrin Fe (III) chloride (3) or thiorate basket porphyrin (4) (Fig. 3.4) (Alenezi et al. 2013). Iron (0) porphyrin (Fig. 3.5) can be used in photochemical processes as a homogeneous catalyst with turnover numbers (TON) of 30 in CO and catalyst selectivity up to 85% (Bonin et al. 2014). The photocatalytic reduction of CO_2 to CO using substituted iron (0) tetraphenylporphyrin has been obtained with high catalytic selectivities (93 and 100%) and TONs of ca. 140 and 60 (Bonin et al. 2014). Phenoxazine-based organic photosensitizer and an iron porphyrin molecular catalyst used as photochemical reduction of CO_2 to CO and CH_4 with TONs of 149 and

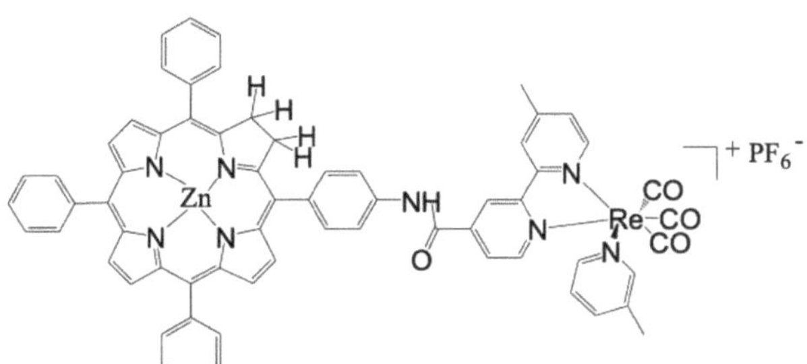

1

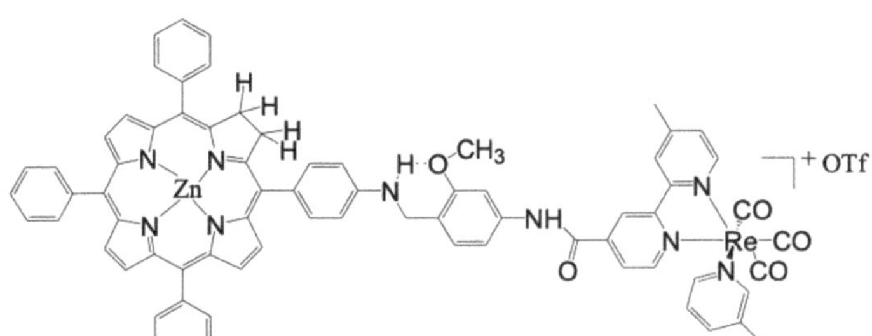

2

Figure 3.3: Structure of dyad 1 and dyad 2 (Windle et al. 2012).

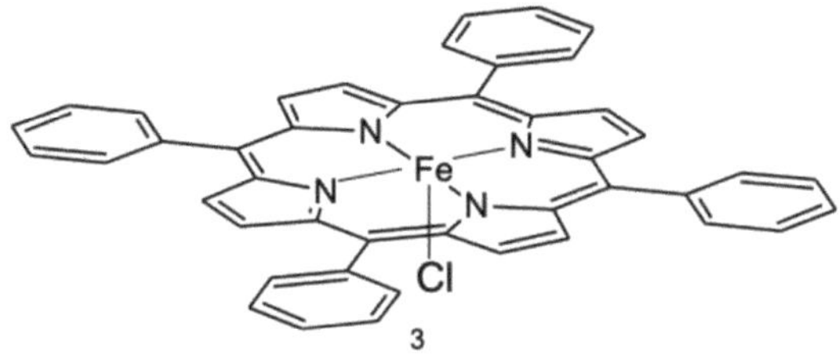

3

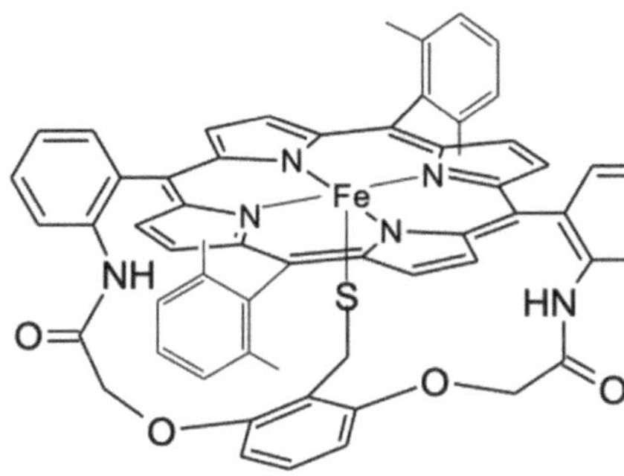

4

Figure 3.4: Structure of meso-tetra phenyl porphyrin Fe (III) chloride (3) or the thiolate basket porphyrin (4) complexes (Alenezi et al. 2013).

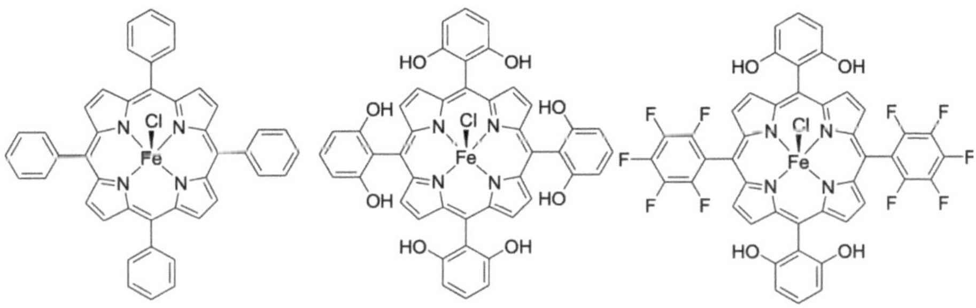

5 6 7

Figure 3.5: Iron porphyrin catalysts (5) Iron(III) 5,10,15,20-tetraphenylporphyrin chloride; (6) chloro iron (III) 5,10,15,20-tetrakis(2',6'-dihydroxyphenyl) porphyrin chloride and (7) pentafluorinated [chloro iron(III) 5,15-bis(2',6'-dihydroxyphenyl)-10,20-bis(pentafluorophenyl)Porphyrin.

29, respectively, under visible-light irradiation with a tertiary amine as sacrificial electron donor (Rao et al. 2018).

Carbon-based materials such as carbon nanotubes (CNTs) have proven to be solid support chains suitable for heterogeneity due to their electron conductivity and ability to form strong, non-covalent π-π interactions with aromatic compounds (Aoi et al. 2015, Zhang et al. 2017). Hu et al. have reported CO_2 to CO reduction with excess efficiency 90% under aqueous conditions using cobalt-tetraphenyl porphyrin (CoTPP) immobilized on CNTs (Hu et al. 2017). Earlier it had been demonstrated that the selective reduction of CO_2 to CO with 90% parabolic efficiency, when immobilizing iron-porphyrin-dimer on Multi-Walled Carbon Nanotubes (MWCNT) (Abdinejad et al. 2020). Very recently, Abdinejad et al. studied the electrocatalytic activity of metallated and non-metallated pyridine-porphyrins Fig. 3.6 deposited onto CNT showed outstanding catalytic performance towards CO_2 electroreduction to CH_4 and CO in aqueous solution (Abdinejad et al. 2020).

Lu et al. demonstrated a combination of experimental and theoretical validation of the Ni porphyrin-based covalent triazine framework (Fig. 3.7) with atomically dispersed NiN_4 centers as an efficient electrocatalyst for CO_2 reduction reactions (Lu et al. 2019). It showed the high stability as well as high selective conversion efficiency of CO_2 to CO toward CO_2 RR with a Faradaic efficiency of > 90% over the range from −0.6 to −0.9 V.

Cheung et al. (Cheung et al. 2019) demonstrated the synthesis of a COF with 5,10,15,20-tetra-(4-aminophenyl)-porphyrin Fe(III) chloride and 2,5-dihydroxyterephthalaldehyde in solvent-free condition for the reduction of CO_2 to CO (Fig. 3.8). It showed turnover frequency was > 600 h^{-1} mol^{-1} of electroactive Fe sites, and Faradaic efficiency 80% for CO at −2.2 V (vs Ag/AgCl) over 3 hours in MeCN with 0.5 M trifluoroethanol. Liu et al. demonstrated the ultrathin imine-linked 2-Dimensional COF-367-Co nanosheets heterogeneous photocatalyst for CO_2 to CO conversion, showing excellent efficiency with a CO production rate as high as 10162 μmol g^{-1} h^{-1} and a selectivity of ca. 78% in aqueous media under visible-light irradiation (Liu et al. 2019). Chen et al. directly synthesized a series of metalloporphyrin-based hypercrosslinked polymers (M-HCPs: M = Al, Co, Fe, Mn) through Friedel-Crafts alkylation reaction for cyclic carbonates from epoxides and CO_2 using tetrabutylammonium bromide as a cocatalyst under mild conditions (Chen et al. 2017). Al-HCP showed a high TOF value of 14880 h^{-1} was succeeded with propylene oxide at 100°C and 3.0 MPa.

The homogeneous metalloporphyrin catalysts are also disadvantageous for the catalytic CO_2 fixation reactions due to low CO_2 adsorption capacity. Xu et al. reported indium based metal–porphyrinic framework, denoted NUPF-3 with high CO_2 adsorption used as a heterogeneous catalyst for cycloaddition of CO_2 and epoxides, under relatively mild conditions, with good recyclability (Xu et al. 2016). Zhang et al. reported a new synthetic approach involving long-term pyrolysis of hemin and melamine molecules in graphene for the manufacture of powerful and

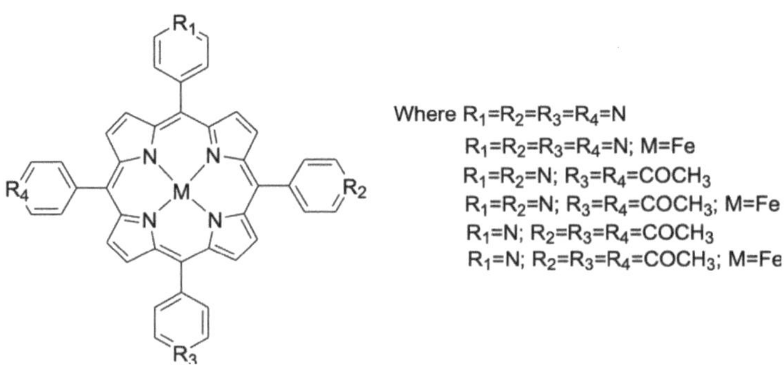

Figure 3.6: Structure of pyridine-porphyrins (Abdinejad et al. 2020).

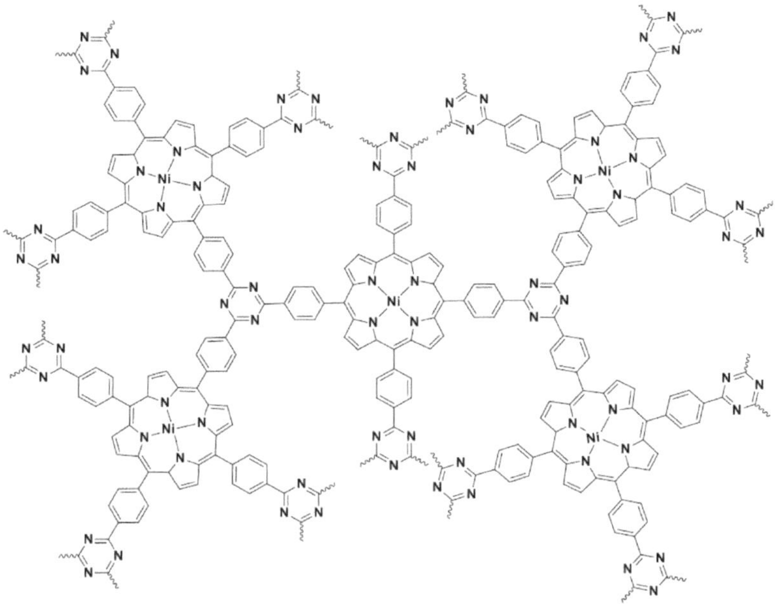

Figure 3.7: Structure of Ni porphyrin-based covalent triazine framework (Lu et al. 2019).

efficient single-iron-atomic electrocatalysts for electrochemical CO_2 reduction with high Faraday efficiency (97.0%) (Zhang et al. 2019). Water is a rich and proton source for CO_2 reduction. However, it can promote catalytic decomposition and off-target H_2 evolution. CO_2 reduction by water-soluble catalysts is rare, but aqueous electrolytes can be used as a catalyst fixation or deposition method on the electrode (Costentin et al. 2015). The porous organic cage composed of six iron tetraphenylporphyrins, which promotes electrochemical CO_2 to CO production with excellent speed, current density and lifetime (Smith et al. 2018). The iron porphyrin complexes produced CO from CO_2 and it could be photochemically reduced to CH_4 (Rao et al. 2017). In addition, Sen et al. reported iron porphyrin complexes for the role of 2nd sphere H-bonding residues in tuning the kinetics of CO_2 reduction to CO (Sen et al. 2019). Maurin et al. demonstrated CNT are modified by attaching Fe-porphyrin via a covalent bond between the peripheral carboxylic acid group and the surface amine group, thereby obtaining high CO selectivity and turnover at a potential of 500 mV (Maurin and Robert 2016). A chemically modified cobalt porphyrins onto the surface of CNT by a substitution reaction reported for CO_2 to CO conversion (Zhu et al. 2019). This provides a total current density of 25.1 $mAcm^{-2}$ and a Faraday efficiency of 98.3% over 490 mV overcharge. Recently, a porphyrin-based metal-free tetrakis (4-carboxyphenyl) porphyrin (H_2TCPP) co-catalyst was developed through solvothermal synthesis route and a metal atom fixed at the center of the porphyrin ring, TCPP-M@TiO_2 (loaded with metal as porphyrin TCPP-M (M = Co, Ni, Cu) into TiO_2 through a bond between the terminal group carboxyl and Ti atoms) (M = Co, Ni, Cu) nanostructured materials (Wang et al. 2020). TCPP-Cu@ TiO_2 was found to exhibit the best activity for photocatalytic CO_2 reduction into CO (13.6 $\mu mol\ g^{-1}\ h^{-1}$) and CH_4 (1.0 $\mu mol\ g^{-1}\ h^{-1}$).

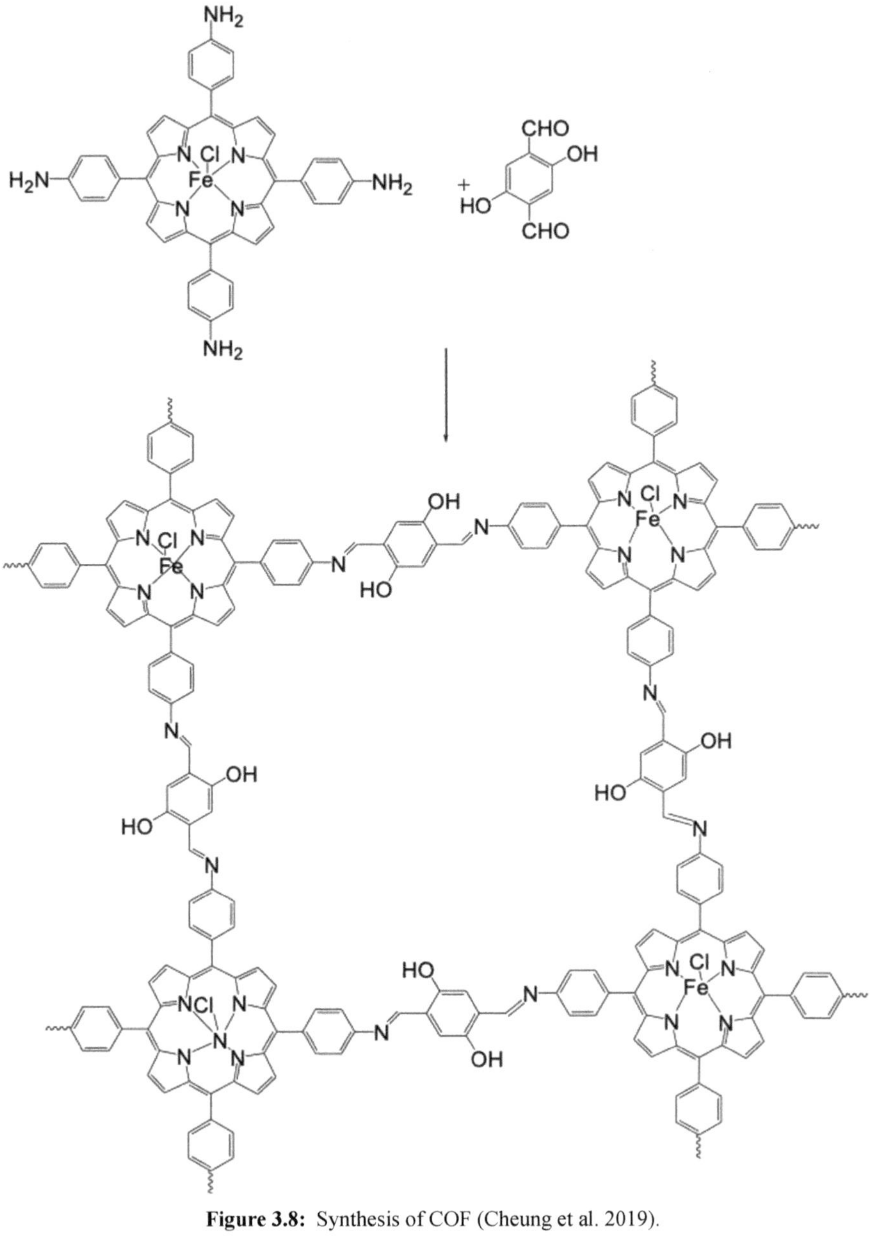

Figure 3.8: Synthesis of COF (Cheung et al. 2019).

The first time cofacial iron tetraphenyl porphyrin dimer with shorter metal-to-metal separation acts as an efficient catalyst to reduce CO_2 to CO with high Faradic efficiency (95%) and TOF (4300 s^{-1}) without the use of external Lewis or Bronsted acids (Mohamed et al. 2015). The presence of two iron centers in dimer where one iron center acts as a Lewis base to push an electron pair into CO_2 molecules, and the second iron center acts as a Lewis acid to promote C-O bond cleavage and

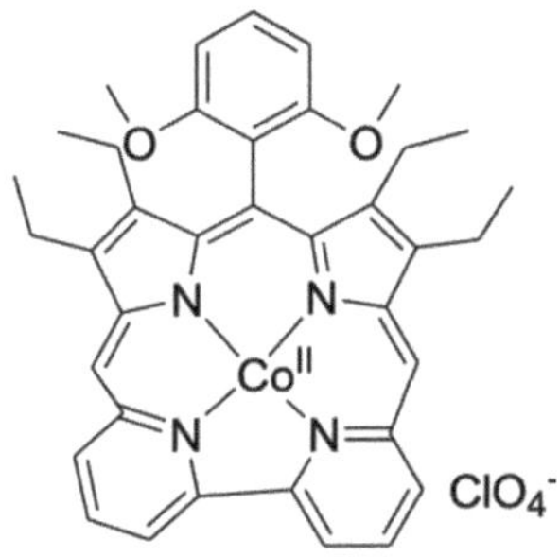

Figure 3.9: Structure of the dianionic tetraphenyl porphyrin ligand and cobalt complex (Ogawa et al. 2019).

CO formation. The cobalt bipyricorrole complex have been found to catalytically promote selective CO_2 electroreduction to CO with a Faraday efficiency of 75% and the overpotential was 0.35 V lower than that of the dianionic tetraphenyl porphyrin ligand and cobalt complex (Ogawa et al. 2019) (Fig. 3.9).

Mao et al. demonstrated a (5,10,15,20-tetrakis(4-carboxyphenyl) porphyrinato)-Fe(III) chloride co-building UiO-66 electrocatalyst for CO_2 reduction in an aqueous solution with the highest Faradaic efficiency of nearly 100% at an overpotential of 450 mV for conversion of CO_2 to CO (Mao et al. 2019).

Methanol

Electrochemical reduction of CO_2 to methanol (CH_3OH) is considered to be the main goal of conversion from fossil fuels to renewable fuels. To catalyze 6-electron-6 proton reduction with CH_3OH of CO_2, it was decided to use metal oxides, metal alloys or chalcogenide-based catalyst electrodes (Sun et al. 2016). The Cu porphyrin (5,10,15,20-tetrakis(4-carboxyphenyl) porphyrin)-based MOF enhanced the photocatalytic conversion of CO_2 to methanol compared with the samples without Cu metal (Liu et al. 2013) (Fig. 3.10). Isatin-porphyrin chromophore chemically bound to the graphene photocatalyst-biocatalyst integrated system has been developed for capable harvesting of sufficient visible light for carrying out the multi electron reduction of CO_2 to methanol at ambient conditions on integration with the sequentially coupled enzymes (Yadav et al. 2014).

Formic acid

Formic acid has the potential to generate alternative energy sources by hydrogenation of CO_2. Formic acid has attracted considerable attention, as it is a practical hydrogen storage material because it is liquid at an ambient temperature and has a high bulk hydrogen density of 53 gL^{-1} (Eppinger and Huang 2017, Bernskoetter and Hazari 2017). Formic acid is also less toxic, making it more environmentally safe compared to petrochemicals. Eder et al. developed benzobisthiazole-linked ruthenium porphyrin-based POP (Fig. 3.11) and utilized for hydrosilylative reduction of CO_2 to potassium formate/formic acid for energy storage applications (Eder et al. 2019).

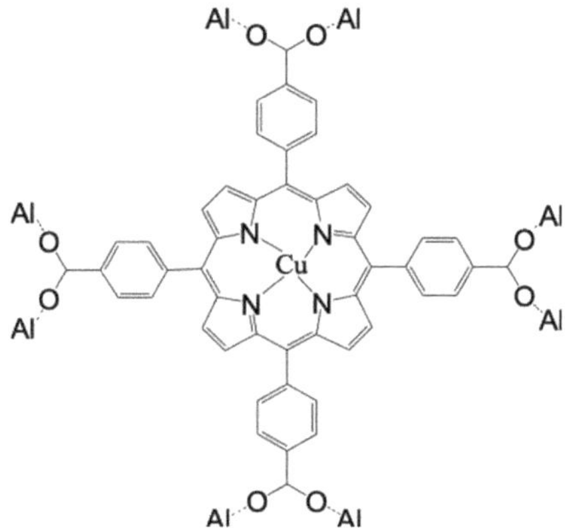

Figure 3.10: Structure of Cu-metallated 5,10,15,20-Tetrakis(4-carboxyphenyl) porphyrin (Liu et al. 2013).

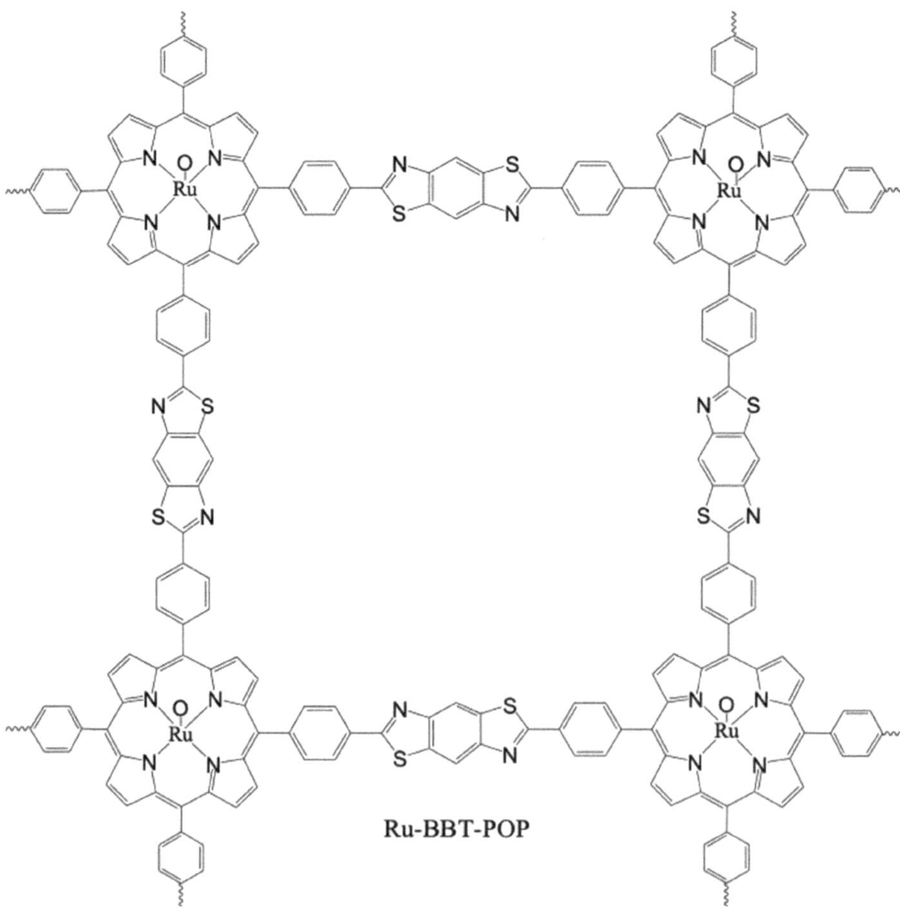

Figure 3.11: Structure of benzobisthiazole-linked ruthenium porphyrin-based porous organic polymer (Ru-BBT-POP) (Eder et al. 2019).

Formic acid generation by CO_2 reduction in water medium on ZnTPPS/,2′-bpy salt/TEOA reached 125 μmol with a TON of 10 and [Mn(phen)(CO)$_3$Br]/ ZnTPP reached TON of 19 (Amao 2018, Hameed et al. 2019). A CoTPP (cobalt meso-tetraphenylporphyrin)/g-C_3N_4 appropriately developed by self-assembly for high efficiency CO_2 reduction in water medium for formic acid production (154.4 μmol with a TON of 137) (Liu et al. 2017). An urchin-like 5,10,15,20-tetrakis(4-aminophenyl) porphyrin/graphene microspheres based photocatalyst have been used for the conversion of formic acid from CO_2 (Li et al. 2017). The microspheres showed enhancement in sunlight absorption and improving light-harvesting efficiency. An effective approach has been adopted for electronic coupling to develop covalent bonding between porphyrin and graphene which increases the rate of the photoinduced charge transfer. Graphene based photocatalyst, covalently bonded with multianthraquinone substituted porphyrin has been used for the artificial photosynthesis system for an efficient photosynthetic production of formic acid from CO_2 (Yadav et al. 2012). Porphyrin adsorbed Nafion membranes have been used for the photocatalytic reduction of CO_2 to formic acid (Premkumar and Ramaraj 1997). Graphene oxide modified cobalt metallated porphyrin photocatalyst for the efficiency of artificial photosynthetic conversion of formic acid from CO_2 was 96.49 μmol for 2 hours (Fig. 3.12) was also developed by us (Kumar et al. 2018).

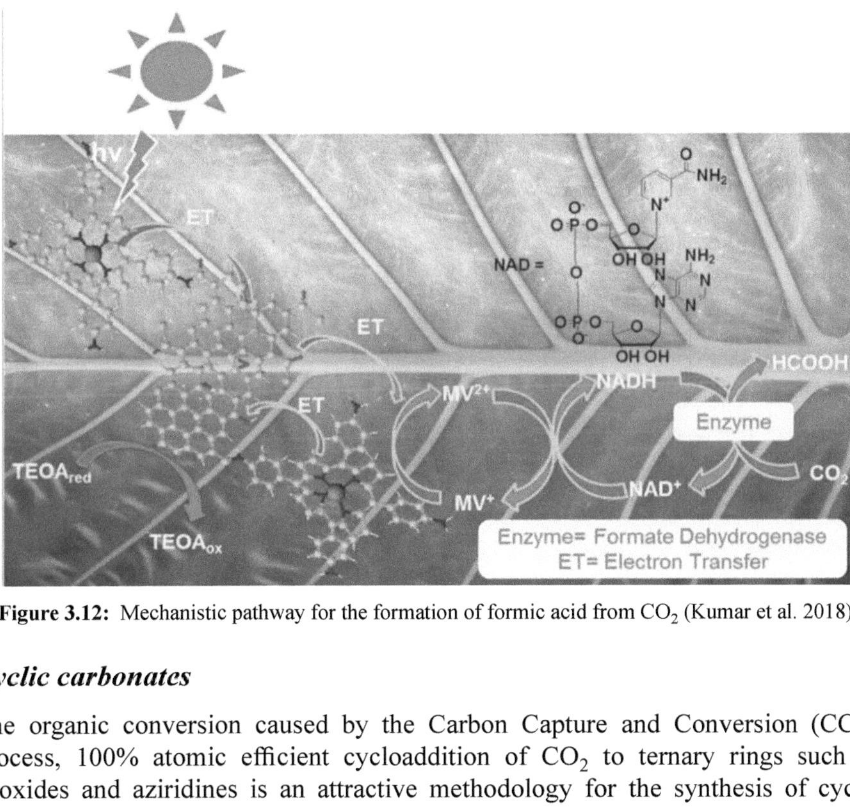

Figure 3.12: Mechanistic pathway for the formation of formic acid from CO_2 (Kumar et al. 2018).

Cyclic carbonates

The organic conversion caused by the Carbon Capture and Conversion (CCC) process, 100% atomic efficient cycloaddition of CO_2 to ternary rings such as epoxides and aziridines is an attractive methodology for the synthesis of cyclic carbonates and oxazolidinone (Shaikh et al. 2018, Liu et al. 2015, Kumar et al. 2019, Kumar et al. 2018, Kumar et al. 2018). They are widely used as raw materials for the manufacture of chemicals and pharmaceutical compounds (Heravi et al. 2016). Synthesis of cyclic carbonates from epoxides and CO_2 is one of the most useful CO_2 fixation methods (Bai et al. 2011, Ema et al. 2014, Robert et al. 2014, Anderson et al. 2012, Qin et al. 2008). Cyclic carbonates have also been used as raw materials for polycarbonates, polar aprotic solvents, fuel additives and electrolytes in lithium-ion secondary batteries. Conversion of cyclic carbonates using porphyrin catalysts (Mg(II) porphyrin) (Fig. 3.13) showed high catalytic activity, high turnover number (TON = 103000) and turnover frequency (TOF = 12000 h^{-1}) (Ema et al. 2012). Maeda et al. synthesized bifunctional metalloporphyrins with quaternary ammonium bromides (nucleophiles) at the meta, para or ortho positions of meso-phenyl groups as catalysts for the formation of cyclic carbonates from epoxides and CO_2 under solvent-free conditions (Maeda et al. 2016). A zinc(II) porphyrin with eight nucleophiles at the meta positions showed very high catalytic activity TON was 240000 at 120°C, TOF was 31500 h^{-1} at 170°C at an initial CO_2 pressure of 1.7 MPa; catalyzed the reaction even at atmospheric CO_2 pressure (balloon) at ambient temperatures (20°C).

Milani et al. demonstrated a series of 5,10,15,20-tetrakis(2,3-dichlorophenyl) porphyrinate complexes of manganese (III) [MnIII(T2,3DCPP)X] (Fig. 3.14) with six different axial ligands (X = NO_3, AcO, IO_3, Br, Cl, HO) used as catalysts in the cycloaddition reactions of CO_2 and styrene oxide, under mild conditions, i.e., atmospheric pressure and 60°C (Milani et al. 2019).

Chen et al. synthesized a Zn-metallated porphyrin-based porous organic framework catalyst (Fig. 3.15) for cycloaddition reactions of CO_2 and propylene oxide to provide a TOF of 433 h^{-1} in the presence of co-catalyst (Chen et al. 2018).

Highly porous metalloporphyrin covalent ionic frameworks developed by coupling 5,10,15,20-tetrakis(4-nitrophenyl)porphyrin with 3,8-diamino-6-phenylphenanithridine or 5,5'-diamino-2,2'-bipyridine, showed high catalyzing efficiency in the cycloaddition of propylene oxide with CO_2 to form propylene carbonate (Liu et al. 2019). Wang et al. reported a high surface area metalporphyrin-based microporous organic polymer (HUST-1-Co) for the cycloaddition of CO_2 with substituted epoxides with various functional groups with > 93% yields at

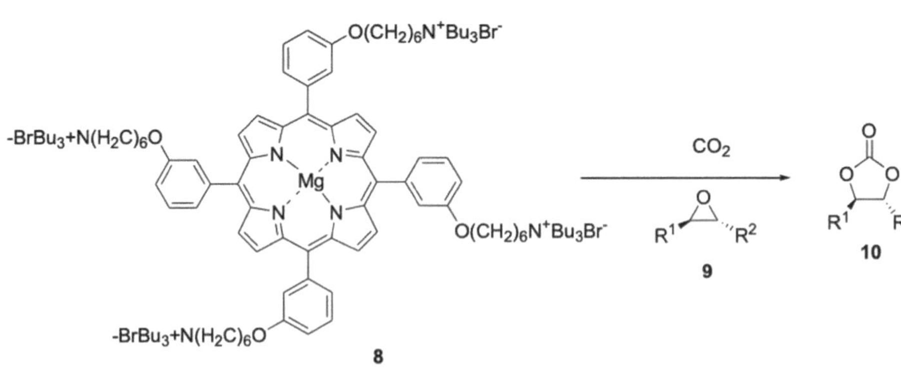

Figure 3.13: Synthesis of cyclic carbonate 10 from epoxide 9 and CO_2 (Ema et al. 2012).

Figure 3.14: Structure of the metalloporphyrin catalysts (Milani et al. 2019).

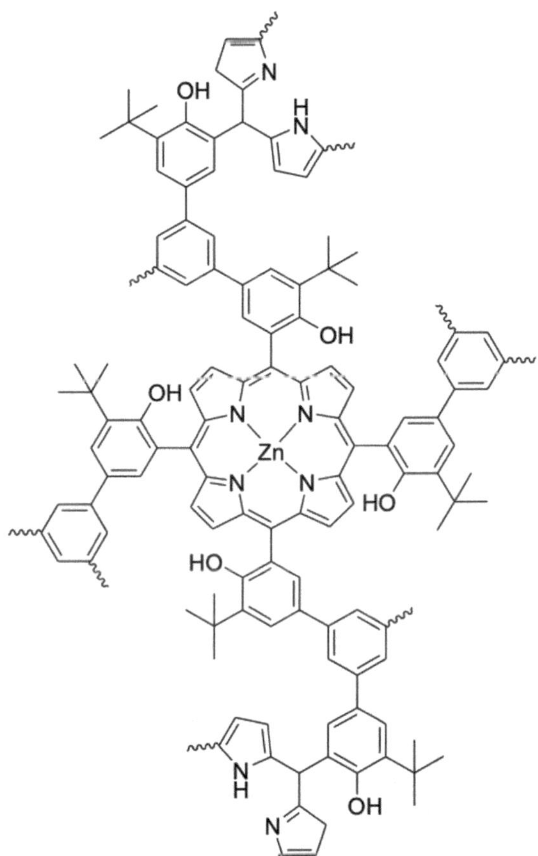

Figure 3.15: Structure of Zn-metallated porphyrin-based porous organic framework catalyst (Chen et al. 2018).

room temperature and atmospheric pressure, with more than 15 times recycling performance (Wang et al. 2017). The highly active and recyclable heterogeneous bifunctional porphyrin based Biogenous Iron Oxide (BIO) catalyst has been reported for formation of cyclic carbonates (Silva et al. 2013) (Fig. 3.16).

Chatterjee and Chisholm (Chatterjee et al. 2013, Chatterjee and Chisholm 2012) demonstrated the reactivities of 5,10,15,20-tetraphenylporphyrin, 5,10,15,20-tetrakis (pentafluorophenyl) porphyrin, 2,3,7,8,12,13,17,18-octaethylporphyirn based aluminum and chromium complexes with respect to their ability to homopolymerize propylene oxide and copolymerize propylene oxide and CO_2 into polypropylene oxide and polypropylene carbonate 31, respectively, with and without the presence of a co-catalyst (4-dimethylaminopyridine or a PPN+ salt where the anion is Cl^- or N^{3-}). In the presence of a co-catalyst, the TFPP complex is the most active in copolymerization to yield PPC. An increase in the PPN^+X^-/[Al] ratio decreases the rate of PPC formation and favors the formation of propylene carbonate, (PC) (30) Fig. 3.17. Manganese-corrole complexes in combination with a co-catalyst [PPN]X ($[PPN]^+$ = bis(triphenylphosphoranylidene) iminium) have been found to be new versatile catalysts for the copolymerization of epoxides with CO_2 to form polycarbonates (31) (Robert et al. 2014). Self-assembly as a burgeoning section of supramolecular chemistry has been a prolific tool for the construction of diverse fascinating hollow 3D structures. Self-assembly of the octatopic porphyrin ligand of tetrakis(3,5-dicarboxybiphenyl) porphine with the *in situ* generated $Cu_2(CO_2)_4$

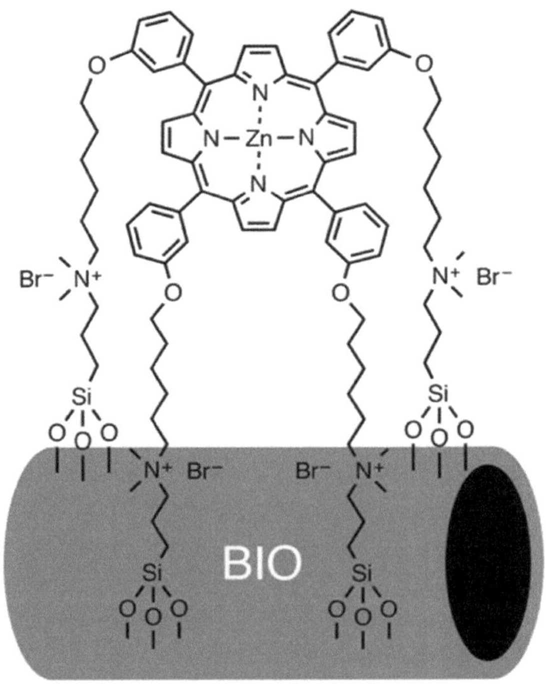

Figure 3.16: Zn II porphyrin linked to BIO via four tetra-alkyl ammonium bromide groups (Silva et al. 2013).

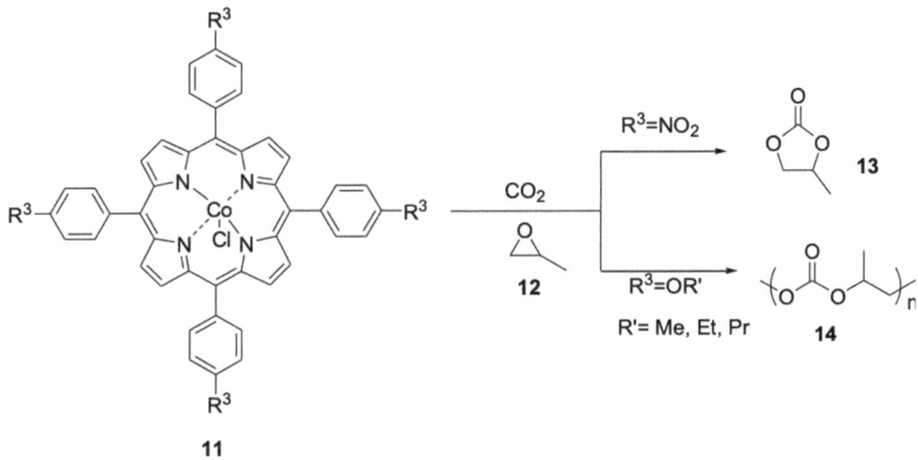

Figure 3.17: Synthesis of cyclic carbonate 13 and poly(propylene carbonate) 14 from epoxide 12 and CO_2.

paddlewheel moieties provided a porous metal-metalloporphyrin framework for chemical fixation of CO_2 to form cyclic carbonates (Gao et al. 2014). Feng et al. demonstrated cobalt porphyrin zirconium MOF, PCN-224 (Co) catalyst for the formation of cyclic carbonate (Feng et al. 2013). Chen et al. demonstrated core-double-shell microsphere [Fe_3O_4@SiO_2@Zn(Por)OP] by coating a core-shell composite of Fe_3O_4 magnetic core and SiO_2 shell (Fe_3O_4@SiO_2) with a zinc porphyrin based organic polymer [Zn(Por)OP] for the cycloaddition of CO_2 and Propylene Oxide (PO) to convert Propylene Carbonate (PC) with a yield of 97% and turnover frequency (TOF) of 13978 mol_{PC} mol_{Zn}^{-1} h^{-1} (Chen et al. 2016). Assembled supramolecular gel from small molecules has been developed as a new form of catalyst due to various catalyst sites and active organic functional groups can be easily introduced into the gelling agent molecule (Chen et al. 2018). Liao et al. synthesized porphyrin/metalloporphyrin imine gels from 5,10,15,20-tetrakis-(p-amino phenyl)porphyrin and metalated porphyrin (MTAPP, M = Zn, Co and Pd) (Liao et al. 2019). H_2TAPP modified with glyoxal (GO) improved its performance. The reduction of the imine bonds of H_2TAPP-GO yielded the H_2TAPP-GO-r gel, and the subsequent incorporation of metal ions into the porphyrin moiety results in MTAPP-GO-r gels. The MTAPP-GO and MTAPP-GO-r gels are active in the heterogeneous cycloaddition of epoxides with CO_2 to cyclic carbonates using wet gels. The ZnTAPP-GO and ZnTAPP-GO-r catalysts show the best catalytic performance with recycling stability during the cycloaddition of epoxides with CO_2 to cyclic carbonates. Kumar et al. developed the facile, environmentally friendly and sustainable chitosan based meso-tetrakis(4-sulfonatophenyl)porphyrin (CS-TPPS) for adsorption and higher conversion of CO_2 and propylene oxide into cyclic carbonate (66%), compared to pure chitosan (31%) (Kumar et al. 2017).

Conclusions

The field of porphyrins based nanostructured for CO_2 conversion is still in its infancy, with only a limited number of reports having appeared that deal with the chemistry and its applications in CO_2 conversion to value added chemicals. It is believed that this chapter provides guidance to design the best porphyrin based nanostructured as, Covalent-Organic Frameworks (COFs), Metal-Organic Frameworks (MOFs), dyads, dimmers or porous materials which would be promising tools to overcome the existing challenges for solving all problems related to implementation of effective sustainable CO_2 conversion technologies.

Acknowledgement

This work was supported by the National Research Foundation of Korea (NRF) grant funded by the Korea government (MSIT) (No. 2018R1D1A1B07045859) and Korea Institute for Advancement of Technology (KIAT) grant funded by the Korea Government (MOTIE) (P0012770).

References

Abdinejad, Maryam, Caitlin Dao, Billy Deng, Filip Dinic, Oleksandr Voznyy, Xiao-an Zhang et al. 2020. Electrocatalytic reduction of CO_2 to CH_4 and CO in aqueous solution using pyridine-porphyrins immobilized onto carbon nanotubes. ACS Sustainable Chemistry & Engineering 8(25): 9549–9557.

Abdinejad, Maryam, Caitlin Dao, Billy Deng, Filip Dinic, Oleksandr Voznyy, Xiao-an Zhang et al. 2020. Enhanced electrochemical reduction of CO_2 to CO upon immobilization onto carbon nanotubes using an iron-porphyrin dimer. Chemistry Select 5(3): 979–984.

Aida, Takazo, Masahide Ishikawa and Shohei Inoue. 1986. Alternating copolymerization of carbon dioxide and epoxide catalyzed by the aluminum porphyrin-quaternary organic salt or -triphenylphosphine system. Synthesis of polycarbonate with well-controlled molecular weight. Macromolecules 19(1): 8–13.

Alenezi, Khalaf, Saad K. Ibrahim, Peiyi Li and Christopher J. Pickett. 2013. Solar fuels: Photoelectrosynthesis of CO from CO_2 at p-type Si using Fe porphyrin electrocatalysts. Chemistry—A European Journal 19(40): 13522–13527.

Amao, Y. 2018. Photoredox systems with biocatalysts for CO_2 utilization. Sustainable Energy & Fuels 2(9): 1928–1950.

Anderson, Carly E., Sergei I. Vagin, Wei Xia, Hanpeng Jin and Bernhard Rieger. 2012. Cobaltoporphyrin-catalyzed CO_2/epoxide copolymerization: Selectivity control by molecular design. Macromolecules 45(17): 6840–6849.

Aoi, Shoko, Kentaro Mase, Kei Ohkubo and Shunichi Fukuzumi. 2015. Selective electrochemical reduction of CO_2 to CO with a cobalt chlorin complex adsorbed on multi-walled carbon nanotubes in water. Chemical Communications 51(50): 10226–10228.

Bai, Dongsheng, Qiong Wang, Yingying Song, Bo Li and Huanwang Jing. 2011. Synthesis of cyclic carbonate from epoxide and CO_2 catalyzed by magnetic nanoparticle-supported porphyrin. Catalysis Communications 12(7): 684–688.

Behar, D., T. Dhanasekaran, P. Neta, C.M. Hosten, D. Ejeh, P. Hambright et al. 1998. Cobalt porphyrin catalyzed reduction of CO_2, radiation chemical, photochemical, and electrochemical studies. The Journal of Physical Chemistry A 102(17): 2870–2877.

Bernskoetter, Wesley H. and Nilay Hazari. 2017. Reversible hydrogenation of carbon dioxide to formic acid and methanol: Lewis acid enhancement of base metal catalysts. Accounts of Chemical Research 50(4): 1049–1058.

Bhugun, Iqbal, Doris Lexa and Jean-Michel Saveant. 1994. Ultraefficient selective homogeneous catalysis of the electrochemical reduction of carbon dioxide by an iron(0) porphyrin associated with a weak Broensted acid cocatalyst. Journal of the American Chemical Society 116(11): 5015–5016.

Bhugun, Iqbal, Doris Lexa and Jean-Michel Savéant. 1996. Catalysis of the electrochemical reduction of carbon dioxide by iron(0) porphyrins: synergistic effect of Lewis acid cations. The Journal of Physical Chemistry 100(51): 19981–19985.

Bhugun, Iqbal, Doris Lexa and Jean-Michel Savéant. 1996. Catalysis of the electrochemical reduction of carbon dioxide by iron(0) porphyrins: Synergystic effect of weak Brönsted acids. Journal of the American Chemical Society 118(7): 1769–1776.

Bonin, Julien, Marie Chaussemier, Marc Robert and Mathilde Routier. 2014. Homogeneous photocatalytic reduction of CO_2 to CO using iron(0) porphyrin catalysts: Mechanism and intrinsic limitations. ChemCatChem. 6(11): 3200–3207.

Bonin, Julien, Marc Robert and Mathilde Routier. 2014. Selective and efficient photocatalytic CO_2 reduction to CO using visible light and an iron-based homogeneous catalyst. Journal of the American Chemical Society 136(48): 16768–16771.

Chatterjee, Chandrani and Malcolm H. Chisholm. 2012. Influence of the metal (Al, Cr, and Co) and the substituents of the porphyrin in controlling the reactions involved in the copolymerization of propylene oxide and carbon dioxide by porphyrin metal(III) complexes. 2. chromium chemistry. Inorganic Chemistry 51(21): 12041–12052.

Chatterjee, Chandrani, Malcolm H. Chisholm, Adnan El-Khaldy, Ruaraidh D. McIntosh, Jeffrey T. Miller and Tianpin Wu. 2013. Influence of the metal (Al, Cr, and Co) and substituents of the porphyrin in controlling reactions involved in copolymerization of propylene oxide and carbon dioxide by porphyrin metal(III) complexes. 3. cobalt chemistry. Inorganic Chemistry 52(8): 4547–4553.

Chen, Aibing, Pengpeng Ju, Yunzhao Zhang, J. Chen, H. Gao, L. Chen et al. 2016. Highly recyclable and magnetic catalyst of a metalloporphyrin-based polymeric composite for cycloaddition of CO_2 to epoxide. RSC Advances 6(99): 96455–96466.

Chen, Jian, M. Zhong, L. Tao, L. Liu, S. Jayakumar, C. Li et al. 2018. The cooperation of porphyrin-based porous polymer and thermal-responsive ionic liquid for efficient CO_2 cycloaddition reaction. Green Chemistry 20(4): 903–911.

Chen, Shoumin, X. Lin, Z. Zhai, R. Lan, J. Li, Y. Wang et al. 2018. Synthesis and characterization of CO_2-sensitive temperature-responsive catalytic poly(ionic liquid) microgels. Polymer Chemistry 9(21): 2887–2896.

Chen, Yaju, Rongchang Luo, Qihang Xu, Wuying Zhang, Xiantai Zhou and Hongbing Ji. 2017. State-of-the-art aluminum porphyrin-based heterogeneous catalysts for the chemical fixation of CO_2 into cyclic carbonates at ambient conditions. ChemCatChem. 9(5): 767–773.

Cheung, Po Ling, Sze Koon Lee and Clifford P. Kubiak. 2019. Facile solvent-free synthesis of thin iron porphyrin COFs on carbon cloth electrodes for CO_2 reduction. Chemistry of Materials 31(6): 1908–1919.

Costentin, Cyrille, Samuel Drouet, Marc Robert and Jean-Michel Savéant. 2012. A local proton source enhances CO₂ electroreduction to CO by a molecular Fe catalyst. Science 338(6103): 90–94.

Costentin, Cyrille, Samuel Drouet, Marc Robert and Jean-Michel Savéant. 2012. Turnover numbers, turnover frequencies, and overpotential in molecular catalysis of electrochemical reactions. Cyclic voltammetry and preparative-scale electrolysis. Journal of the American Chemical Society 134(27): 11235–11242.

Costentin, Cyrille, Marc Robert, Jean-Michel Savéant and Arnaud Tatin. 2015. Efficient and selective molecular catalyst for the CO₂-to-CO electrochemical conversion in water. Proceedings of the National Academy of Sciences 112(22): 6882.

Darensbourg, Donald J. 2007. Making plastics from carbon dioxide: Salen metal complexes as catalysts for the production of polycarbonates from epoxides and CO_2. Chemical Reviews 107(6): 2388–2410.

Dhanasekaran, T., J. Grodkowski, P. Neta, P. Hambright and Etsuko Fujita. 1999. p-Terphenyl-sensitized photoreduction of CO_2 with cobalt and iron porphyrins. Interaction between CO and reduced metalloporphyrins. The Journal of Physical Chemistry A 103(38): 7742–7748.

Eder, Grace M., David A. Pyles, Eric R. Wolfson and Psaras L. McGrier. 2019. A ruthenium porphyrin-based porous organic polymer for the hydrosilylative reduction of CO_2 to formate. Chemical Communications 55(50): 7195–7198.
Ema, Tadashi, Yuki Miyazaki, Shohei Koyama, Yuya Yano and Takashi Sakai. 2012. A bifunctional catalyst for carbon dioxide fixation: cooperative double activation of epoxides for the synthesis of cyclic carbonates. Chemical Communications 48(37): 4489–4491.
Ema, Tadashi, Yuki Miyazaki, Junta Shimonishi, Chihiro Maeda and Jun-ya Hasegawa. 2014. Bifunctional porphyrin catalysts for the synthesis of cyclic carbonates from epoxides and CO_2: Structural optimization and mechanistic study. Journal of the American Chemical Society 136(43): 15270–15279.
Eppinger, Jörg and Kuo-Wei Huang. 2017. Formic acid as a hydrogen energy carrier. ACS Energy Letters 2(1): 188–195.
Feng, Dawei, W.-C. Chung, Z. Wei, Z.-Y. Gu, H.-L. Jiang, Y.-P. Chen et al. 2013. Construction of ultrastable porphyrin Zr metal–organic frameworks through linker elimination. Journal of the American Chemical Society 135(45): 17105–17110.
Gao, Wen-Yang, Lukasz Wojtas and Shengqian Ma. 2014. A porous metal–metalloporphyrin framework featuring high-density active sites for chemical fixation of CO_2 under ambient conditions. Chemical Communications 50(40): 5316–5318.
Grodkowski, J., D. Behar, P. Neta and P. Hambright. 1997. Iron porphyrin-catalyzed reduction of CO_2. Photochemical and radiation chemical studies. The Journal of Physical Chemistry A 101(3): 248–254.
Grodkowski, J. and P. Neta. 2000. Ferrous ions as catalysts for photochemical reduction of CO_2 in homogeneous solutions. The Journal of Physical Chemistry A 104(19): 4475–4479.
Hameed, Yasmeen, Patrick Berro, Bulat Gabidullin and Darrin Richeson. 2019. An integrated Re(i) photocatalyst/sensitizer that activates the formation of formic acid from reduction of CO_2. Chemical Communications 55(74): 11041–11044.
Haszeldine, R. Stuart. 2009. Carbon capture and storage: How green can black be? Science 325(5948): 1647–1652.
Heravi, Majid M., Vahideh Zadsirjan and Behnaz Farajpour. 2016. Applications of oxazolidinones as chiral auxiliaries in the asymmetric alkylation reaction applied to total synthesis. RSC Advances 6(36): 30498–30551.
Hu, Xin-Ming, Z. Salmi, M. Lillethorup, E.B. Pedersen, M. Robert, S.U. Pedersen et al. 2016. Controlled electropolymerisation of a carbazole-functionalised iron porphyrin electrocatalyst for CO_2 reduction. Chemical Communications 52(34): 5864–5867.
Hu, Xin-Ming, Magnus H. Rønne, Steen U. Pedersen, Troels Skrydstrup and Kim Daasbjerg. 2017. Enhanced catalytic activity of cobalt porphyrin in CO_2 electroreduction upon immobilization on carbon materials. Angewandte Chemie International Edition 56(23): 6468–6472.
Klaus, Stephan, Maximilian W. Lehenmeier, Carly E. Anderson and Bernhard Rieger. 2011. Recent advances in CO_2/epoxide copolymerization—New strategies and cooperative mechanisms. Coordination Chemistry Reviews 255(13): 1460–1479.
Kumar, Santosh, Mohmmad Y. Wani, Cláudia T. Arranja, Joana de A. e Silva, B. Avula and Abilio J. F. N. Sobral. 2015. Porphyrins as nanoreactors in the carbon dioxide capture and conversion: a review. Journal of Materials Chemistry A 3(39): 19615–19637.
Kumar, Santosh, Joana de A. e Silva, Mohmmad Y. Wani, João M. Gil and Abilio J. F. N. Sobral. 2017. Carbon dioxide capture and conversion by an environmentally friendly chitosan based meso-tetrakis(4-sulfonatophenyl) porphyrin. Carbohydrate Polymers 175: 575–583.
Kumar, Santosh, Mohmmad Y. Wani, Joonseok Koh, João M. Gil and Abilio J. F. N. Sobral. 2018. Carbon dioxide adsorption and cycloaddition reaction of epoxides using chitosan–graphene oxide nanocomposite as a catalyst. Journal of Environmental Sciences 69: 77–84.
Kumar, Santosh, Rajesh K. Yadav, Kirpa Ram, António Aguiar, Joonseok Koh and Abilio J. F. N. Sobral. 2018. Graphene oxide modified cobalt metallated porphyrin photocatalyst for conversion of formic acid from carbon dioxide. Journal of CO_2 Utilization 27: 107–114.
Kumar, Santosh, K. Prasad, João M. Gil, Abilio J. F. N. Sobral and Joonseok Koh. 2018. Mesoporous zeolite-chitosan composite for enhanced capture and catalytic activity in chemical fixation of CO_2. Carbohydrate Polymers 198: 401–406.

Kumar, Santosh, Dinesh Kumar Mishra, Abilio J. F. N. Sobral and Joonseok Koh. 2019. CO_2 adsorption and conversion of epoxides catalyzed by inexpensive and active mesoporous structured mixed-phase (anatase/brookite) TiO_2. Journal of CO_2 Utilization 34: 386–394.
Leung, Kevin, Ida M. B. Nielsen, Na Sai, Craig Medforth and John A. Shelnutt. 2010. Cobalt–porphyrin catalyzed electrochemical reduction of carbon dioxide in water. 2. Mechanism from first principles. The Journal of Physical Chemistry A 114(37): 10174–10184.
Li, Qi, Qi Zeng, Lina Gao, Z. Ullah, H. Li, Y. Guo et al. 2017. Self-assembly of urchin-like porphyrin/graphene microspheres for artificial photosynthetic production of formic acid from CO_2. Journal of Materials Chemistry A 5(1): 155–164.
Liao, Peisen, Guangmei Cai, Jianying Shi and Jianyong Zhang. 2019. Post-modified porphyrin imine gels with improved chemical stability and efficient heterogeneous activity in CO_2 transformation. New Journal of Chemistry 43(25): 10017–10024.
Liu, Jiahui, Guoying Zhao, Ocean Cheung, Lina Jia, Zhenyu Sun and Suojiang Zhang. 2019. Highly porous metalloporphyrin covalent ionic frameworks with well-defined cooperative functional groups as excellent catalysts for CO_2 cycloaddition. Chemistry—A European Journal 25(38): 9052–9059.
Liu, Jibo, Huijie Shi, Qi Shen, Chenyan Guo and Guohua Zhao. 2017. A biomimetic photoelectrocatalyst of Co–porphyrin combined with a g-C_3N_4 nanosheet based on π–π supramolecular interaction for high-efficiency CO_2 reduction in water medium. Green Chemistry 19(24): 5900–5910.
Liu, Qiang, Lipeng Wu, Ralf Jackstell and Matthias Beller. 2015. Using carbon dioxide as a building block in organic synthesis. Nature Communications 6(1): 5933.
Liu, Wenbo, Xiaokang Li, C. Wang, H. Pan, W. Liu, K. Wang, Q. Zeng et al. 2019. A scalable general synthetic approach toward ultrathin imine-linked two-dimensional covalent organic framework nanosheets for photocatalytic CO_2 reduction. Journal of the American Chemical Society 141(43): 17431–17440.
Liu, Yuanyuan, Yanmei Yang, Qilong Sun, Z. Wang, B. Huang, Y. Dai et al. 2013. Chemical adsorption enhanced CO_2 capture and photoreduction over a copper porphyrin based metal organic framework. ACS Applied Materials & Interfaces 5(15): 7654–7658.
Lu, Chenbao, Jian Yang, Shice Wei, S. Bi, Y. Xia, M. Chen, Y. Hou et al. 2019. Atomic Ni anchored covalent triazine framework as high efficient electrocatalyst for carbon dioxide conversion. Advanced Functional Materials 29(10): 1806884.
Maeda, Chihiro, Junta Shimonishi, Ray Miyazaki, Jun-ya Hasegawa and Tadashi Ema. 2016. Highly active and robust metalloporphyrin catalysts for the synthesis of cyclic carbonates from a broad range of epoxides and carbon dioxide. Chemistry—A European Journal 22(19): 6556–6563.
Magdesieva, T. V., T. Yamamoto, D. A. Tryk and A. Fujishima. 2002. Electrochemical reduction of CO[sub 2] with transition metal phthalocyanine and porphyrin complexes supported on activated carbon fibers. Journal of The Electrochemical Society 149(6): D89.
Mao, Fangxin, Yan-Huan Jin, Peng Fei Liu, P. Yang, L. Zhang, L. Chen et al. 2019. Accelerated proton transmission in metal–organic frameworks for the efficient reduction of CO_2 in aqueous solutions. Journal of Materials Chemistry A 7(40): 23055–23063.
Maurin, Antoine and Marc Robert. 2016. Catalytic CO_2-to-CO conversion in water by covalently functionalized carbon nanotubes with a molecular iron catalyst. Chemical Communications 52(81): 12084–12087.
Mikkelsen, Mette, Mikkel Jørgensen and Frederik C. Krebs. 2010. The teraton challenge. A review of fixation and transformation of carbon dioxide. Energy & Environmental Science 3(1): 43–81.
Milani, Jorge L. S., Alexandre M. Meireles, Werberson A. Bezerra, Dayse. C. S. Martins, Danielle Cangussu and Rafael P. das Chagas. 2019. MnIII porphyrins: Catalytic coupling of epoxides with CO_2 under mild conditions and mechanistic considerations. ChemCatChem. 11(17): 4393–4402.
Mohamed, Eman A., Zaki N. Zahran and Yoshinori Naruta. 2015. Efficient electrocatalytic CO_2 reduction with a molecular cofacial iron porphyrin dimer. Chemical Communications 51(95): 16900–16903.
Morris, Amanda J., Gerald J. Meyer and Etsuko Fujita. 2009. Molecular approaches to the photocatalytic reduction of carbon dioxide for solar fuels. Accounts of Chemical Research 42(12): 1983–1994.
Nielsen, Ida M. B. and Kevin Leung. 2010. Cobalt–porphyrin catalyzed electrochemical reduction of carbon dioxide in water. 1. a density functional study of intermediates. The Journal of Physical Chemistry A 114(37): 10166–10173.

Ogawa, Ayumu, Koji Oohora, Wenting Gu and Takashi Hayashi. 2019. Electrochemical CO_2 reduction by a cobalt bipyricorrole complex: decrease of an overpotential value derived from monoanionic ligand character of the porphyrinoid species. Chemical Communications 55(4): 493–496.

Ogura, Kotaro and Ichiro Yoshida. 1988. Electrocatalytic reduction of CO_2 to methanol: Part 9: Mediation with metal porphyrins. Journal of Molecular Catalysis 47(1): 51–57.

Premkumar, J. and R. Ramaraj. 1997. Photocatalytic reduction of carbon dioxide to formic acid at porphyrin and phthalocyanine adsorbed Nafion membranes. Journal of Photochemistry and Photobiology A: Chemistry 110(1): 53–58.

Qin, Yusheng, Xianhong Wang, Suobo Zhang, Xiaojiang Zhao and Fosong Wang. 2008. Fixation of carbon dioxide into aliphatic polycarbonate, cobalt porphyrin catalyzed regio-specific poly(propylene carbonate) with high molecular weight. Journal of Polymer Science Part A: Polymer Chemistry 46(17): 5959–5967.

Ramírez, G., G. Ferraudi, Y. Y. Chen, E. Trollund and D. Villagra. 2009. Enhanced photoelectrochemical catalysis of CO_2 reduction mediated by a supramolecular electrode of packed CoII(tetrabenzoporphyrin). Inorganica Chimica Acta 362(1): 5–10.

Rao, Heng, Luciana C. Schmidt, Julien Bonin and Marc Robert. 2017. Visible-light-driven methane formation from CO_2 with a molecular iron catalyst. Nature 548(7665): 74–77.

Rao, Heng, Chern-Hooi Lim, Julien Bonin, Garret M. Miyake and Marc Robert. 2018. Visible-light-driven conversion of CO_2 to CH_4 with an organic sensitizer and an iron porphyrin catalyst. Journal of the American Chemical Society 140(51): 17830–17834.

Robert, Carine, Takahiro Ohkawara and Kyoko Nozaki. 2014. Manganese-corrole complexes as versatile catalysts for the ring-opening homo- and co-polymerization of epoxide. Chemistry—A European Journal 20(16): 4789–4795.

Sakakura, Toshiyasu, Jun-Chul Choi and Hiroyuki Yasuda. 2007. Transformation of carbon dioxide. Chemical Reviews 107(6): 2365–2387.

Sakakura, Toshiyasu and Kazufumi Kohno. 2009. The synthesis of organic carbonates from carbon dioxide. Chemical Communications (11): 1312–1330.

Savéant, Jean-Michel. 2008. Molecular catalysis of electrochemical reactions. Mechanistic aspects. Chemical Reviews 108(7): 2348–2378.

Sen, Pritha, Biswajit Mondal, Dibyajyoti Saha, Atanu Rana and Abhishek Dey. 2019. Role of 2nd sphere H-bonding residues in tuning the kinetics of CO_2 reduction to CO by iron porphyrin complexes. Dalton Transactions 48(18): 5965–5977.

Shaikh, Rafik Rajjak, Suriyaporn Pornpraprom and Valerio D'Elia. 2018. Catalytic strategies for the cycloaddition of pure, diluted, and waste CO_2 to epoxides under ambient conditions. ACS Catalysis 8(1): 419–450.

Silva, Joana de A. e., Valdemar F. Domingos, Daniela Marto, Leticia D. Costa, Mariana Marcos, Manuela R. Silva et al. 2013. Reversible sequestering of CO_2 on a multiporous crystalline framework of 2-quinolyl-porphyrin. Tetrahedron Letters 54(20): 2449–2451.

Smith, Peter T., Bahiru Punja Benke, Zhi Cao, Younghoon Kim, Eva M. Nichols, Kimoon Kim et al. 2018. Iron porphyrins embedded into a supramolecular porous organic cage for electrochemical CO_2 reduction in water. Angewandte Chemie International Edition 57(31): 9684–9688.

Sun, Xiaofu, Qinggong Zhu, Xinchen Kang, Huizhen Liu, Qingli Qian, Zhaofu Zhang et al. 2016. Molybdenum–bismuth bimetallic chalcogenide nanosheets for highly efficient electrocatalytic reduction of carbon dioxide to methanol. Angewandte Chemie International Edition 55(23): 6771–6775.

Tripkovic, Vladimir, Marco Vanin, Mohammedreza Karamad, Mårten E. Björketun, Karsten W. Jacobsen, Kristian S. Thygesen et al. 2013. Electrochemical CO_2 and CO reduction on metal-functionalized porphyrin-like graphene. The Journal of Physical Chemistry C 117(18): 9187–9195.

Wang, Shaolei, Kunpeng Song, Chengxin Zhang, Yu Shu, Tao Li and Bien Tan. 2017. A novel metalporphyrin-based microporous organic polymer with high CO_2 uptake and efficient chemical conversion of CO_2 under ambient conditions. Journal of Materials Chemistry A 5(4): 1509–1515.

Wang, Ziyi, Wei Zhou, Xin Wang, Xueliang Zhang, Huayu Chen, Huilin Hu et al. 2020. Enhanced photocatalytic CO_2 reduction over TiO_2 using metalloporphyrin as the cocatalyst. Catalysts 10(6): 654.

Windle, Christopher D., Marius V. Câmpian, Anne- K. Duhme-Klair, Elizabeth A. Gibson, Robin N. Perutz and Jacob Schneider. 2012. CO_2 photoreduction with long-wavelength light: dyads and monomers of zinc porphyrin and rhenium bipyridine. Chemical Communications 48(66): 8189–8191.

Wu, Chaoyong, Jinyao Wang, Pingjing Chang, Haiyang Cheng, Yancun Yu, Zhijian Wu et al. 2012. Polyureas from diamines and carbon dioxide: synthesis, structures and properties. Physical Chemistry Chemical Physics 14(2): 464–468.

Xu, Lei, Meng-Ke Zhai, Xin-Chao Lu and Hong-Bin Du. 2016. A robust indium–porphyrin framework for CO_2 capture and chemical transformation. Dalton Transactions 45(46): 18730–18736.

Xu, Xiaochun, Chunshan Song, John M. Andrésen, Bruce G. Miller and Alan W. Scaroni. 2003. Preparation and characterization of novel CO_2 "molecular basket" adsorbents based on polymer-modified mesoporous molecular sieve MCM-41. Microporous and Mesoporous Materials 62(1): 29–45.

Yadav, Rajrsh K., J.-O. Baeg, G.H. Oh, N.-J. Park, K.-j. Kong, J. Kim et al. 2012. A photocatalyst–enzyme coupled artificial photosynthesis system for solar energy in production of formic acid from CO_2. Journal of the American Chemical Society 134(28): 11455–11461.

Yadav, Rajesh K., Gyu Hwan Oh, No-Joong Park, Abhishek Kumar, Ki-jeong Kong and Jin-Ook Baeg. 2014. Highly selective solar-driven methanol from CO_2 by a photocatalyst/biocatalyst integrated system. Journal of the American Chemical Society 136(48): 16728–16731.

Younus Wani, Mohmmad, Avula Balakrishna, Santosh Kumar and Abilio J. F. N. Sobral. 2015. Covalently linked free-base and metallo-bis-porphyrins: Chemistry and diversity. Current Organic Chemistry 19(7): 599–651.

Zhang, Huinian, Jing Li, Shibo Xi, Yonghua Du, Xiao Hai, Junying Wang et al. 2019. A graphene-supported single-atom FeN_5 catalytic site for efficient electrochemical CO_2 reduction. Angewandte Chemie International Edition 58(42): 14871–14876.

Zhang,Xing, Zishan Wu, Xiao Zhang, Liewu Li, Yanyan Li, Haomin Xu et al. 2017. Highly selective and active CO_2 reduction electrocatalysts based on cobalt phthalocyanine/carbon nanotube hybrid structures. Nature Communications 8(1): 14675.

Zhu, Minghui, Jiacheng Chen, Libei Huang, Ruquan Ye, Jing Xu and Yi-Fan Han. 2019. Covalently grafting cobalt porphyrin onto carbon nanotubes for efficient CO_2 electroreduction. Angewandte Chemie International Edition 58(20): 6595–6599.

CHAPTER 4

Application of Metallic Foam in Solar Power System

Ajay Kumar P.[1,2,*] and *Vishnu Namboodiri V.*[3]

Introduction

Energy harnessing strategies have exhibited a remarkable development in past few decades. The development of renewable energy was considered as a bigger step towards sustainable and ecofriendly power generations. At the sametime, the evolution of solar power generation systems were placed amongst the highest in development. The solar power generation systems can be broadly classified into two systems, they are the photovoltaic system and solar thermal systems. The photovoltaic systems can be further classified into non concentrating and concentrating types. The solar thermal systems have various classifications which depend on their applications but most of them are solar concentrating types. The solar based power systems are designed based on the heat transfer mechanisms. In general, it is understood that the higher heat transfer rates enable greater efficiency and better performance. The non-conventional design routes are of greater interest for researchers. In this context, one of the emerging areas is to improve the efficiency and performance by using special kinds of materials which offer greater thermal conductivities. Furthermore, the evolution of Phase Change Materials (PCM) and various advancements in the solar power systems pushed research interest to another level in performance improvements. But there are some challenges that are observed for *in situ* PCM, they have low thermal conductivity, fire safety and ecofriendly aspects. This scenario catalyzed evolution of Metal Foams (MF). The MF is a cellular structured form of a metal. The common fabrication routes involved for MF is through injection of gas into a molten metal which forms a cellular structure with open or closed pores. These

[1] Department of Mechanical Engineering, Indian Institute of Technology Tirupati (A.P.) India-517506.
[2] Department of Materials Science and Engineering, University of Wisconsin-Milwaukee USA-53201.
[3] National Institute of Construction Management and Research (NICMAR), Hyderabad (Telangana) India.
* Corresponding author: ajaymits85@gmail.com

MF exhibit the base metal properties to some extent. As the foam is of metal this helps in the improvement of fire safety aspects. The basic use of MF in solar based power generations systems are required to define better understandings. Past studies have revealed that the MF can be used along with PCM in Thermal Energy Storage Systems (TES), in stand alone solar thermal systems, and in thermal management systems of allied equipments. Considering all the possible applications, the PCM with MF in TES have more research interests. The detailed features and applications are established later. The metal foam is often known as a catalyst for heat transfer enhancement along with the PCM in thermal energy storages. This may be due to the unique properties such as, high porosity, large specific surface area and considerable thermal and mechanical stability.

Fundamentals of metal foams

Metal foams are new kinds of materials which have a cellular structure of a metal. The metal foams have significant applications in manufacturing of lightweight structures, energy and thermal management and high temperature filters are a few of them. In general, the properties of the metal foam are based on the properties of base material, relative density and cell topology. There are various methods available for the fabrication of metal foams and are discussed next.

Fabrication of metal foams

There are several techniques that can be adopted for the manufacture of metal foams. They are, melt gas injection method, gas releasing particle decomposition in melt and semi solids method, casting using a precursor, metal deposition and entrapped gas expansion are few of them. A few of them are discussed in detail next.

Melt Gas Injection (MGI) method

The melt gas injection methods are normally adopted for low density alloys. The aluminum alloys exhibit a low-density property and have a controlled oxidation nature. Thus, aluminum alloy-based metal foam can be routed through the MGI method. One of the methods has been illustrated in the Fig. 4.1. According to this method, the aluminum or aluminum-based alloys are initially heated and ceramic particles are added for stabilization which is normally 5–15 wt%. These ceramic particles are of 0.5–25 μm in diameter and made of alumina, zirconia, silicon carbide or titanium diboride. The following gases can be considered for bubble production: air, carbon dioxide, oxygen and inert gases. In some cases, even water can be used to produce bubbles. The process starts with the melting of the metal and stabilizer, then a stirrer with provision of gas injection is used to stir and produce bubbles in the liquid metal. The bubbles will float and move towards the surface, then they are drained and start the solidification and produce the metal foam. The low density and closed cell foams can be produced especially by optimization of the gas injection process and foam cooling rate (Ashby et al. 2000).

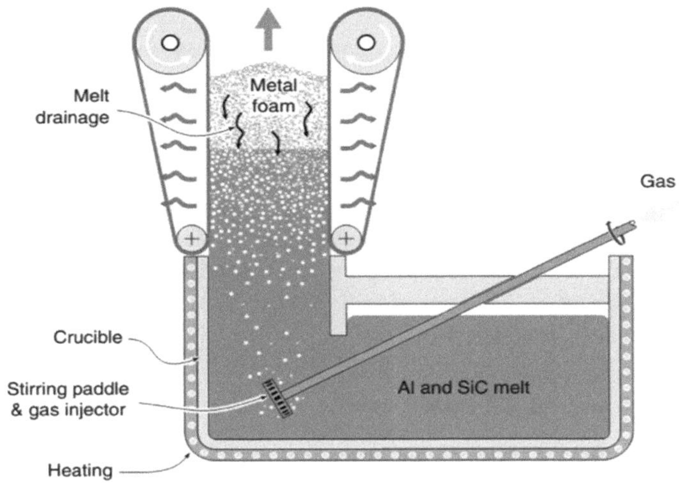

Figure 4.1: Melt gas injection setup for an aluminum foam (Reprinted with permission from Ashby et al. 2000).

Gas-releasing particle decomposition method in semi-solid

The gas releasing particle decomposition method uses a foaming agent in the process. The titanium hydride is widely used as the foaming agent. The forming agent decomposes at a lower temperature than the melting temperature of the base material. For this method aluminum based alloys also have greater interest for the fabrication of the foam. Initially the foaming agent is mixed with the aluminum powder and subsequently increases the temperature in a mold. About 465°C, the Ti based foaming agent starts decomposing and releases the gas. At this time, the aluminum powders are in a semi solid form. The gases create the bubble and the bubble grows to a limiting point. Then the semi solid mixture cools and forms the Al metal foam (Fig. 4.2). This method has been developed by IFAM in Bremen, Germany, LKR in Randshofen, Austria, and Neuman-Alu in Marktl, Austria. Closed cell foams can also be fabricated through this method successfully (Ashby et al. 2000).

Metal deposition method on cellular preforms

In this method, an open cell polymer template/substrate are chosen and placed inside the chemical vapor decomposition reactor. The nickel foam is synthesized by the decomposition of the nickel carbonyl. The nickel carbonyl was introduced to the chemical vapor decomposition reactor and the nickel carbonyl decomposes about 100°C to nickel and carbon monoxide. During this process, the nickel is coated in the polymer template. Once the coatings get over, the nickel with polymer template will be removed. The polymer template is then heated, destroyed and forms the nickel foam. A subsequent sintering is carried out to densify the ligaments. On the other hand, this method is very challenging due to the high toxic nature of nickel carbonyl. This method is often used only for the pure metals like nickel and titanium. It also gives the lowest relative density (0.02–0.05) foams available today (Ashby et al. 2000).

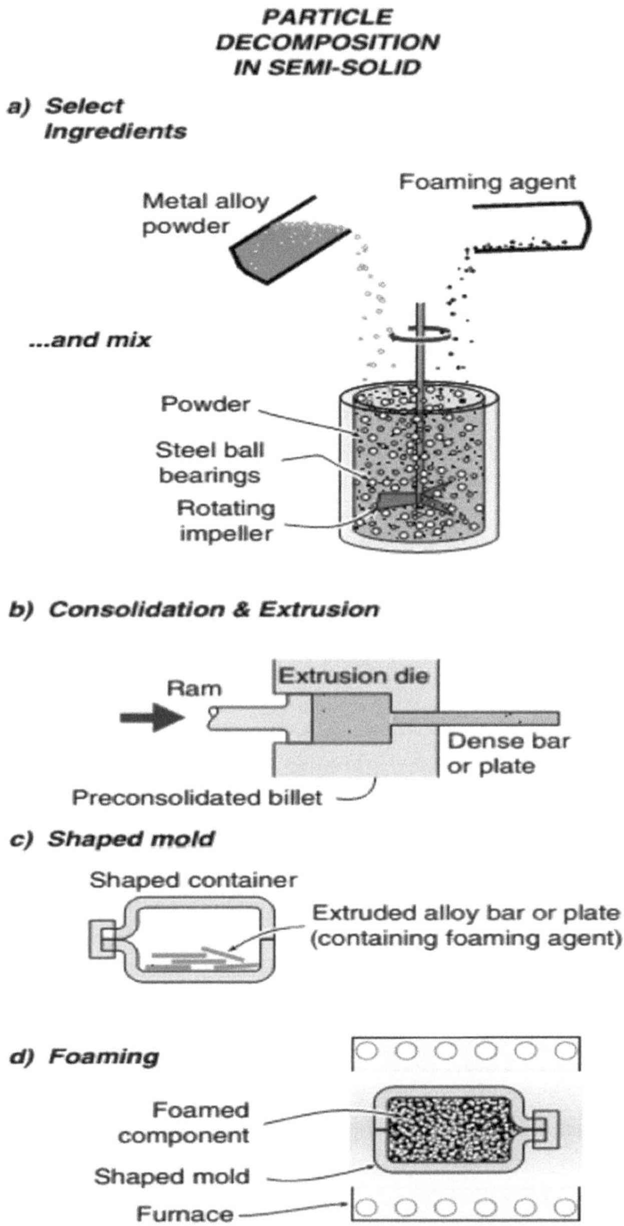

Figure 4.2: Gas releasing particle decomposition in a semi-solid method with various processing steps (Reprinted with permission from Ashby et al. 2000).

Characterization and properties of metal foams

The cell topology, relative density, cell size and shape and anisotropy are the structural characterization entities of metal foams. The cell topology can be divided into an open and closed cell. The structural characterization features of the metal foams can be studied with the help of various techniques such as, Scanning Electron Microscopy (SEM), optical microscopy, and X-ray Computed Tomography (CT) (Ashby et al. 2000). Along with the structural features, the properties of the metal foams also need to be studied. Some of the mechanical properties of various metal foams are demonstrated next.

Mechanical properties

The mechanical properties of Al metal foam were established by (Papadopoulos et al. 2004). As discussed earlier, melt foaming methods were adopted for the fabrication of the Al metal foam. An experimental and numerical study is established for the calculation of the Young's Modulus (E) and the plateau stress for the Al metal foam. Initially a stress strain curve was obtained for the Al foam (Fig. 4.3) through compressive testing for a deformation rate of 3×10^{-3}. It has been observed that there is a crack initiation at the cell wall during the compressive testing and original pore starts gradual elongation. This infers the ductile behavior of Al metal foam.

The numerical results for the Young's Modulus (E) and the plateau stress (σ_{pl}) are obtained based on the Gibson and Ashby (Gibson and Ashby 1997) theory for two cases (fraction of solid contained in cell edge $\phi = 0.30$ and 0.50). For $\phi = 0.30$ the E = 27.1 GPa and $\sigma_{pl} = 9.5$ MPa and the experimental results are E = 23 GPa and $\sigma_{pl} = 9.2$ MPa. Another theory by (Onck et al. 2001), the numerical value for $\phi = 0.30$ the E = 19.4 GPa and $\sigma_{pl} = 8.9$ MPa and the experimental results are E = 23 GPa and $\sigma_{pl} = 9.2$ MPa. In general, the numerical and experimental results are in good agreement and it is observed that the E value for Al form is between 19–23 GPa.

Anisotropy and strain rate sensitivity of open cell aluminum foam was analyzed numerically and experimentally by (Vesenjak et al. 2012). The results exhibited that the mechanical properties are in anisotropic nature. Furthermore, the yield

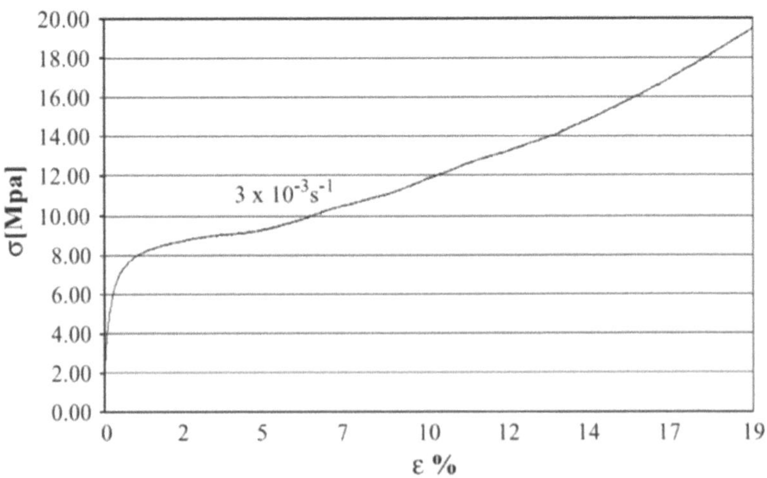

Figure 4.3: Stress-strain curve for Al metal foam obtained from compressive testing (Reprinted with permission from Papadopoulos et al. 2004).

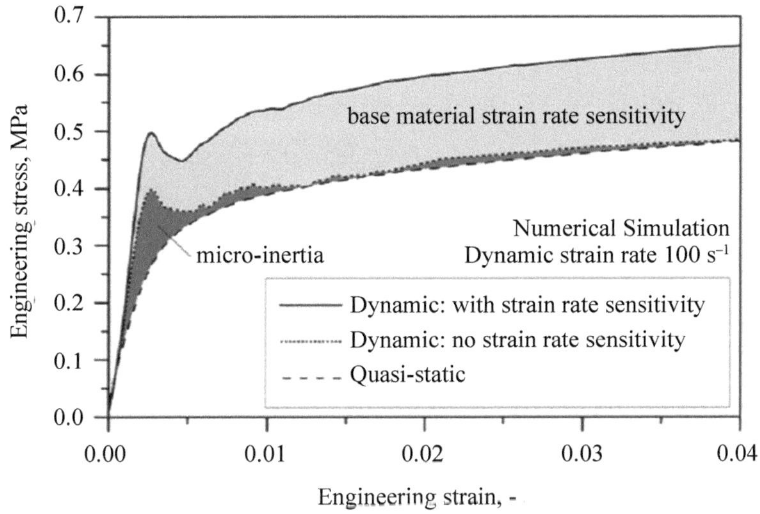

Figure 4.4: Influence of the strain rate and mass inertia (Reprinted with permission from Vesenjak et al. 2012).

and plateau stress grows with increase in strain rates. The studies are related to the influence of strain rates and mass inertia. The results indicate that the dynamic simulations are influenced by micro inertia (dark region) and a subsequent calculation of strain rate energy (area of the dark region which includes both quasi static and dynamic simulation) shows that micro-inertia has a less significant material stiffness (Fig. 4.4). In general, the strain rates are sensitive to influencing various basic stress in the materials (Vesenjak et al. 2012).

Thermophysical properties

The numerical and experimental (Al and Reticulated Vitreous Carbon (RCV) foams used for the studies) analysis of thermophysical properties like effective thermal conductivity, permeability, and inertial coefficient of various metal foams were established by (Bhattacharya et al. 2002). The results indicate that the effective thermal conductivity is significantly influenced by porosity. However, a non-systematic influence was observed for pore density for effective thermal conductivity. In the case of permeability as porosity and pore diameter progress, the permeability also increases. This indicated a strong influence of porosity and pore diameter in the improvement of permeability. Furthermore, the inertial coefficient is only influenced by porosity. Both numerical and experimental analysis shows good agreement (Bhattacharya et al. 2002).

Applications and performance of metal foams in solar power systems

The use of the MF in the solar based systems can be classified broadly into two. They are (1) direct applications in the receiver (as absorbent), and in Thermal Energy

Storages (TES) (2) as a component in thermal management systems of solar based power systems.

Direct applications

The performance of a solar based power generation system which works based on solar thermo chemical process of different catalytically activated metal foam (CA-MF) absorber in the reactor section has been established by Kodama. Furthermore, the objective of these types of systems to produce the syngas through a solar thermo chemical process and further can be compressed and used for power generation. A Ni-Cr-Al metal foam disk was considered for the fabrication of CA-MF with various materials like Ru, Rh and Ni. The two different CA-MF absorber types were synthesized: (1) with alumina coatings and (2) noncoated. The CA-MF absorber with alumina coating and noncoated are placed in the reactor and the irradiation are simulated artificially. The Ru, Rh and Ni absorber 7.3 and 14.3 μmol-metal/cm^2 in the surface of Ni-Cr-Al metal foam disk with alumina coated and noncoated. The results show that Ru-Alumina-Ni-Cr-Al metal foam disk absorber exhibited better yield than other configurations. In addition, it has been observed that the metal foams can be a good candidate for the fabrication of CA-MF absorber for a solar based power generation system (Kodama et al. 2003).

Latent heat thermal energy storage systems in which the thermal energy is stored in the PCM material and is further utilized for power generation. A concentric heat pipe (foil/foam) and copper disk based latent heat thermal energy storage system with PCM and metal foam or combination of both in a copper disk and heat pipe (foil/metal foam) in various configurations has been experimentally studied by (Allen et al. 2017) with respect to the influences of the inclination in the solidification and melting of the PCM. Figure 4.5 illustrates the experimental setup. The n-octadecane selected as the PCM and Duocelaluminum foam selected as the metal foam for the studies.

The influence of the inclination (0°, 30°, 60° and 90°) in solidification and melting of the concentric heat pipe and foil based latent heat thermal energy storage setup has been established. The results show that the orientation has very limited influences in the solidification. In addition, the melting process are highly influenced in the alteration of melting rates and improve the melting rates for various orientations especially found better in horizontal orientation. Also, the heat pipe with foam configuration can be a good candidate for PCM based latent heat thermal energy storage systems (Figs. 4.6 and 4.7) (Allen et al. 2017).

The energy storage efficiency of a solar based CO_2 reforming of the methane system in which the reactor is embedded with the metal foam was established by Fuqiang et al. and numerical studies were established. Furthermore, the experimental studies were validated with results from (Gokon et al. 2009). Ru/c-Al_2O_3/Ni-Cr-Al metallic foams are considered for the experimental studies by (Gokon et al. 2009). The comparison results indicate good agreement between the numerical and experimental analysis. It is clear that the increasing temperature augments the reaction rates. Furthermore, the temperature distribution along the centerline of the

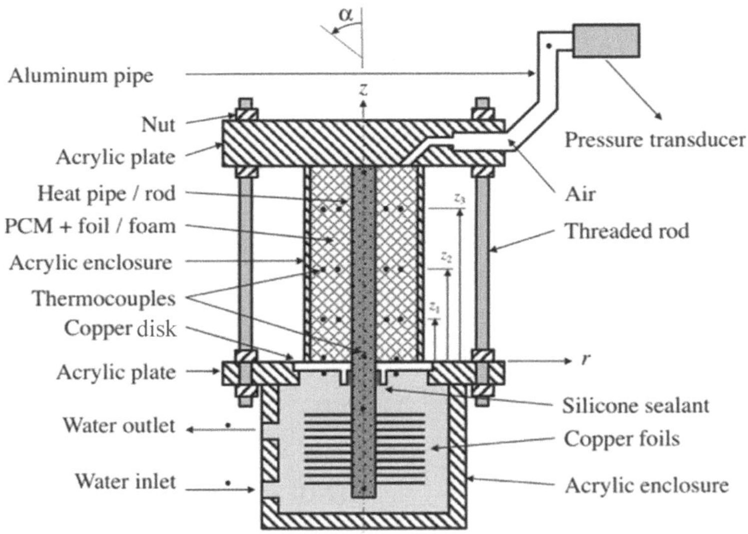

Figure 4.5: Latent heat thermal energy storage setup with various configurations (Reprinted with permission from Allen et al. 2017).

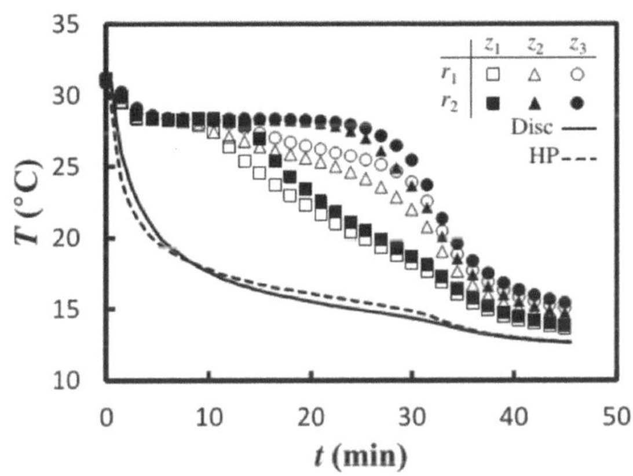

Figure 4.6: Solidification temperature distribution for heat pipe-foam-PCM at 0° (Reprinted with permission from Allen et al. 2017).

metal foam was obtained through numerical simulations for various gas mixture velocities at specific porosity (0.90).

The results indicate that as gas mixture velocity increases the fluid phase temperature decreases (Fig. 4.8). In addition, as porosity progresses the fluid phase temperature decreases (Fig. 4.9) and the porosity has low significance in the fluid entrance surface.

The increase in porosity does not significantly improve the methane conversion efficiency and energy storage efficiency for metal foam at specific gas mixture velocity of 0.20 m/s (Fig. 4.10). A slight improvement in both the efficiencies are also observed after a porosity value of 0.90 but was not significant enough (Fuqiang

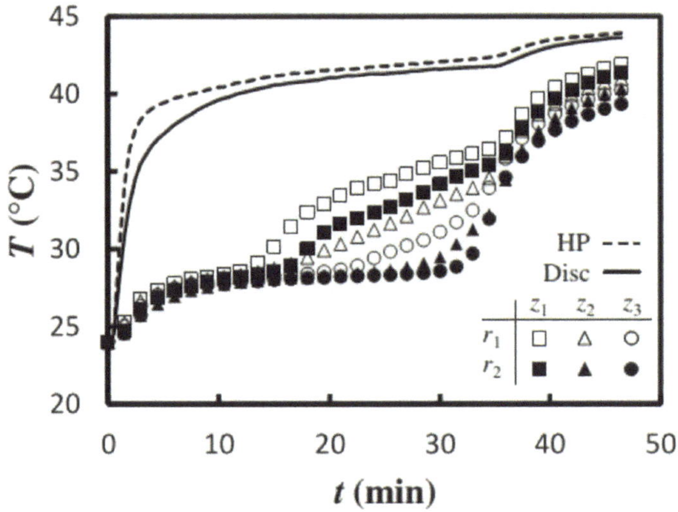

Figure 4.7: Melting temperature distribution for heat pipe-foam-PCM at 0° (Reprinted with permission from Allen et al. 2017).

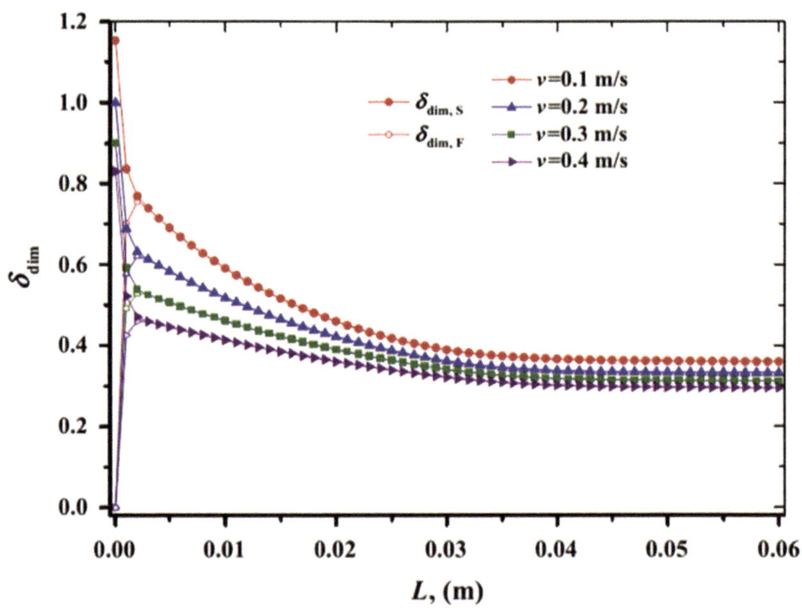

Figure 4.8: Centerline temperature distribution in metal foam for various gas mixture velocities (Reprinted with permission from Fuqiang et al. 2017).

et al. 2017). In general, the temperature increases with the gas mixture velocity and increases with porosity. Furthermore, in terms of the storage and methane efficiency the porosity progress decreases the efficiencies (Fuqiang et al. 2017).

An experimental analysis of open cell porous metal foams and nanofluids for solar absorbent applications were established by (Valizade et al. 2019). The optical

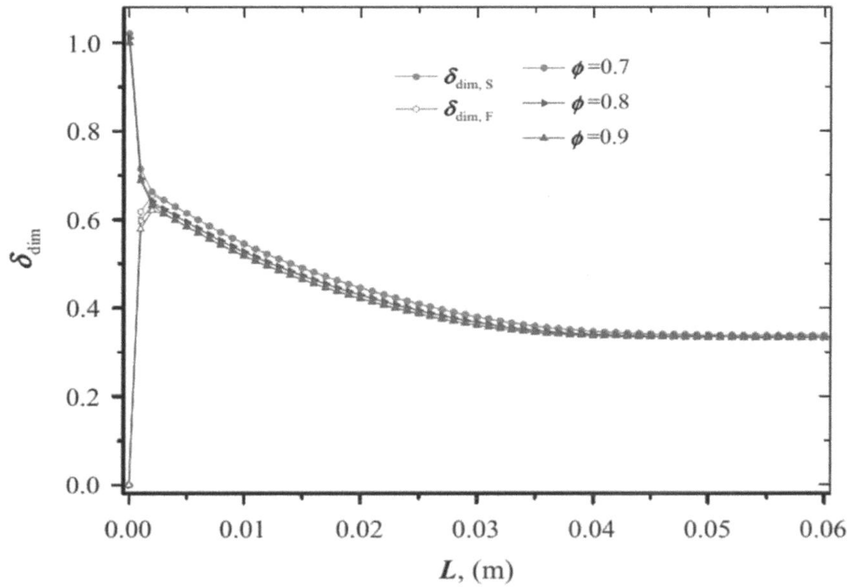

Figure 4.9: Centerline Temperature Distribution in metal foam for various porosity values (Reprinted with permission from Fuqiang et al. 2017).

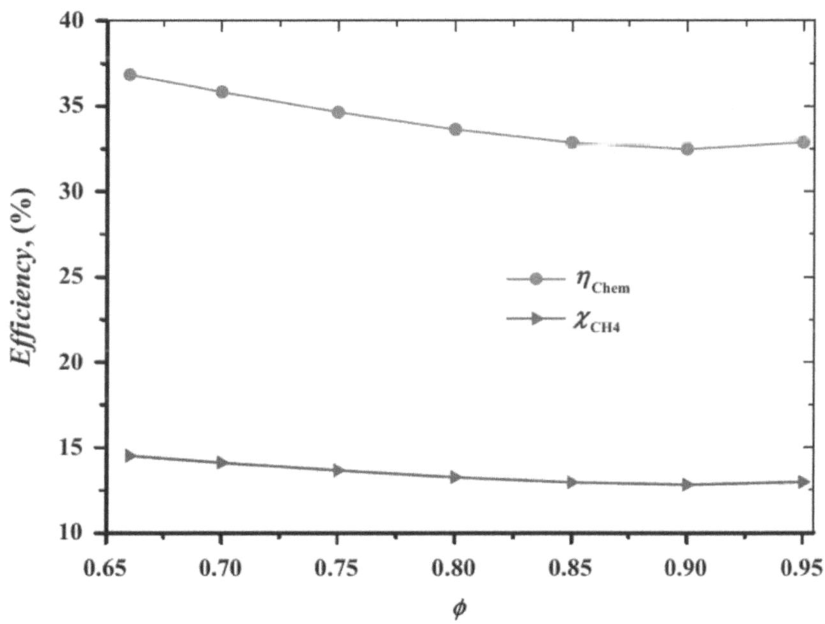

Figure 4.10: The methane conversion efficiency (χ_{CH4}) and energy storage efficiency (η_{Chem}) for metal foam have different porosity values (Reprinted with permission from Fuqiang et al. 2017).

properties of these porous and nanofluids were the objectives of the studies. Better thermal performance is the basic requirement for the solar based absorbers. To understand the thermal performance, one has to understand the optical properties of the foam material. The optical properties greatly influence the thermal behavior of the porous foams as an absorbent in the solar based power systems. For better understanding the variable porosity and pore diameter metal foam was considered for the experimental study. The CuO and SiC based porous foam (10 mm thick) with porosity (90–95%), and pore diameter (10PPI (Pores per Inch)-20PPI-30PPI) were chosen for the studies. The same materials are considered for the preparation of the nanofluids. The optical properties of porous foams are determined from the spectroscopy analysis. The results are obtained for CuO with water and air as the reference for the transmittance. The results show that increasing number of pores will reduce the rate of transmittance. By increasing the solid surface area, the rate of absorption also increases. If the porosity increases, the amount of absorption decreases and transmittance coefficient increases (Fig. 4.11). If the porosity increases from 90 to 95% for a fixed cell diameter exhibits an increase in transmittance by four times. Furthermore, the foam material has little effect in the transmittance (Valizade et al. 2019).

The performance of copper foam in an absorber for solar parabolic trough collector was established by (Jamal-Abad et al. 2017). The porosity of the copper foam is 0.9 and pore density is 30 PPI. The experiments were executed with various flow rate regimes (0.5 to 1.5 Ltrs/min). The ASHRAE standard 93 was used for the performance calculation of the collector. The absorber was filled with copper foam that was used for experimental analysis. The results show that as the flow rate progresses, the collector efficiency is enhanced. This may due to reduction of the energy loss and subsequent increase in the efficiency (Jamal-Abad et al. 2017).

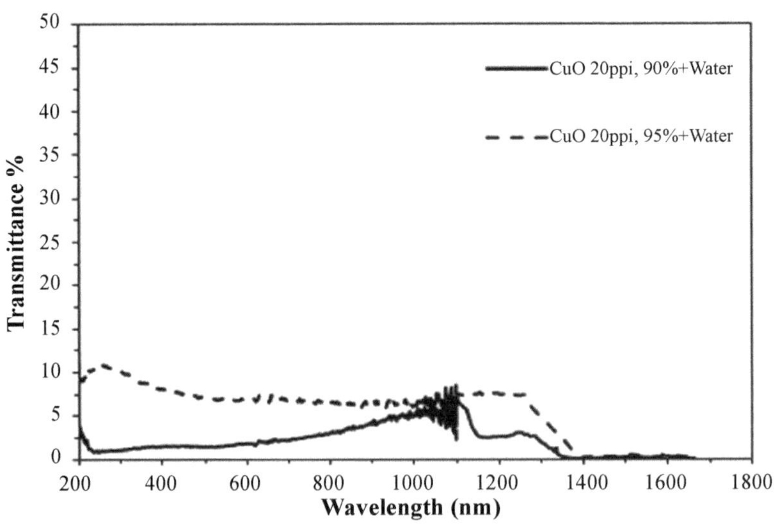

Figure 4.11: Influence of different porosity in the transmittance for CuO foam with fixed pore diameter (Reprinted with permission from Valizade et al. 2019).

Valizade et al. established the performance of copper foam in an absorber for a solar parabolic trough collector. The porosity of the copper foam is 0.95 and pore density is 10 PPI in different regimes (full porous, semi-porous and without porous). The experiments are established with various flow rate conditions and different inlet temperatures. The ASHRAE standard 93 was used for the performance calculation of the collector. The results show that the pore foams increase the friction factor. Furthermore, the collector thermal efficiency can be improved by increasing the flow rate and decreasing the inlet temperature (Fig. 4.12). The full porous and semi-porous foams exhibited maximum efficiency (Valizade et al. 2020).

Kim et al. incorporated graphite foam along with PCM for a latent heat thermal energy storage application. The number of heat pipes required in the PCM-graphite foam tank and PCM tank were evaluated and the results indicated that the number of heat transfer pipes can be reduced by the incorporation of the graphite foam in the PCM tank (Kim et al. 2014). The performance of the latent heat thermal energy storage works on molten salt (eutectic salt (50 wt% $NaNO_3$, 50 wt% KNO_3)) and metal foam/eutectic salt composite as PCM materials were studied and it was found that the metal foam/salt composite exhibits significant enhancement in the heat retrieval regime (Zhang et al. 2015). Zhao et al. investigated the influences of the heat flux, porosity and pore density of a metal foam embedded in the paraffin wax RT58 PCM for a thermal energy storage system. The results show that the heat transfer performance can be significantly enhanced with metal foam in addition to PCM. Better results are obtained for the configuration having smaller porosity and pore density (Zhao et al. 2010). Chen et al. exhibited the numerical studies related to the performance of a solar flat-plate collector in which the collector is embedded with the aluminum foam and paraffin. The temperature variation results show that

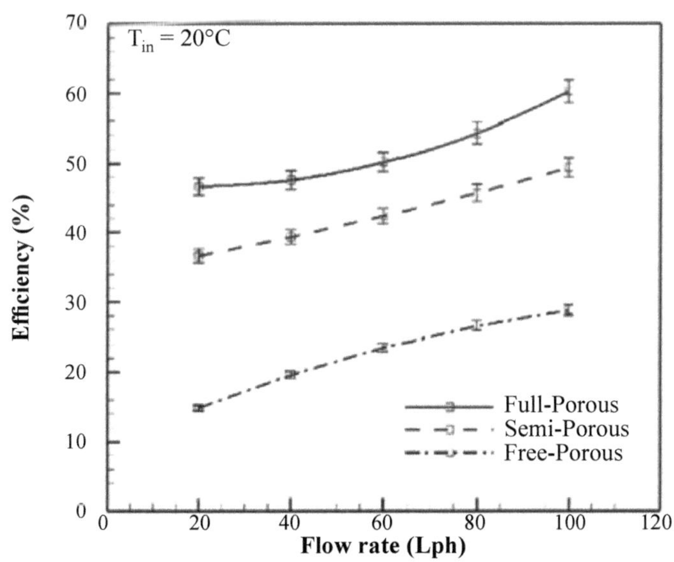

Figure 4.12: The efficiency of various porous foam regimes at temperature which is low as compared to other evaluation (Reprinted with permission from Valizade et al. 2020).

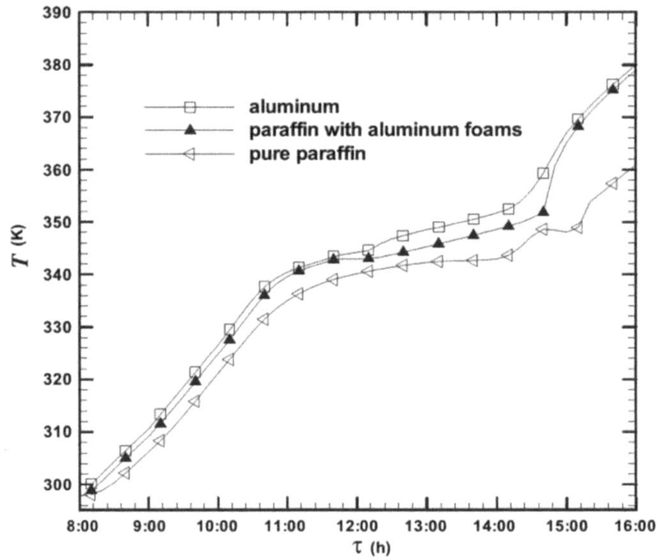

Figure 4.13: The temperature variations for different materials solarflat-plate collector (Reprinted with permission from Chen et al. 2010).

the significant temperature improvement with paraffin with aluminum metal foam configurations (Fig. 4.13). In general, a significant heat transfer enhancement was observed for paraffin with aluminum metal foam (Chen et al. 2010).

Indirect applications

The solar photovoltaic power systems are widely used as a reliable and renewable power source which ranges from few kW to MW capacities. There are different allied electrical systems that are required along with the PV module. It is clear that most of the sites experience high temperature conditions. The electrical components are to be cooled and should be maintained at a particular temperature range to exhibit considerable efficiencies. The uses of PCM that are attained are of greater interest in these applications, but PCM offers low temperature conductivity. Thus, the uses of PCM materials in thermal management systems are very limited. To overcome these challenges one remarkable consideration was established with metal foam filled with PCM. Engeda et al. conducted numerical and experimental studies of latent cooling system for an electrical allied system in a solar PV system with metal foam PCM composite. This composite will act as a heat dissipator for a cooling system. Paraffin-metal foam composite has been considered for numerical studies. The paraffin-metal foam composite material is filled in a cylindrical module which is attached to the cooling module. A CFD numerical simulation with k-epsilon model has been used for the analysis. The CFD results of inlet and outlet temperature show that paraffin-metal foam composite material has a greater role in the thermal management and significantly reduces the outlet temperature Fig. 4.14. The prototype of the cooling module was also fabricated and evaluated. Both the numerical and experimental

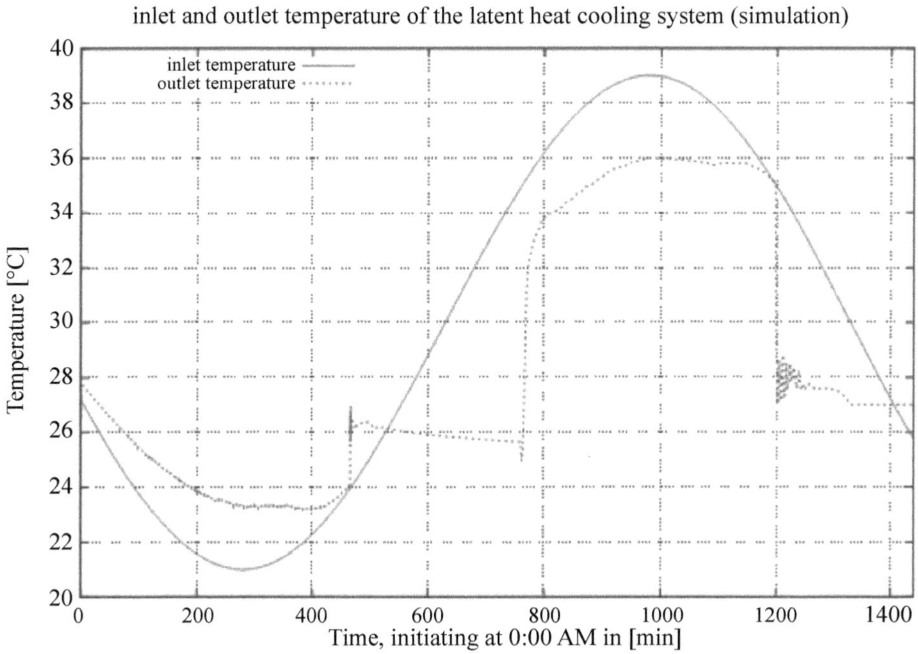

Figure 4.14: Inlet and Outlet Temperature Results from CFD Simulation of Cooling Module with paraffin-metal foam composite material (Reprinted with permission from Engeda et al. 2010).

results show that the paraffin-metal foam composite material are a good candidate in the thermal management systems for allied electrical components in PV module power generation systems (Engeda et al. 2010).

Concluding remarks

Metal foams have considerable mechanical and thermophysical properties. This enables the wide applications of metal foams. The selection of fabrication methods of metal foams is dependent on the material which one requires to covert as foam. This may increase the fabrication cost of the metal foam. In general, aluminum, aluminum alloy and copper-based foams are widely used in the industry. It could be because the fabrication routes are affordable. Furthermore, the better thermal property of metal foams gained the attention for use in various solar based power generations.

The direct application of the metal foam can be defined for use as an absorber in the solar receiver, along with PCM materials in the thermal energy storage systems, and also in the absorber of solar flat-plate collector. In addition, it is clear that the heat transfer can be enhanced by use of metal foam. But there are some parameters that need to be optimized to yield maximum efficiencies for the solar based power generation systems. The indirect applications of the metal foams can be defined for thermal management for solar based power systems, especially for cooling of electrical systems. It should also be noted that the use of metal foam as components in thermal management systems are much less in research numbers. Finally, the metal foams are promising in various applications and continuous improvements are further needed for cost effective and highly reliable foam which can be used for a wide range of applications.

Future prospects

1. Development of new metal foams and fabrication routes are to be undertaken.
2. The use of metal foam in thermal management systems need rigorous studies to understand the potentiality and influences.
3. The influence of corrosion in metal foam and their effects in the performances of various solar based power generation systems.

References

Allen, M. J., N. Sharifi, A. Faghri and T. L. Bergman. 2015. Effect of inclination angle during melting and solidification of a phase change material using a combined heat pipe-metal foam or foil configuration. International Journal of Heat and Mass Transfer 80: 767–780.

Ashby, M. F., A. G. Evans, N. A. Fleck, L. J. Gibson, J. W. Hutchinson and H. N. G. Wadley. 2000. Metal foams—A design guide. Butterworth-Heinemann, Boston.

Bhattacharya, A., V. V. Calmidi and R. L. Mahajan. 2002. Thermophysical properties of high porosity metal foams. International Journal of Heat and Mass Transfer 45(5): 1017–1031.

Chen, Z., M. Gu and D. Peng. 2010. Heat transfer performance analysis of a solar flat-plate collector with an integrated metal foam porous structure filled with paraffin. Applied Thermal Engineering 30(14-15): 1967–1973.

Engeda, S., D. Girlich, A. Kneer, E. Martens and B. Nestler. 2010. Development of a Metal Foam Based Latent Heat Cooling System in the Field of Solar Power Generation. CELLMAT2010 International Conference.

Fuqiang, W., C. Ziming, T. Jianyu, Z. Jiaqi, L. Yu and L. Linhua. 2017. Energy storage efficiency analyses of CO_2 reforming of methane in metal foam solar thermochemical reactor. Applied Thermal Engineering 111: 1091–1100.

Gibson, L. J. and M. F. Ashby. 1997. Cellular Solids. Second edition, Cambridge Univ. Press, Cambridge, UK.

Gokon, N., Y. Osawa, D. Nakazawa and T. Kodama. 2009. Kinetics of CO_2 reforming of methane by catalytically activated metallic foam absorber for solar receiver reactors. Int. J. Hydrogen. Energy 34: 1787–1800.

Jamal-Abad, M. T., S. Saedodin and M. Aminy. 2017. Experimental investigation on a solar parabolic trough collector for absorber tube filled with porous media. Renewable Energy 107: 156–163.

Kim, T., D. M. France, W. Yu, W. Zhao and D. Singh. 2014. Heat transfer analysis of a latent heat thermal energy storage system using graphite foam for concentrated solar power. Solar Energy 103: 438–447.

Kodama, T., A. Kiyama and K. -I. Shimizu. 2003. Catalytically activated metal foam absorber for light-to-chemical energy conversion via solar reforming of methane. Energy & Fuels 17(1): 13–17.

Onck, P. R., E. W. Andrews and L. J. Gibson. 2001. Size effects in ductile cellular solids: Part I. Modeling. International Journal of Mechanical Sciences 43: 681–699.

Papadopoulos, D., I. C. Konstantinidis, N. Papanastasiou, S. Skolianos, H. Lefakis and D. Tsipas. 2004. Mechanical properties of Al metal foams. Materials Letters 58(21): 2574–2578.

Shang, B., J. Hu, R. Hu, J. Cheng and X. Luo. 2018. Modularized thermal storage unit of metal foam/paraffin composite. International Journal of Heat and Mass Transfer 125: 596–603.

Valizade, M., M. M. Heyhat and M. Maerefat. 2019. Experimental comparison of optical properties of nanofluid and metal foam for using in direct absorption solar collectors. Solar Energy Materials and Solar Cells 195: 71–80.

Valizade, M., M. M. Heyhat and M. Maerefat. 2020. Experimental study of the thermal behavior of direct absorption parabolic trough collector by applying copper metal foam as volumetric solar absorption. Renewable Energy 145: 261–269.

Vesenjak, M., C. Veyhl and T. Fiedler. 2012. Analysis of anisotropy and strain rate sensitivity of open-cell metal foam. Materials Science and Engineering: A 541: 105–109.

Zhang, P., X. Xiao, Z. N. Meng and M. Li. 2015. Heat transfer characteristics of a molten-salt thermal energy storage unit with and without heat transfer enhancement. Applied Energy 137: 758–772.

Zhao, C. Y., W. Lu and Y. Tian. 2010. Heat transfer enhancement for thermal energy storage using metal foams embedded within phase change materials (PCMs). Solar Energy 84(8): 1402–1412.

CHAPTER 5

Valorization Chemistry

A Compendium on Photoreduction of CO_2 to Biofuels Over Nano TiO_2

Sounak Roy

Introduction

The primary energy feedstock of today's world is non-renewable finite carbon-based fossil fuels such as, petroleum, coal and natural gas. According to the Statistical Review published by British Petroleum in 2018, the world's main primary energy sources consist of 34% petroleum, 27% coal and 24% natural gas, amounting to an 85% share for fossil fuels in primary energy consumption in the world (BP Statistical Review 2019). Interestingly, primary energy consumption grew at a rate of 2.9% in 2018, which was almost double its 10-year average of 1.5% per year. This dependency would continuously grow as the increase in energy demand is inevitable. Energy Information Administration (EIA) predicts that world energy consumption will grow by nearly 50% between 2018 and 2050 (EIA 2020). This large reliance on fossil fuel has two major problems. First the primary energy stock is finite, and is quickly depleting. As a result, the annual average oil price rose to US$71.31 per barrel in 2018, up from $54.19/barrel in 2017. Secondly, the uncontrolled combustion accumulated a significant amount of CO_2 in the environment, and as of 2019, its concentration was almost 48% above pre-industrial levels (ESRL 2019).

Both the problems, depletion of fossil fuel as well as accumulation of greenhouse gas in atmosphere, can be handled well by valorization chemistry. Valorization of anthropogenic CO_2 into useful platform chemicals and biofuels like CH_4, CO, HCOOH, HCHO, and CH_3OH would not only reduce the accumulated

Department of Chemistry, Birla Institute of Technology and Science (BITS) Pilani, Hyderabad Campus, Jawahar Nagar, Shameerpet Mandal, Hyderabad-500078, India.
Email: sounak.roy@hyderabad.bits-pilani.ac.in

CO_2 from the atmosphere, but would also develop renewable alternative fuels as a replacement of existing fossil fuels. Sun being a inexhaustible primary energy source, would be an ideal and optimum solution to convert CO_2 to biofuels with the help of solar illumination. This artificial photosynthesis would be a completely carbon neutral cycle with positive energy balance. Solar reduction of CO_2 requires visible light active semiconductor materials, which can provide electrons and protons to surface adsorbed CO_2. In spite of a wide variety of reported solar-active catalysts for CO_2 photoreduction, a significant amount of them suffer from low energy conversion efficiency, poor product selectivity and inability in suppressing the competing hydrogen generation. Therefore, the design and fabrication of highly active photocatalytic systems with high conversion efficiency and selectivity for CO_2 reduction remains a great challenge.

Water photocatalysis was invented by Fujishima and Honda in 1972 (Fujishima and Honda 1972). TiO_2 semiconductor has emerged as the catalyst of choice due to excellent features such as high surface area, non-toxicity, high thermal, chemical and economic stability, and due to abundant availability (Roy et al. 2007, Challagulla and Roy 2017a, Soman et al. 2018, Roy et al. 2018, Daghrir et al. 2013, Low et al. 2017). In the recent past, TiO_2-based nanostructured photocatalysts such as 1-D, 2-D, 3-D or hierarchical structures have attracted significant attention in the area of CO_2 photoreduction (Kočí et al. 2008, Razzaq and Su-II 2019, Al Jitan et al. 2020, Habisreutinger et al. 2013, Ola and Maroto-Valer 2015). The distinct properties of high surface area and aspect ratio, well separation of photogenerated charges made nano TiO_2 valuable and worthwhile to use in the photocatalysis research domain. Despite the eminent benefits of nanostructured TiO_2, the material possesses limitations in absorbing solar radiation due to their wider band gap (E_g = ~ 3.0–3.2 eV). Therefore, to overcome this, strategies including metal and non-metal doping, noble metal loading, hetero-junctioning TiO_2 nanostructures by anchoring low band gap materials are popularly adopted. In this chapter, the recent advances on CO_2 photoreduction over TiO_2-based catalysts are critically discussed. At first, photocatalytic reduction mechanism of CO_2 is explained. Later, various modification techniques of TiO_2, including morphological and size modifications, metal/non-metal doping, formation of semiconductor heterostructures and dispersion on high surface area supports are summarized. At the end, a future outlook has been portrayed.

Photocatalytic reduction of CO_2 over TiO_2: Fundamentals and mechanism

In photocatalytic reduction of CO_2, TiO_2 plays the most important role. Once light is irradiated, the energy absorbed by TiO_2 produces electron–hole pairs with electrons (e^-) in the Conduction Band (CB) and holes (h^+) at the Valence Band (VB). The electrons are transferred to surface adsorbed CO_2, while holes are transferred to the interface between TiO_2 and water present in the reaction medium in order to participate in the red-ox process. These holes from the VB oxidize water to form oxygen and protons, which in turn along with the electrons from CB react with the surface-adsorbed CO_2 to produce the desired products by the proton-assisted multi-electron photoreduction process. The selectivity of the products depends on the number of available electrons and protons as can be seen in Eqn. 1–7. The adsorbed CO_2 on reaction with two electrons and protons form a HCOOH and gaseous CO, whereas, with 6 electrons/protons reaction lead to formation of CH_3OH. All the reduction potentials are in Normal Hydrogen Electrode (NHE) at pH 7.

$$CO_2 + 2H^+ + 2e^- \rightarrow HCO_2H; E^o_{red} = -0.61 \text{ V} \tag{1}$$

$$CO_2 + 2H^+ + 2e^- \rightarrow CO + H_2O; E^o_{red} = -0.53 \text{ V} \tag{2}$$

$$CO_2 + 4H^+ + 4e^- \rightarrow HCHO + H_2O; E^o_{red} = -0.48 \text{ V} \tag{3}$$

$$CO_2 + 6H^+ + 6e^- \rightarrow CH_3OH + H_2O; E^o_{red} = -0.38 \text{ V} \tag{4}$$

$$CO_2 + 8H^+ + 8e^- \rightarrow CH_4 + 2H_2O; E^o_{red} = -0.24 \text{ V} \tag{5}$$

The photoreduction efficacy and selectivity are primarily determined by the semiconductor TiO_2. For efficient photoreduction of CO_2 with the high selectivity of the desired product, the CB and the VB positions of TiO_2 with respect to the reduction potentials are very important. The CB minima of TiO_2 should be positioned energetically higher (more negative) compared to the corresponding CO_2 reduction potential. The VB maxima should be well below (more positive than) the water oxidation level to produce protons and oxygen ($2H_2O + 4h^+ \rightarrow O_2 + 4H^+ + 4e^-$, $E^o_{red} = 0.81$ V). In most cases, water is used as a reductant and solvent.

A pictorial representation in Fig. 5.1 describes the energy levels of TiO_2 and the corresponding redox potentials in the CO_2 reduction and water oxidation. The CB and VB levels of TiO_2 vary with its polymorphic phase (Challagulla et al. 2017b), morphology and dimensions (Challagulla et al. 2017a), heterojunctions with other suitable semiconductors (Challagulla et al. 2016, Nagarjuna et al. 2015) and cations and anions doping (Roy et al. 2007, Challagulla et al. 2017c, Roy et al. 2008a). The band positions of TiO_2 depends on its polymorphic phases: anatase, brookite and rutile. Anatase TiO_2 is the ambient-temperature phase with a band gap of

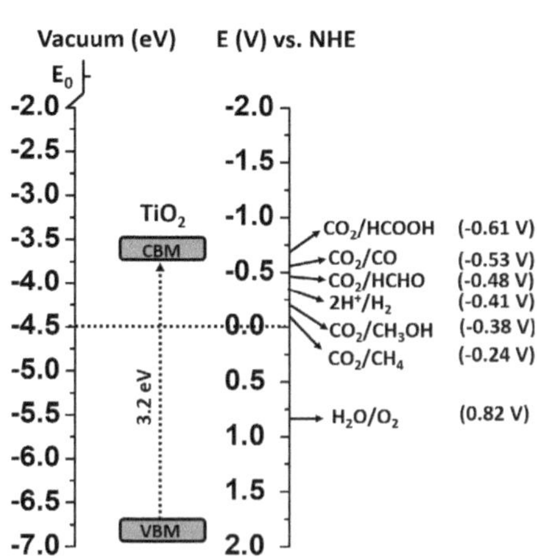

Figure 5.1: VB and CB levels of TiO_2 with respect to the redox potentials in the CO_2 reduction and water oxidation.

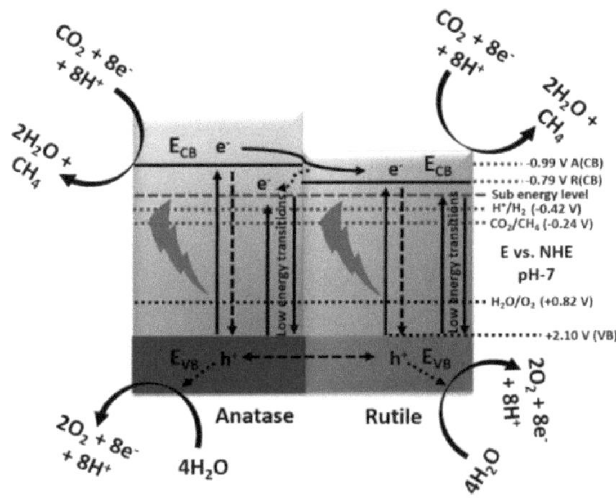

Figure 5.2: CO_2 reduction potentials and electron movement in mixed phase TiO_2. Reproduced with permission from (Kar et al. 2019).

3.2 eV, whereas the high temperature rutile phase has a slightly lower band gap of 3.0 eV. On the other hand, the brookite phase is cumbersome to synthesize. Another school of thought claims that the use of mixed-phase TiO_2 to maximize the visible light harvesting ability (Liu et al. 2012, Kar et al. 2019). The proposed biphasic CO_2 photoreduction mechanism is shown in Fig. 5.2. In addition to the VB and CB position, the other two challenging factors over TiO_2 are (i) prohibiting the photogenerated electron–hole recombination, and (ii) restricting the photoproduction of H_2 ($2H^+ + 2e^- \rightarrow H_2$, $E^{o}_{red} = -0.42$ V), which is a competitive reaction to CO_2 photoreduction.

Morphology and dimension alteration

The surface morphology and the size of TiO_2 play a significant role in determining adsorption of gaseous CO_2, longer life of photogenerated electrons and the efficacy of electron–hole recombination in order to achieve the best performance of CO_2 photoreduction. Adsorption is a primary step before photocatalytic reduction and different exposed facets, such as (110) of rutile and (101) of anatase surfaces exhibit significantly different Lewis acidity and polarizing power, possibly resulting in different reactivity toward CO_2 (Mino et al. 2011). Studies have shown that CO_2 is mainly adsorbed in a linear form at the anatase (101) surface, while the formation of a variety of surface carbonates occur at the (001) surface (Mino et al. 2014, Sorescu et al. 2011). Various morphologies of TiO_2 nanostructures, such as 1 D rods, wires, tubes and 2D sheets, etc., exhibited excellent photocatalytic reduction of CO_2 (Meryem et al. 2018). The nanostructures provide high surface to bulk ratio and facilitate the photogenerated charge movement. Yu et al. changed the ratio of (001) and (101) facets of TiO_2 nanosheets with HF leaching and found that the two facets with a ratio of 45:55 showed the highest CO_2 photoreduction with CH_4 yield at a rate of 1.35 μmol h^{-1} g^{-1} (Yu et al. 2014). The co-exposed facets of (101) and (001) formed a homojunction within TiO_2 that helped in charge separation and transportation.

H_2SO_4 leaching on TiO_2 nanosheets also enhanced $CO_2 \rightarrow CH_4$ yield (He et al. 2016). However, in this study acidification did not change the ratios of the exposed facets of TiO_2. Tunable microspheres TiO_2 exhibited enhanced storage and diffusion of gaseous CO_2 that efficiently converted $CO_2 \rightarrow$ reduced the product in the presence of simulated solar illumination (Wang et al. 2019). Among the three different types of TiO_2 microspheres such as solid, yolk/shell and hollow microspheres, the solid one showed the maximum UV light assisted catalytic CO yield of 17.7 μmol $g^{-1}\,h^{-1}$, was 1.2 and 1.6 times higher than that obtained with yolk/shell and hollow microspheres, respectively. However, under visible light illumination, the hollow microspheres outperformed the other structures. The amplified photocatalytic performance was ascribed to the increased pore size of the hollow microspheres, leading to quick diffusion of CO_2 towards catalytic active sites.

Creations of heterojunctions and dispersion of nano-TiO_2 on conducting supports

The primary limitation of pristine TiO_2 to be recognized as the most efficient catalyst for CO_2 photoreduction is primarily attributed to the lower position (i.e., more positive) of the electrons in the CB (–0.5 V) than the single electron reduction potential of $CO_2 \rightarrow CO_2^-$ (–1.90 V). Additionally, the co-presence of reducing electrons and free protons from water oxidation is detrimental as they interact to generate H_2. Therefore, several modification strategies are incorporated to help overcome these drawbacks of pure TiO_2.

Sensitizing TiO_2 with other semiconductors to form heterostructures is a popular method used to improve the photoactivity. The nature of photogenerated charge separation yields four different types of heterostructures: conventional type-II, p-n, Z-scheme and surface heterojunctions. Bi_2WO_6 (Yuan et al. 2018), CdS (Beigi et al. 2014), Bi_2S_3 (Li et al. 2012), CdSe (Wang et al. 2010), PbS (Wang et al. 2011), AgBr (Asi et al. 2011), etc., are the most common semiconductors to couple with TiO_2. CdS/TiO_2 composite enhanced the visible light absorption and the photocatalytic performance of the TiO_2 owing to the lower recombination of photogenerated excitons. However, different products like CH_4 and CH_3OH were observed from different studies over CdS/TiO_2 (Beigi et al. 2014, Li et al. 2012). The photocatalytic activity of CO_2 to methyl formate was greatly enhanced in CuO/TiO_2 due to the heterojunction between the pristine materials. The surface-junction improved the transfer of charges by enabling the migration of electrons from CB of TiO_2 to VB of CuO, and as a result the charge recombination probability was lowered and the photocatalytic efficiency was enhanced (Qin et al. 2011).

The metal free 2D semiconductors like g-C_3N_4 is also a popular choice to form composite with TiO_2 for efficient photoreduction reaction (Zhou et al. 2018, Reli et al. 2016, Wu et al. 2019a, Wada et al. 2017, He et al. 2020, Wu et al. 2019b, Wang et al. 2020). The recent surge in artificial photosynthesis over g-C_3N_4 is due to its medium band gap (2.7 eV) appearing from 2D π layers, a particular tectonic structure with the tri-s-triazine building blocks, and specific Lewis and Brønsted basic sites on the surface that promotes the acidic CO_2 adsorption. TiO_2/g-C_3N_4 layered nanosheets, prepared with urea as a precursor of g-C_3N_4, yielded CO from visible light reduction of CO_2 (Zhou et al. 2014). The urea precursor not only produced g-C_3N_4, but also helped to dope N in TiO_2, which aligned the band edges with respect to the redox potentials of prospective products. Figure 5.3 describes the band alignment and charge separation in the composite. The CB of TiO_2 lies below the CB of g-C_3N_4. Therefore, the photogenerated electrons in g-C_3N_4 are easily transferred to CB of TiO_2, and the combined electrons reduced CO_2 molecules to C1 intermediates. The holes formed in TiO_2 due to N-doping easily moved to the VB of g-C_3N_4 and oxidized H_2O molecules absorbed on the surface of g-C_3N_4 to generate O_2 and protons. This Z-scheme suppressed the recombination rate to a large extent. The TiO_2 nanosheets on g-C_3N_4 nanosheets has been used for photocatalytic CO_2 conversion and the 2D composite showed 12 times higher CO yield than pristine TiO_2 (Crake et al. 2019). A defect-rich TiO_2 quantum dots embedded within g-C_3N_4 nanosheets exhibited a five times superior photocatalytic conversion of CO_2 into CO compared to pristine g-C_3N_4 (Shi et al. 2019). Interestingly, the $TiO_{2-\delta}$/g-C_3N_4 nanostructure was synthesized by *in situ* pyrolysis of melamine with Ti-MOF (MIL-125-NH_2). Carbon coated C-TiO_{2-x}@g-C_3N_4 was synthesized by calcining polyoxotitanium $[Ti_{17}O_{24}(OPr^i)_{20}]$ and urea showed a very high activity of 12.30 mmol g^{-1} (204.96 mmol $g_{TiO2}{}^{-1}$) CO generation within 60 hours visible-light irradiation (Zhou et al. 2018). The wide variety of synthesis strategies impacted in morphology and band alignment of the composite, which in turn showed a variation of photo CO_2 reduction performance.

The incorporation of conducting carbon-based materials, such as carbon nanotubes (CNT) or graphene (or reduced graphene oxide, rGO) with TiO_2 is attracting attention in the area of photocatalysis. These conducting large surface area materials enhance the electron–hole separation, help accumulating electrons on TiO_2 surface and also facilitate the CO_2 adsorption on its surface through π–π conjugations. The multi walled CNT/TiO_2 nanocomposite under visible light, yielded as high as 0.17 μmol methane/g-catalyst/h (Gui et al. 2014). Xia et al. explored the role of

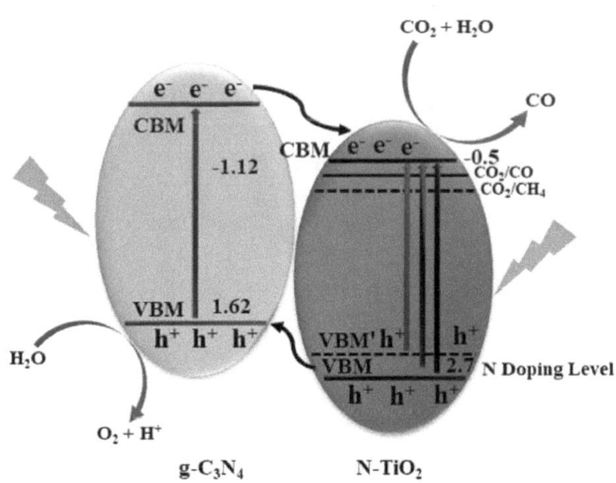

Figure 5.3: Charge separation mechanism with Z-scheme over N-doped TiO_2/g-C_3N_4 composite. Reproduced with permission from (Zhou et al. 2014).

two different synthesis methods sol-gel and hydrothermal to produce MWCNT/TiO_2 (Xia et al. 2007). The sol–gel method formed anatase TiO_2, whereas rutile-phase was formed by the hydrothermal method. The anatase MWCNT/TiO_2 produced C_2H_5OH (29.87 µmol/g-cat/h) on photoreduction of CO_2, whereas the rutile composite reduced CO_2 to formic acid (25.02 µmol/g-cat/h). The MWCNTs helped reduce the agglomeration of TiO_2 nanoparticles and helped increase the separation of the photogenerated electron–hole pairs. A recent study showed the formation of Ti–C bond between CNT and TiO_2 nanoparticles (Olowoyo et al. 2018). The computational exploration revealed that decahedral anatase nanoparticles was weakly attached to CNT with the (001) surface and strongly attached to CNT with the (101) surface. The binding with (101) resulted in efficient charge transfer between TiO_2 and CNT, and consequently the composite exhibited a very high photocatalytic CO_2 reduction activity. Graphene based oxides also serve as a high electrically conducting material to facilitate the charge separation and migration owing to the oxygen functioning groups in the basal plane of rGO. Composite of rGO/TiO_2 improved the photocatalytic CO_2 reduction by seven-fold compared to pristine TiO_2 under visible light illumination (Liang et al. 2011). The CB of TiO_2, which is constructed from 3*d*-orbitals of Ti energetically match well with the π orbital of rGO to form d-π electron orbital overlap that facilitated CO_2 activation before adsorption (Liang et al. 2012, Tan et al. 2015, Gillespie et al. 2019). Further improvements in CO_2 adsorption and photoreduction were achieved by N doping in rGO or creating oxygen rich $TiO_{2+\delta}$ (Lin et al. 2017, Tan et al. 2017). The composite of rGO/TiO_2 showed Langmuir-Hinshelwood adsorption, where both CO_2 and H_2O were adsorbed simultaneously through the π-π conjugated electrons interaction. An optimum limit of graphene content was reported over TiO_2 (Jiang et al. 2011). rGO supported TiO_2 with co-exposed specific facets of {001} and {101} demonstrated that on irradiation, the photogenerated electrons were collected in the {101} facet, while the holes flowed towards {001} facet, and the rGO junction with TiO_2 promoted the efficient electron extraction for enhanced CO_2 photoreduction to produce CH_4 and CO as shown in Fig. 5.4 (Xiong et al. 2016). Zhang et al. did a comprehensive study to probe the reaction mechanism of photocatalytic CO_2 reduction over TiO_2/rGO, where they proposed formation of as the rate-determining step of this reductive reaction (Zhang et al. 2015).

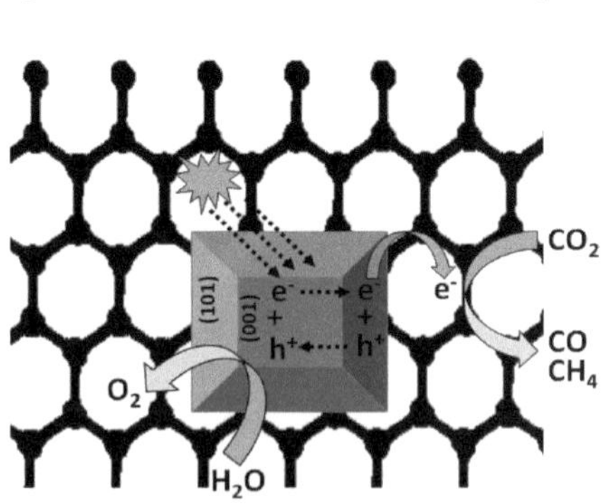

Figure 5.4: Mechanism of photocatalytic CO_2 reduction over {001} and {101} facets of TiO_2 supported on rGO. Reproduced with permission from (Xiong et al. 2016).

Impurity doping and metal deposition

One of the most popular strategies that enhance the photocatalytic activity of TiO_2 photocatalysts is cationic and anionic doping. Owing to noble and transition metal doping in TiO_2, the formed solid solution introduces the point defects in the lattice that extend the light absorption window by narrowing the band gap and also suppresses the recombination rate of excitons (Roy et al. 2007, Roy et al. 2008a, Roy et al. 2008b). The doped noble metals create a Schottky barrier at the interface that act as an electron trap, facilitating the separation of photo-generated charges (Challagulla et al. 2017c). Further, the doped plasmonic nano metals enable the photocatalysts to utilize a wide range of visible light and enhance their photocatalytic performance.

One of the most commonly used dopant metal in TiO_2 is Cu. 1.2 wt.% of Cu-doped TiO_2 synthesized by hydrothermal method exhibited CO_2 conversion to CH_3OH with a yield of 0.45 μmol/g-cat/h under UV irradiation (Wu et al. 2005). Sol-gel synthesized with 2 wt. Cu/TiO_2 showed higher CH_3OH yield of 12.5 μmol g-cat^{-1}h^{-1} (Tseng et al. 2002). The CH_3OH yield from photoreduction of CO_2 was further enhanced to 194 μmol g-cat^{-1}h^{-1} over 3 wt.% of Cu/TiO_2 (Evonik P-25) prepared by the impregnation method (Slamet et al. 2005). Copper doping helped in lowering the chances of electron–hole recombination, and thus boosted the photocatalytic efficiency. Additionally, owing to Localized Surface Plasmon Resonance (LSPR) the effect of Cu nanoparticles, Cu/TiO_2 also exhibited enhanced photocatalytic activity under visible light (Liu et al. 2015). LSPR was more prominent over Au nano-particles doped in TiO_2 as evidenced by the visible light photoconversion of $CO_2 \rightarrow CH_3OH$ with a yield of 12.65 μmole g cat^{-1} h^{-1} over 0.5 wt.% Au/TiO_2 (Tahir et al. 2015). The LSPR assisted mechanism is pictorially represented in Fig. 5.5. Under visible light irradiations, the LSPR effect of Au on TiO_2 surface promotes the electron to the CB of TiO_2. Next, the photogenerated CB electrons of TiO_2 are trapped by Au, and enhance charges separation. The energy of the trapped electrons is enhanced by the LSPR effect of Au, and reduces CO_2 to CH_3OH. The Ag doped TiO_2 core-shell hetero-junction on visible light reduction of CO_2 produced CO with a rate of 983 μmole g cat^{-1} h^{-1} at selectivity 98% and apparent quantum yield of 0.15%, which was 109 times higher than CO produced over pure TiO_2 (Tahir et al. 2017a, Low et al. 2018). Ag/TiO_2 exhibited strong absorption of visible light due to LSPR excitation, trapped electrons and hindered charges recombination rate. Researchers have also explored the role of LSPR in nano-alloys of AgAu doped in TiO_2 over visible light photoreduction, where the materials exhibited CO yield of 1813 μmole g cat^{-1} h^{-1} (Tahir et al. 2017b). Several other bimetallic catalysts also showed effective photocatalytic activities in different reactions (Wang et al. 2012, Jiao et al. 2015, Collado et al. 2013, Yui et al. 2011, Singhal et al. 2017). AuCu loaded on nanotubes of $SrTiO_3$/TiO_2 composite showed efficient CO_2 photoreduction with hydrazine hydrate that produced CO and hydrocarbons (Kang et al. 2015). Bimetallic $Au_{0.25}Pt_{0.75}$ over nanofibers of TiO_2 also demonstrated high CH_4 formation on photoreduction of CO_2 (Zhang et al. 2013). The electron-trapping effect of Pt along with the SPR effect of Au nanoparticles contributed to the high catalytic efficacy. Bimetallic nanoalloys of AgPt and core–shell Ag@SiO_2 particles on TiO_2 showed notable enhancement in the photoreduction of CO_2 to CH_4 (Mankidy et al. 2013).

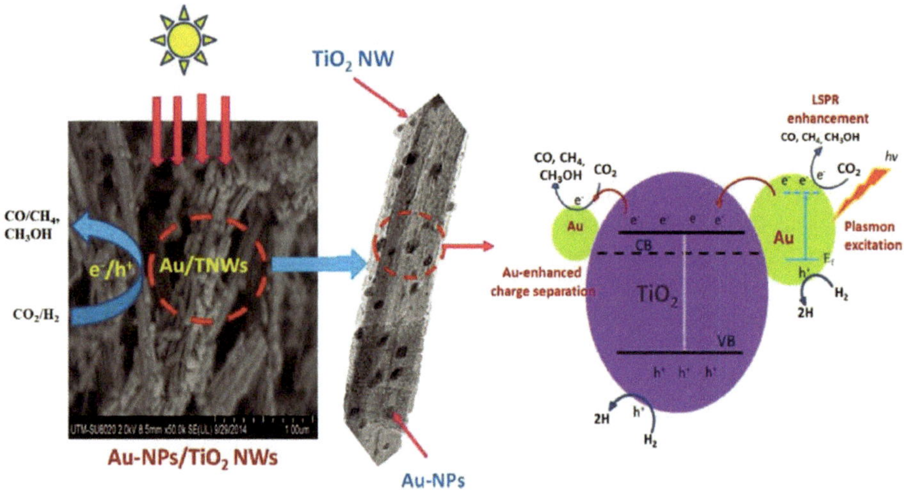

Figure 5.5: Schematic illustration of the LSPR mechanism for photocatalytic CO_2 reduction over Au/TiO_2. Reproduced with permission from (Tahir et al. 2015).

The transition metals are also widely doped in TiO_2 to achieve better photocatalytic CO_2 reduction. Kwak et al. studied CO_2 photoreduction over solvothermally synthesized Ni-doped TiO_2 (Kwak et al. 2015). 0.1 mol % Ni doping yielded the highest amount of CH_4 with a rate of 14 μmol g cat^{-1} h^{-1}. CNT@Ni/TiO_2 nanocomposite was able to absorb a high quantity of visible light with a reduction in the band gap of 2.22 eV (Ong et al. 2013). The high catalytic performance of CNT@Ni/TiO_2, producing a maximum CH_4 product yield of 0.145 μmol g cat^{-1} h^{-1}, was attributed to synergistic combination of CNTs and TiO_2, extended light absorption range and improved transportation of photogenerated charge carriers. Composites of V-, Cr- and Co doped TiO_2 have not only shown a 'red shift' in the absorption spectra in the visible light region, but also enhanced the photocatalytic reduction of CO_2 (Ola et al. 2015). Sol–gel synthesized rare-earth Ce-doped TiO_2 reduced the band gap for effective absorption of visible light and also enhanced charge separation (Matějová et al. 2014). About 0.28 mol % cerium doping in TiO_2 yielded methane with rate of 0.889 μmol g cat^{-1} h^{-1}.

Substitutional anionic non-metals, such as N, C, and S are usually doped to introduce defect energy states localized at the impurity site that consequently reduce the TiO_2 optical band gap and increase the absorption of visible light. N-doped mesoporous anatase TiO_2 retained the mesoporosity after nitrogen doping and exhibited excellent CO_2 photoreduction by water in gas phase (Li et al. 2012). With increasing nitridation temperature, the mean pore diameter increased while the surface area decreased. The optimum nitridation temperature was 525°C, and the amount of doped N was 0.84% on the basis of the lattice oxygen atoms to achieve the best visible light CO_2 reduction activity. Hexamethylene tetramine was used as N source for hydrothermally synthesized N-doped TiO_2 nanotubes that produced HCO_2H (1039 μmol g cat^{-1} h^{-1}), CH_3OH (94.4 μmol g cat^{-1} h^{-1}), and HCHO (76.8 μmol g cat^{-1} h^{-1}) on CO_2 photoreduction (Zhao et al. 2012). C substituted TiO_2

exhibited lowering of the band gap, and improved the charge separation efficiency that consequently photoreduced CO_2 to HCO_2H at the rate of 439 μmol g cat^{-1} h^{-1} (Xue et al. 2011).

Conclusion and outlook

This chapter briefly reviews the importance of CO_2 abatement and its valorization, the mechanistic aspect of photoreduction of CO_2 and imperatives of nanostructured TiO_2 in photocatalytic CO_2 conversion. The important aspects to be a material of choice for photocatalytic CO_2 reduction are: (i) the appropriate position of VB and CB with respect to the reduction potentials of CO_2, (ii) CO_2 activation and its efficient adsorption on the surface, (iii) poor electron-hole recombination, (iv) low H_2 generation, and last but not the least (v) the ability to absorb solar light to generate photoelectrons. In spite of having several advantages, the pristine TiO_2 suffers from some inherent issues, which are overcome by varying the morphology and dimensions, creating heterojunctions with other suitable semiconductors and doping impurity levels cations and anions. This chapter encompasses these recent developments in TiO_2 to make it a better and efficient photocatalyst for CO_2 reduction. The studies have demonstrated the substantial development of the CO_2 photoreduction process with modified TiO_2 based catalysts.

However in spite of the notable laboratory achievement, the practical outdoor application of TiO_2 for the visible light of CO_2 is still far from realization. The major hurdles are that the minute modifications in TiO_2 are not technically and economically feasible, particularly in a large industrial scale. An encouraging solution could be to associate photocatalysis with other CO_2 reduction technologies like electrochemical catalytic reduction. Photoelectrochemical conversion of CO_2 over modified TiO_2 with solar light and renewable electrical energy could be a new direction of efficient CO_2 valorization technology for large scale practical application.

References

Al Jitan, S., G. Palmisano and C. Garlisi. 2020. Synthesis and surface modification of TiO_2-based Phpotocatalysts for the conversion of CO_2. Catalysts 10: 227–256.

Asi, M. A., C. He, M. Su, D. Xia, L. Lin, H. Deng et al. 2011. Photocatalytic reduction of CO_2 to hydrocarbons using AgBr/TiO_2 nanocomposites under visible light. Catalysis Today 175: 256–263.

Beigi, A., S. Fatemi and Z. Salehi. 2014. Synthesis of nanocomposite CdS/TiO_2 and investigation of its photocatalytic activity for CO_2 reduction to CO and CH_4 under visible light irradiation. Journal of CO_2 Utilization 7: 23–29.

BP Statistical Review of World Energy. 2019. London, B. P. C. Primary energy: consumption by fuel. Statistical Review of World Energy 2019, Retrieved, 9.

Challagulla, S., R. Nagarjuna, R. Ganesan and S. Roy. 2016. Acrylate-based polymerizable sol–gel synthesis of magnetically recoverable TiO_2 supported Fe_3O_4 for Cr (VI) photoreduction in aerobic atmosphere. ACS Sustainable Chemistry and Engineering 4: 974–982.

Challagulla, S. and S. Roy. 2017a. The role of fuel to oxidizer ratio in solution combustion synthesis of TiO_2 and its influence on photocatalysis. Journal Materials Research 32: 2764–2772.

Challagulla, S., K. Tarafder, R. Ganesan and S. Roy. 2017b. Structure sensitive photocatalytic reduction of nitroarenes over TiO_2. Scientific Reports 7: 1–11.

Challagulla, S., K. Tarafder, R. Ganesan and S. Roy. 2017c. All that glitters is not gold: a probe into photocatalytic nitrate reduction mechanism over noble metal doped and undoped TiO_2. The Journal of Physical Chemistry C 121: 27406–27416.

Collado, L., P. Jana, B. Sierra, J. M. Coronado, P. Pizarro, D. P. Serrano et al. 2013. Enhancement of hydrocarbon production via artificial photosynthesis due to synergetic effect of Ag supported on TiO_2 and ZnO semiconductors. Chemical Engineering Journal 224: 128–135.

Crake, A., K. C. Christoforidis, R. Godin, B. Moss, A. Kafizas, S. Zafeiratos et al. 2019. Titanium dioxide/carbon nitride nanosheet nanocomposites for gas phase CO_2 photoreduction under UV-visible irradiation. Applied Catalysis B: Environmental 242: 369–378.

Daghrir, R., P. Drogui and D. Robert. 2013. Modified TiO_2 for environmental photocatalytic applications: a review. Journal of Industrial and Engineering Chemistry 52: 3581–3599.

Fujishima, A. and K. Honda. 1972. Photolysis-decomposition of water at the surface of an irradiated semiconductor. Nature 238: 37–38.

Gui, M. M., S. P. Chai, B. Q. Xu and A. R. Mohamed. 2014. Enhanced visible light responsive MWCNT/TiO_2 core-shell nanocomposites as the potential photocatalyst for reduction of CO_2 into methane. Solar Energy Materials and Solar Cells 122: 183–189.

Gillespie, P. N. O. and N. Martsinovich. 2019. Origin of charge trapping in TiO_2/reduced graphene oxide photocatalytic composites: insights from theory. ACS Applied Materials & Interfaces 11: 31909–31922.

Habisreutinger, S. N., L. Schmidt-Mende and J. K. Stolarczyk. 2013. Photocatalytic reduction of CO_2 on TiO_2 and other semiconductors. Angewandte Chemie International Edition 52: 7372–7408.

He, Z., J. Tang, J. Shen, J. Chen and S. Song. 2016. Enhancement of photocatalytic reduction of CO_2 to CH_4 over TiO_2 nanosheets by modifying with sulfuric acid. Applied Surface Science 364: 416–427.

He, F., B. Zhu, B. Cheng, J. Yu, W. Ho and W. Macyk. 2020. 2D/2D/0D $TiO_2/C_3N_4/Ti_3C_2$ MXene composite S-scheme photocatalyst with enhanced CO_2 reduction activity. Applied Catalysis B: Environmental 272: 119006–119017.

https://www.Eia.Gov/Todayinenergy.

https://www.Esrl.Noaa.Gov/Gmd/Ccgg/Trends/Weekly.html.

Kang, Q., T. Wang, P. Li, L. Liu, K. Chang, M. Li et al. 2015. Photocatalytic reduction of carbon dioxide by hydrous hydrazine over Au-Cu alloy nanoparticles supported on $SrTiO_3/TiO_2$ coaxial nanotube arrays. Angewandte Chemie International Edition 54: 841–845.

Kar, P., S. Zeng, Y. Zhang, E. Vahidzadeh, A. Manuel, R. Kisslinger et al. 2019. High rate CO_2 photoreduction using flame annealed TiO_2 nanotubes. Applied Catalysis B: Environmental 243: 522–536.

Kočí, K., L. Obalová and Z. Lacný. 2008. Photocatalytic reduction of CO_2 over TiO_2 based catalysts. Chemical Papers 62: 1–9.

Kwak, B. S., K. Vignesh, N. K. Park, H. J. Ryu, J. I. Baek and M. Kang. 2015. Methane formation from photoreduction of CO_2 with water using TiO_2 including Ni ingredient. Fuel 143: 570–576.

Jiang, B., C. Tian, W. Zhou, J. Wang, Y. Xie, Q. Pan et al. 2011. *In situ* growth of TiO_2 in interlayers of expanded graphite for the fabrication of TiO_2-graphene with enhanced photocatalytic activity. Chemistry—A European Journal 17: 8379–8387.

Jiao, J., Y. Wei, Z. Zhao, W. Zhong, J. Liu, J. Li et al. 2015. Synthesis of 3D ordered macroporous TiO_2-supported Au nanoparticle photocatalysts and their photocatalytic performances for the reduction of CO_2 to methane. Catalysis Today 258: 319–326.

Liang, Y. T., B. K. Vijayan, K. A. Gray and M. C. Hersam. 2011. Minimizing graphene defects enhances titania nanocomposite-based photocatalytic reduction of CO_2 for improved solar fuel production. Nano Letters 11: 2865–2870.

Liang, Y. T., B. K. Vijayan, O. Lyandres, K. A. Gray and M. C. Hersam. 2012. Effect of dimensionality on the photocatalytic behavior of carbon-titania nanosheet composites: charge transfer at nanomaterial interfaces. The Journal of Physical Chemistry Letters 3: 1760–1765.

Li, X., H. Liu, D. Luo, J. Li, Y. Huang, H. Li et al. 2012. Adsorption of CO_2 on heterostructure $CdS(Bi_2S_3)$/TiO_2 nanotube photocatalysts and their photocatalytic activities in the reduction of CO_2 to methanol under visible light irradiation. Chemical Engineering Journal 180: 151–158.

Lin, L. Y., Y. Nie, S. Kavadiya, T. Soundappan and P. Biswas. 2017. N-doped reduced graphene oxide promoted nano TiO_2 as a bifunctional adsorbent/photocatalyst for CO_2 photoreduction: effect of N species. Chemical Engineering Journal 316: 449–460.

Liu, L., H. Zhao, J. M. Andino and Y. Li. 2012. Photocatalytic CO_2 reduction with H_2O on TiO_2 nanocrystals: comparison of anatase, rutile, and brookite polymorphs and exploration of surface chemistry. ACS Catalysis 2: 1817–1828.

Liu, E., L. Qi, J. Bian, Y. Chen, X. Hu, J. Fan et al. 2015. A facile strategy to fabricate plasmonic Cu modified TiO_2 nano-flower films for photocatalytic reduction of CO_2 to methanol. Materials Research Bulletin 68: 203–209.

Low, J., B. Cheng and J. Yu. 2017. Surface modification and enhanced photocatalytic CO_2 reduction performance of TiO_2: a review. Applied Surface Science 392: 658–686.

Mankidy, B. D., B. Joseph and V. K. Gupta. 2013. Photo-conversion of CO_2 using titanium dioxide: enhancements by plasmonic and co-catalytic nanoparticles. Nanotechnology 24: 405402–405409.

Matějová, L., K. Kočí, M. Reli, L. Čapek, A. Hospodková, P. Peikertová et al. 2014. Preparation, characterization and photocatalytic properties of cerium doped TiO_2: on the effect of Ce loading on the photocatalytic reduction of carbon dioxide. Applied Catalysis B: Environmental 152: 172–183.

Meryem, S. S., S. Nasreen, M. Siddique and R. Khan. 2018. An overview of the reaction conditions for an efficient photoconversion of CO_2. Reviews in Chemical Engineering 34: 409–425.

Mino, L., A. M. Ferrari, V. Lacivita, G. Spoto, S. Bordiga and A. Zecchina. 2011. CO adsorption on anatase nanocrystals: a combined experimental and periodic DFT study. The Journal of Physical Chemistry C 115: 7694–7700.

Mino, L., G. Spoto and A. M. Ferrari. 2014. CO_2 capture by TiO_2 anatase surfaces: a combined DFT and FTIR study. The Journal of Physical Chemistry C 118: 25016–25026.

Nagarjuna, R., S. Challagulla, N. Alla, R. Ganesan and S. Roy. 2015. Synthesis and characterization of reduced-graphene oxide/TiO_2/Zeolite-4A: a bifunctional nanocomposite for abatement of methylene blue. Materials & Design 86: 621–626.

Ola, O. and M. M. Maroto-Valer. 2015. Review of material design and reactor engineering on TiO_2 photocatalysis for CO_2 reduction. Journal of Photochemistry and Photobiology C: Photochemistry Reviews 24: 16–42.

Olowoyo, J. O., M. Kumar, S. L. Jain, J. O. Babalola, A. V. Vorontsov and U. Kumar. 2018. Insights into reinforced photocatalytic activity of the CNT-TiO_2 nanocomposite for CO_2 reduction and water splitting. The Journal of Physical Chemistry C 123: 367–378.

Ong, W. J., M. M. Gui, S. P. Chai and A. R. Mohamed. 2013. Direct growth of carbon nanotubes on Ni/ TiO_2 as next generation catalysts for photoreduction of CO_2 to methane by water under visible light irradiation. RSC Advances 3: 4505–4509.

Qin, S., F. Xin, Y. Liu, X. Yin and W. Ma. 2011. Photocatalytic reduction of CO_2 in methanol to methyl formate over CuO-TiO_2 composite catalysts. Journal of Colloid and Interface Science 356: 257–261.

Razzaq, A. and In. Su-Il. 2019. TiO_2 based nanostructures for photocatalytic CO_2 conversion to valuable chemicals. Micromachines 10: 326–350.

Reli, M., P. Huo, M. Šihor, N. Ambrožová, I. Troppová, L. Matějová et al. 2016. Novel TiO_2/C_3N_4 photocatalysts for photocatalytic reduction of CO_2 and for photocatalytic decomposition of N_2O. The Journal of Physical Chemistry A 120: 8564–8573.

Roy, S., M. S. Hegde, N. Ravishankar and G. Madras. 2007. Creation of redox adsorption sites by Pd^{2+} ion substitution in nanoTiO_2 for high photocatalytic activity of CO oxidation, NO reduction, and NO decomposition. The Journal of Physical Chemistry C 111: 8153–8160.

Roy, S., B. Viswanath, M. S. Hegde and G. Madras. 2008a. Low-temperature selective catalytic reduction of NO with NH_3 over $Ti_{0.9}M_{0.1}O_{2-\delta}$ (M = Cr, Mn, Fe, Co, Cu). The Journal of Physical Chemistry C 112: 6002–6012.

Roy, S., A. Marimuthu, P. A. Deshpande, M. S. Hegde and G. Madras. 2008b. Selective catalytic reduction of NO_x: mechanistic perspectives on the role of base metal and noble metal ion substitution. Industrial & Engineering Chemistry Research 47: 9240–9247.

Roy, S., S. Payra, S. Challagulla, R. Arora, S. Roy and C. Chakraborty. 2018. Enhanced photoinduced electrocatalytic oxidation of methanol using Pt nanoparticle-decorated TiO_2-polyaniline ternary nanofibers. ACS Omega 3: 17778–17788.

Shi, H., S. Long, S. Hu, J. Hou, W. Ni, C. Song et al. 2019. Interfacial charge transfer in 0D/2D defect-rich heterostructures for efficient solar-driven CO_2 reduction. Applied Catalysis B: Environmental 245: 760–769.

Singhal, N. and U. Kumar. 2017. Noble metal modified TiO_2: selective photoreduction of CO_2 to hydrocarbons. Molecular Catalysis 439: 91–99.

Slamet, H., W. Nasution, E. Purnama, S. Kosela and J. Gunlazuardi. 2005. Photocatalytic reduction of CO_2 on copper-doped titania catalysts prepared by improved-impregnation method. Catalysis Communications 6: 313–319.

Soman, B., S. Challagulla, S. Payra, S. Dinda and S. Roy. 2018. Surface morphology and active sites of TiO_2 for photoassisted catalysis. Research on Chemical Intermediates 44: 2261–2273.

Sorescu, D. C., W. A. Al-Saidi and K. D. Jordan. 2011. CO_2 adsorption on TiO_2 (101) anatase: a dispersion-corrected density functional theory study. The Journal of Chemical Physics 135: 124701–124717.

Tahir, M., B. Tahir and N. A. S. Amin. 2015. Gold-nanoparticle-modified TiO_2 nanowires for plasmon-enhanced photocatalytic CO_2 reduction with H_2 under visible light irradiation. Applied Surface Science 356: 1289–1299.

Tahir, M., B. Tahir, N. A. S. Amin and Z. Y. Zakaria. 2017a. Photo-induced reduction of CO_2 to CO with hydrogen over plasmonic Ag-NPs/TiO_2 NWs core/shell hetero-Junction under UV and visible light. Journal of CO_2 Utilization 18: 250–260.

Tahir, M., B. Tahir and N. A. S. Amin. 2017b. Synergistic effect in plasmonic Au/Ag alloy NPs co-coated TiO_2 NWs toward visible-light enhanced CO_2 photoreduction to fuels. Applied Catalysis B: Environmental 204: 548–560.

Tan, L. L., W. J. Ong, S. P. Chai, B. T. Goh and A. R. Mohamed. 2015. Visible-light-active oxygen-rich TiO_2 decorated 2D graphene oxide with enhanced photocatalytic activity toward carbon dioxide reduction. Applied Catalysis B: Environmental 179: 160–170.

Tan, L. L., W. J. Ong, S. P. Chai and A. R. Mohamed. 2017. Photocatalytic reduction of CO_2 with H_2O over graphene oxide-supported oxygen-rich TiO_2 hybrid photocatalyst under visible light irradiation: process and kinetic studies. Chemical Engineering Journal 308: 248–255.

Tseng, I., W. Chang and J. C. S. Wu. 2002. Photoreduction of CO_2 using sol–gel derived titania and titania-supported copper catalysts. Applied Catalysis B: Environmental 37: 37–48.

Wada, K., C. S. K. Ranasinghe, R. Kuriki, A. Yamakata, O. Ishitani and K. Maeda. 2017. Interfacial manipulation by rutile TiO_2 nanoparticles to boost CO_2 reduction into CO on a metal-complex/ semiconductor hybrid photocatalyst. ACS Applied Materials & Interfaces 9: 23869–23877.

Wang, C., R. L. Thompson, J. Baltrus and C. Matranga. 2010. Visible light photoreduction of CO_2 using CdSe/Pt/TiO_2 heterostructured catalysts. The Journal of Physical Chemistry Letters 1: 48–53.

Wang, C., R. L. Thompson, P. Ohodnicki, J. Baltrus and C. Matranga. 2011. Size-dependent photocatalytic reduction of CO_2 with PbS quantum dot sensitized TiO_2 heterostructured photocatalysts. Journal of Materials Chemistry 21: 13452–13457.

Wang, H., D. Wu, W. Wu, D. Wang, Z. Gao, F. Xu et al. 2019. Preparation of TiO_2 microspheres with tunable pore and chamber size for fast gaseous diffusion in photoreduction of CO_2 under simulated sunlight. Journal of Colloid and Interface Science 539: 194–202.

Wu, J. C. S., H. M. Lin and C. L. Lai. 2005. Photo reduction of CO_2 to methanol using optical-fiber photoreactor. Applied Catalysis A: General 296: 194–200.

Wu, J., Y. Feng, B. E. Logan, C. Dai, X. Han, D. Li et al. 2019a. Preparation of Al-O-linked porous-g-C_3N_4/TiO_2-nanotube Z-scheme composites for efficient photocatalytic CO_2 conversion and 2,4-dichlorophenol decomposition and mechanism. ACS Sustainable Chemistry & Engineering 7: 15289–15296.

Wu, J., Y. Feng, D. Li, X. Han and J. Liu. 2019b. Efficient photocatalytic CO_2 reduction by P–O linked g-C_3N_4/TiO_2-nanotubes Z-scheme composites. Energy 178: 168–175.

Yu, J., J. Low, W. Xiao, P. Zhou and M. Jaroniec. 2014. Enhanced photocatalytic CO_2-reduction activity of anatase TiO_2 by coexposed {001} and {101} facets. Journal of the American Chemical Society 136: 8839–8842.

Yuan, L., K. Q. Lu, F. Zhang, X. Fu and Y. J. Xu. 2018. Unveiling the interplay between light-driven CO_2 photocatalytic reduction and carbonaceous residues decomposition: a case study of Bi_2WO_6-TiO_2 binanosheets. Applied Catalysis B: Environmental 237: 424–431.

Yui, T., A. Kan, C. Saitoh, K. Koike, T. Ibusuki and O. Ishitani. 2011. Photochemical reduction of CO_2 using TiO_2: effects of organic adsorbates on TiO_2 and deposition of Pd onto TiO_2. ACS Applied Materials & Interfaces 3: 2594–2600.

Xia, X. H., Z. J. Jia, Y. Yu, Y. Liang, Z. Wang and L. L. Ma. 2007. Preparation of multi-walled carbon nanotube supported TiO_2 and its photocatalytic activity in the reduction of CO_2 with H_2O. Carbon 45: 717–721.

Xue, L. M., F. H. Zhang, H. J. Fan and X. F. Bai. 2011. Preparation of C doped TiO_2 photocatalysts and their photocatalytic reduction of carbon dioxide. Advanced Materials Research 183: 1842–1846.

Xiong, Z., Y. Luo, Y. Zhao, J. Zhang, C. Zheng and J. C. S. Wu. 2016. Synthesis, characterization and enhanced photocatalytic CO_2 reduction activity of graphene supported TiO_2 nanocrystals with coexposed {001} and {101} Facets. Physical Chemistry Chemical Physics 18: 13186–13195.

Zhang, Z., Z. Wang, S. W. Cao. and C. Xue. 2013. Au/Pt nanoparticle-decorated TiO_2 nanofibers with plasmon-enhanced photocatalytic activities for solar-to-fuel conversion. The Journal of Physical Chemistry C 117: 25939–25947.

Zhang, Q., C. F. Lin, B. Y. Chen, T. Ouyang and C. T. Chang. 2015. Deciphering visible light photoreductive conversion of CO_2 to formic acid and methanol using waste prepared material. Environmental Science & Technology 49: 2405–2417.

Zhao, Z., J. Fan, J. Wang and R. Li. 2012. Effect of heating temperature on photocatalytic reduction of CO_2 by N-TiO_2 nanotube catalyst. Catalysis Communications 21: 32–37.

Zhou, S., Y. Liu, J. Li, Y. Wang, G. Jiang, Z. Zhao et al. 2014. Facile in situ synthesis of graphitic carbon nitride (g-C_3N_4)-N-TiO_2 heterojunction as an efficient photocatalyst for the selective photoreduction of CO_2 to CO. Applied Catalysis B: Environmental 158: 20–29.

Zhou, J., H. Wu, C. Y. Sun, C. Y. Hu, X. L. Wang, Z. H. Kang et al. 2018. Ultrasmall C-TiO_{2-x} nanoparticle g-C_3N_4 composite for CO_2 photoreduction with high efficiency and selectivity. Journal of Materials Chemistry A 6: 21596–21604.

CHAPTER 6

Concept of Nanocatalyst and Its Application in the Energy Domain

Anand Sharma,[1] *Aparna Shukla*[2] and *Jay Singh*[3,*]

Introduction

Catalysts and nanocatalysts

Catalysts occupy a vital place in chemistry and this has led to numerous chemical reactions and patterns in academics and industrialized laboratory level. TA catalytic reagents can diminish the transformation temperature, reagent-based waste and enhance the reaction rate and effectively avoid the unwanted co-reaction leading to green chemistry. Typically, a catalyst is an entity that speeds up a chemical reaction without involving the chemical reaction (Li and Somorjai 2010). This capacity is typically referred to as the activity of a catalyst. Two scientists Anastas and Warner gave 12 principles to the chemical organization, which is the key idea for green chemistry's world to mitigate or remove excess chemicals and chemical reactions that have some drawbacks and other negative impacts on the environment. Making blueprints and developing a well-defined catalyst is the central idea of green chemistry. Product variations, i.e., perfect chemicals, polymers, composites lubricants and other essential products in the organization would not to be possible in the absence of the catalyst. Followed by catalytic protocols one can make a more eco-friendly, green and sustainable future (Anastas and Warner 1998, Li and Somorjai 2010, Singh and Tandon 2014).

[1] Solid State and Structural Chemistry Unit, Indian Institute of Science, Bengaluru, 560012, India. Email: anandbhu12191@gmail.com

[2] Department of Materials Engineering, Indian Institute of Science, Bengaluru, 560012, India.

[3] Department of Chemistry, Institute of Science, Banaras Hindu University, Varanasi-221005, UP, India.

* Corresponding author: jaimnnit@gmail.com

Nanomaterials and its synthesis

First the essential nanomaterials before the nanocatalyst will be discussed. Therefore a brief introduction to nanomaterials is described here. Nanotechnology or nanoscience is the field which studies nanomaterials and concerned devices like the smallest transistor, light source, laser, sensors, memory and carbon nanotube devices, etc., these all are nanometers in size. Nanomaterials are those which have at least one dimension in nanometers. They are mainly divided into three types based on their dimensions. First, thin-film or layers (surface coating), all these types of materials are limited in one direction of the nanometer size and the other two extended in size. Second, those are extended in only one direction and limit the other two in nanometers range, generally known as nanowire, e.g., nanotubes. And in the end, these materials in which all dimension restrictions in nanometer size are called quantum dots, for example precipitates, colloids and quantum dots (tiny particles of semiconductor materials). Nanocrystalline materials, made up of nanometer-sized grains, also fall into this category (Vollath 2008, Singh and Tandon 2014). Two main characteristics make nanomaterials different from other materials. One is the increased surface area and the second is quantum effects. These two factors change or amplify the properties such as electrical performance, reactivity and strength. The particle size reduces, the enhanced proportion of atoms are noticed at the surface compared to the inside. For example a particle size of 40 nm at 5% of its atoms on its surface, 10 nm 20% of its atoms and 4 nm 50% of its atoms. Hence nanoparticles have more surface area than larger particles. As growth and catalytic chemical reactions occur at surfaces, this means that a given mass of material in the nanoparticulate form will be much more reactive than the same mass of material made up of larger particles (Chaturvedi et al. 2012).

As mentioned above the size of nanoparticle is an order of 10^{-9} meters. Over the last few years, novel synthesis approaches/methods for nanomaterials (e.g., metal nanoparticles, Quantum Dots (QDs), carbon nanotubes (CNTs), graphene and their composites) have been an attractive area in nanoscience. To control the size of nanoparticles between 1–100 nm is the key task of nanochemistry or synthesis. To synthesize nanoparticles of required sizes, shape and functionalities, two different fundamental principles of synthesis first top-down second bottom-up methods have already been investigated in the reported literature (Fig. 6.1) (Singh et al. 2018).

Why nanocatalyst?

Catalysts are developed in three directions homogeneous, heterogeneous and enzymatic. Molecular and solid state-communities supported well known homogeneous and heterogeneous catalysis in two different domains. While both search for a similar goal that is the breakthrough of the best catalytic performance. Types of catalyst materials such as solid-state particles versus molecular complexes in solution (often granted onto the support) plus applied conditions of chemical reactions (e.g., liquid-phase reactions versus gas-phase) are the possible reasons present in dissimilarity among homogenous and heterogeneous catalysis. Homogeneous and heterogeneous catalysis both have its advantages and disadvantages. From the

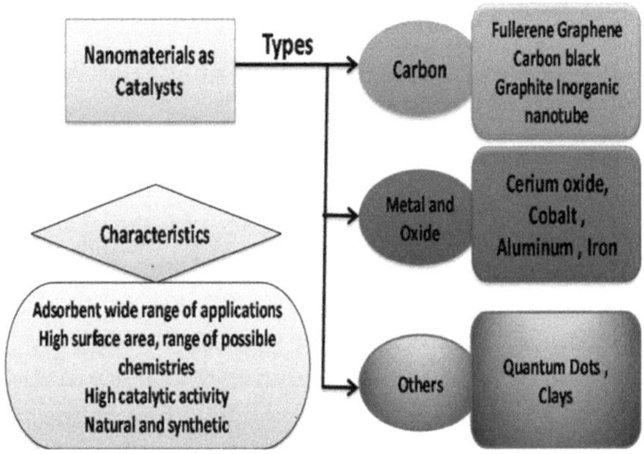

Figure 6.1: Various synthesis routes existing for the production of metal nanoparticles. Reprinted with permission from (Singh et al. 2018).

heterogeneous catalyst point of view, these are revived easily but some demerits exist, for example radical conditions that can lead to mass transport problems. Besides this in homogeneous catalysts, these have superior activity and selectivity, however in negative aspects, the dissociation of expensive transition metal catalysts from substrates and products remains a major issue for industrial applications (Gawande et al. 2013). Due to these disadvantages, there is a need to develop a new catalytic system, which should be a combination of homogenous and heterogeneous catalysts, i.e., it should be active like the homogenous catalyst and also simply be able to restore like a heterogeneous catalyst (Dunworth and Nord 1954). At the end of the 90th century, nanomaterials, nanoscience and nanotechnology were the emerging field as a domain at the interface between homogeneous and heterogeneous catalysis, which resolve these problems and improves the challenging conditions for catalyst development. Therefore, with the help of these nanoscience outcomes, the primary objective is to design the ideal type of catalyst namely called nanocatalyst. These nanocatalysts have highly efficient selectivity, stability and a high recovery rate. These good quality parameters such as specific reactivity can be expected because of the nano dimension (Thomas 2010). On the other hand, regular materials or non-nano materials cannot attempt these specific properties.

Quantum and surface area effects dominate the properties of materials as the size is reduced to the nanoscale. This can alter the electrical, optical and magnetic properties of materials. Hence, the size is the important factor in nanomaterials, this can be mainly controlled by using various stabilizing or capping substances, e.g., ligands, surfactants, polymers, dendrimers ion, polyoxoanions, etc. For this purpose, a suitable agent is also a major component due to this, the surface properties of the nanoparticles (NPs) can vary, and it can also amend the nature of morphology (active sites) and the surface chemical environment (electronic effects). Two parameters affect the selectivity and reactivity of NPs. First, control of crystal structure and morphology and second is control of surface composition (Tandon 2019).

Nanomaterials as catalyst or nanocatalyst

The catalyst is one of the best applications of nanomaterials which is used in wide applications. The types of the nanomaterials used as a catalyst are listed in Fig. 6.2 (Sharma et al. 2015). For many years now, alumina, titanium dioxide, iron, silica and clays are broadly used as a catalyst in various applications, for example, electronics, magnetic, chemical and industrial applications. Catalytic activity strongly depends on the nanorange size such as structure and the shape of materials. Better selectivity is attained because of composition, i.e., fine-tuning of nanocatalysts. To resolve all these issues, nanocatalysts are designed by scientists, which are highly active, highly selective and highly resilient. Nanomaterials are used very frequently as a catalyst in a broad area from chemical manufacturing to energy conversion and storage (Rajendran et al. 2019). The variable and particle-specific catalytic activity of nanoparticles are due to its heterogeneity and their differences in size and shape. Figure 6.3 represents the basic difference in bulk catalysis and catalysis shown by nanoscale materials.

In catalysis, chemical reactions in solids, gases or liquids are accelerated by introducing a solid phase that ideally contains large amounts of the right kind of site for chemical reactants to adsorb, react and desorbs.

The effect of the fundamental properties of the nanomaterials on catalysts can be understood by using the theory behind nanocatalysis. Catalytic activity is vitally affected by the fundamental properties of the nanomaterials and can be categorized as

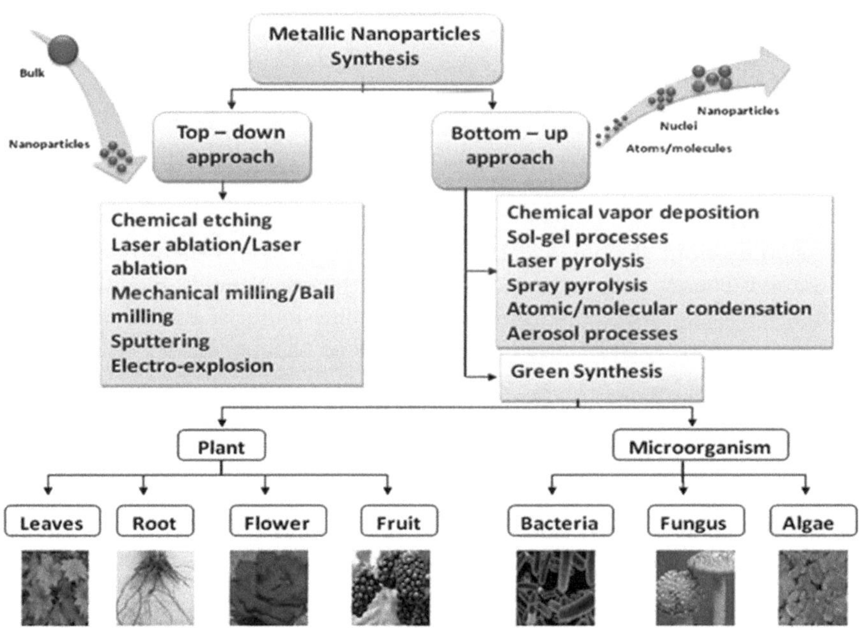

Figure 6.2: Categories of catalytic nanomaterials and their characteristics.

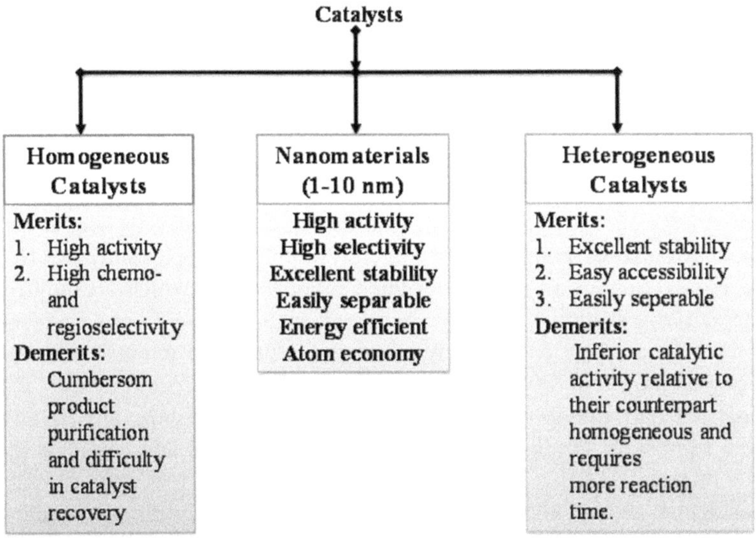

Figure 6.3: Comparative efficiency of homogeneous, heterogeneous and nanocatalysis.

i) Quantities that are apparently associated to bond lengths, for example, the mean lattice constant, atomic density and binding energy. Densification and surface relaxation was induced by lattice contraction in nano solid,

ii) Quantities that also depend on cohesive energy per discrete atom, for example self-organization growth; thermal stability; Coulomb blockade and evaporation in a nano-solid; and the activation energy for atomic dislocation, diffusion and chemical reactions,

iii) Properties that differ with the binding energy density in the relaxed continuum region for example the Hamiltonian that establish the band structure and connected properties such as band gap, core level energy, photoabsorption and photoemission and

iv) Properties from the joint effect of the binding energy density and atomic cohesive energy such as the mechanical strength Young's modulus, surface energy, surface stress, extensibility and compressibility of a nano-solid, as well as the magnetic performance of a ferromagnetic nano-solid.

The catalyst is affected by the fundamental properties of nanomaterials and their size dependency will induce, directly or not (Fig. 6.4), this part is discussed in detail throughout this chapter.

Applications of nano-catalyst in various useful areas

Plasmonic catalyst for hydrogen generation

In the last few years, nanomaterials have attracted immense propensity due to their great properties such as mechanical, electronics, optical, electronic and surface properties. The high surface area to volume area ratio of these materials has

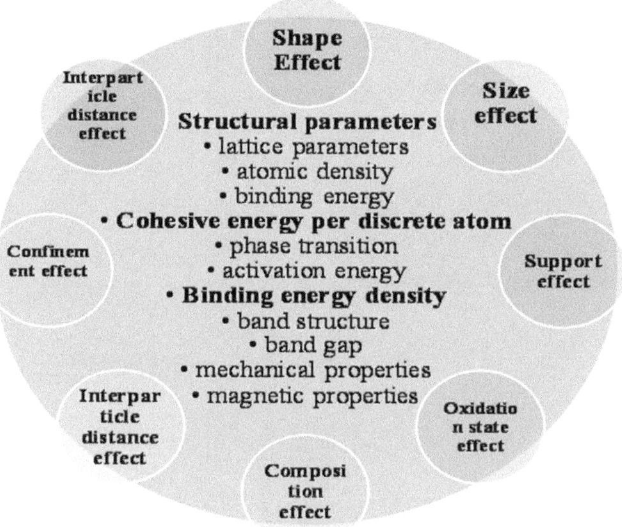

Figure 6.4: The effect of the fundamental properties of nanomaterials on catalysis.

significant implications for hydrogen storage devices. A large quantity of hydrogen is stored by some nanostructured based potential candidates, e.g., carbon nanotubes, nano magnesium-based hydrides, complex carbon nanocomposites, boron nitride nanotubes and metal-organic frameworks (Seayad and Antonelli 2004, Goswami and Kreith 2007). Here the focus is on the application of nanostructured catalyst materials for storing atomic or molecular hydrogen. The synergistic effects of nanocrystalinity and nanocatalyst doping on the metal or complex hydrides for improving the thermodynamics and hydrogen reaction kinetics are discussed. Besides, various carbonaceous nanomaterials and novel sorbent systems (e.g., carbon nanotubes, fullerenes, nanofibers, polyaniline nanospheres and metal-organic frameworks, etc.), and their hydrogen storage characteristics are outlined (Fichtner 2005, Niemann et al. 2008).

Hydrogen has generated a lot of interest because of its clean character as a fuel. It can be produced by splitting of water, dissociation of ammonia-based complex, e.g., NH_3BH_3, ABC, etc. Plasmonic catalyst also uses a nanocatalyst for various applications, Halas et al. reported a lowering of the activation barrier for ammonia decomposition by using a Cu–Ru based plasmonic photocatalyst (Fig. 6.5) (Zhou et al. 2018). Cu-Ru catalyst exhibited approximately 20 and 180 times superior activity than pristine Cu and Ru nanoparticles because of the synergistic effect on the catalytic performance. Notably, the catalyst could be reused up to five times and the maximum TOF based on Ru loading was as high as 15 s^{-1} (Fig. 6.5B and C). In hydrogen storage application, ABC is widely used as a potential candidate and also demonstrates a hydrogen capacity of 20 wt%. C-isolated Ag-C-Co fabricated sandwiched structure through synchronous growth and assembly process by a research group. Comparing the catalytic assessment of the Ag–C–Co with Ag-Co and C-Co for H_2 generation is shown in Fig. 6.6 (Zhou et al. 2013, Bao and Carter 2019).

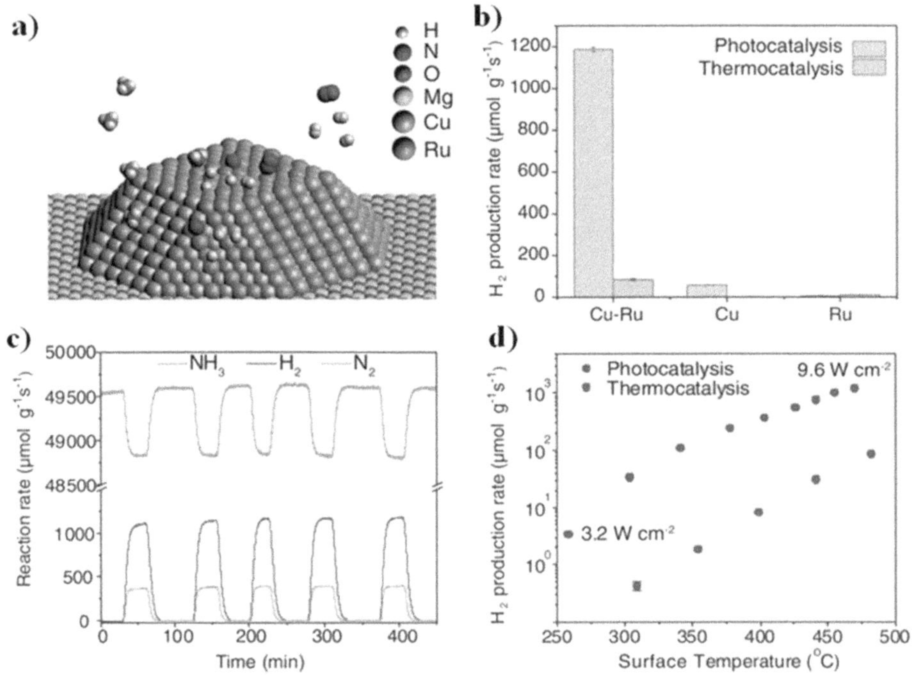

Figure 6.5: (a) Schematic of the structure of Cu–Ru-AR consisting of a Cu NP antenna with a Cu–Ru surface alloy. (b) The H_2 formation rate of photocatalysis and thermocatalysis on Cu–Ru-AR, Cu, and Ru NPs. (c) Multiple-hour measurement of photocatalytic rates on Cu–Ru-AR under light without external heating. (d) Comparison of photocatalytic and thermocatalytic rates on Cu–Ru-AR. Reprinted with permission from (Dhiman 2020).

From the graph is it clear that all of the ABC have been decomposed by Ag-C-O within 11 minutes while in curve e–g (Fig. 6.6a) isolated Ag and C, Ag-C core-shell did not exhibit any catalytic activity. Figure 6.6 d shows activation energy 34.87 kJ mol^{-1} for the Ag–C–Co catalyst which is calculated by the Arrhenius equation. Yamashita and co-workers utilized SBA-15 endorsed spherical silver NPs and nanorods for H_2 made from ABC. Under dark conditions, silver NPs displayed higher catalytic activity in the production of H_2 from ABC. Even though, in the presence of light Ag nanorods exhibited maximum enhancement (Mori et al. 2015).

To elaborate the impact of maintaining the preparation of catalysts for hydrogen production by water splitting was studied in detail by Bruckner and his team. Their outcomes revealed that the support phase (e.g., anatase, rutile, brookite or composites) and structural properties emphatically affected the activity of plasmonic Au-TiO_2 catalyst using the method of gold deposition such as sol-immobilization, photodeposition (Fig. 6.7).

Mesoporous silica nanomaterials (MSNs)

The zeolite opened various new opportunities in the area of the catalyst due to its unique shape and execution. Catalytic cracking of heavy oil macromolecules and

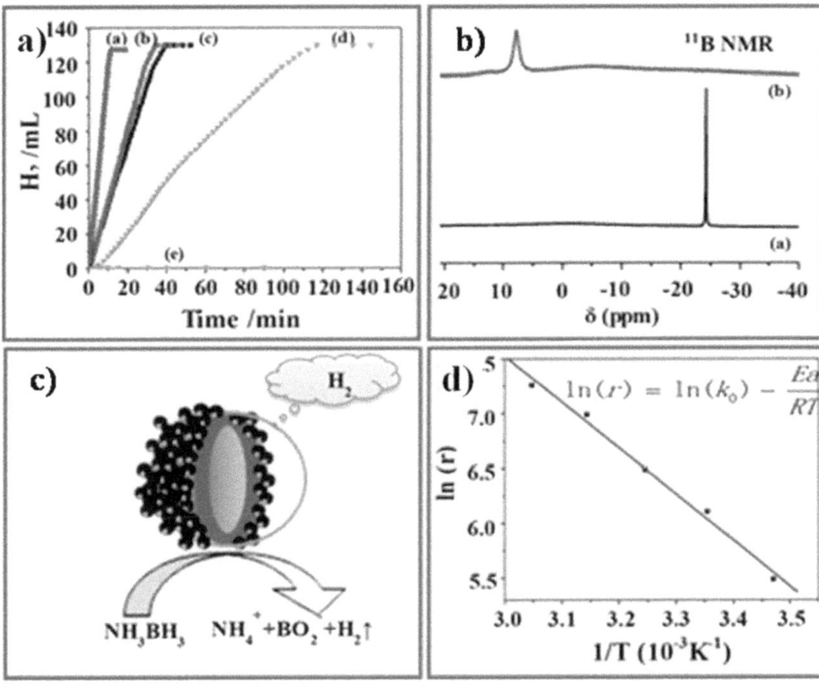

Figure 6.6: (a) Production of hydrogen from ABC aqueous solution by using different catalyst systems such as (a) Ag–C–Co nanospheres, (b) Ag–Co nanospheres, (c) C–Co nanospheres, (d) Co nanospheres, (e) Ag–C nanospheres, (f) isolated Ag nanospheres, and (g) isolated C nanospheres at r.t.; (b) ^{11}B NMR spectra of different aqueous solution: (a) freshly prepared ABC, and (b) at the end of reaction in the presence of Ag–C–Co sandwich sphere structures; (c) Scheme showing the catalytic reaction mechanism; (d) Arrhenius plot for the hydrolysis of ABC. Reprinted with permission from (Dhiman 2020).

immobilization of the macrocyclic complexes are some of the difficulties that have been faced in the processing of zeolite. Scientists introduced first-time mesoporous materials in the early part of the 18th century. These materials have been attracting more interest due to their materials, chemical, physical and other properties. Mesoporous silica nanocatalyst has a wide range of novel applications (Table 6.1) in the catalytic field because of its appealing nanostructured mesoporous materials, that has a large surface area, high activity and great adsorption capacity (Fig. 6.8) (Della Pina et al. 2008, Wang et al. 2008).

The tunable property of pore size makes this material comfortable to install in the nanocatalysts. This variable uniform size can be used as a potential component for the nanoparticles size effect, surface effect and quantum effect to provide additional features in it. Mesoporus materials are those materials which have regular pore diameter in the range of 2 nm to 50 nm and large surface area as well as distinct absorption capacity. Catalytic cracking and manufacturing of fine materials are the new novel opportunities that are presented by these characteristics. Multifunctionalization and therefore, multiple catalyst immobilizations hydrogenation catalysis, magnetic acid catalysis and nano noble-metal catalysis, etc., are some of the remarkable

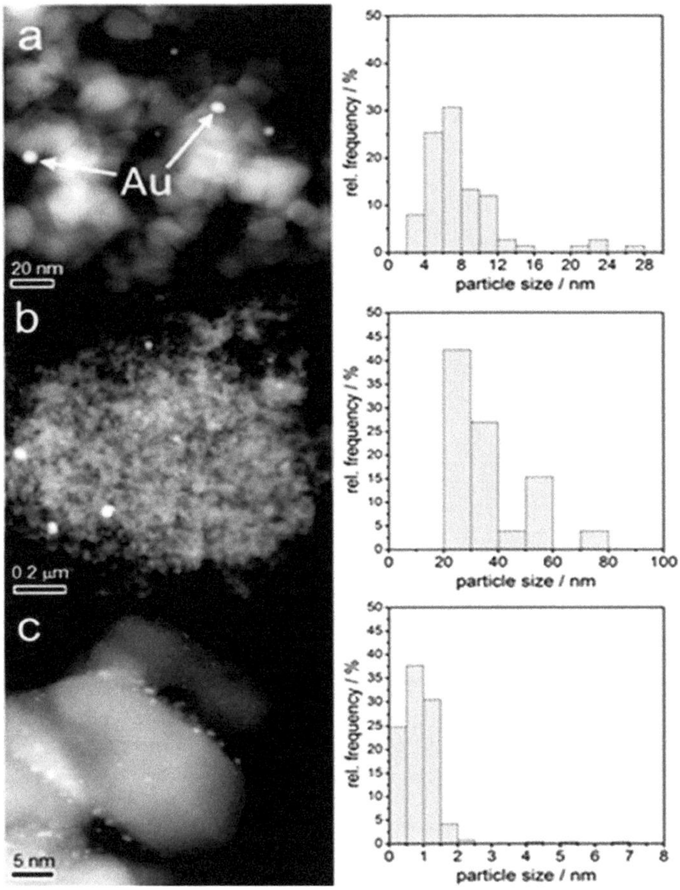

Figure 6.7: TEM images and distribution of (a) sol-immobilization, (b) photodeposition and (c) deposition–precipitation, prepared sample gold particle. Reprint permission from (Dhiman 2020).

Table 6.1: Applications of MSNs materials for catalysis (Anpo et al. 1998).

Uses	Materials
Olefin epoxidation	Ti/V-MS
Olefin oligomerization	Cr-MS
Aromatics oxidation	Ti-MS
Photocatalysis	Ti-MS
MPV reduction	Al-MS
Olefin metathesis	Mo-MS
Olefin polymerization	MAO-MS
Hydrocracking	Ni–Mo–Al-M
Base catalysis catalytic alkylation	$(CH_2)xNR_2$-MS

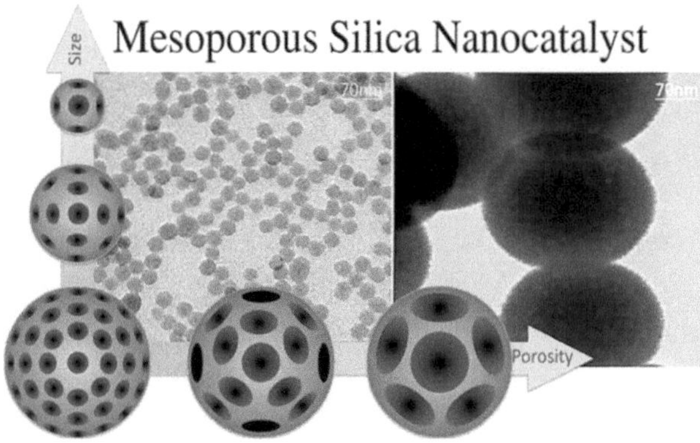

Figure 6.8: Variation in the pore size and particle size of the mesoporous silica nanocatalyst. Reproduced with the permission from (Sharma et al. 2015).

applications of mesoporous silica shell nanospheres in catalysis (Chen et al. 2005, Wang et al. 2008, Liu et al. 2010).

Synthesis of solid magnetic mesoporous silica nanospheres

We and his co-authors suggested the sol-gel synthesis of a mesoporous magnetic composite and is illustrated schematically in Fig. 6.9 (Wu et al. 2004). In this sol-gel method sodium silicate was used as the silica source. Thin layer formation of silicon oxide on the top of the magnetic particle surface was used by slowly adding H_2SO_4. Moreover this layer of silicon oxide is helpful for the self assembly of the surfactant. This magnetic silicon layer wrapped in silicon oxide coated Fe_3O_4 particles was prepared using the sol gel method and a molecular template. A thin layer of silicon

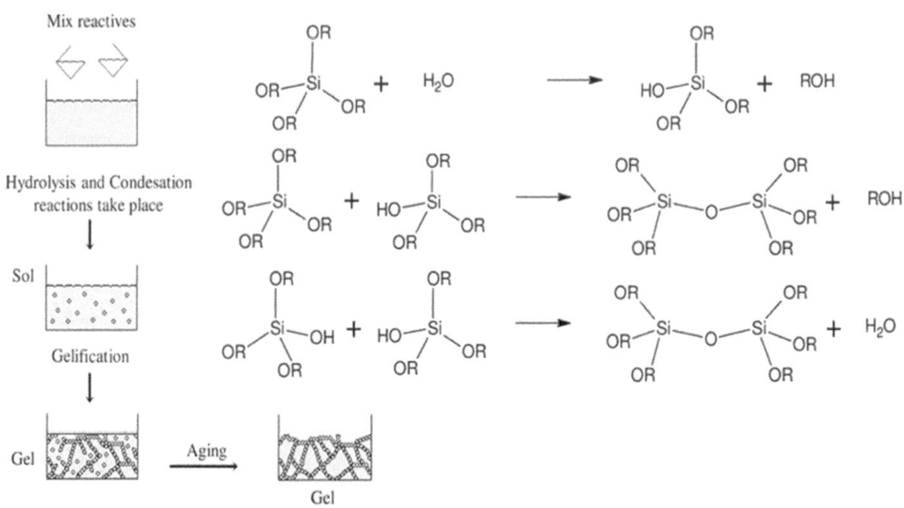

Figure 6.9: Different typical steps using the synthesize of MSNs. Reprinted with permission from (Sharma et al. 2015).

oxide was formed on the top of magnetic particles after dropwise addition of dilute sulfuric acid. The purpose of the mother liquor is to protect against nuclear magnetic immersion under industrial conditions TEOS (tetraethylorthosilicate) was used as a silica source and cationic surfactant molecules through ad-hoc loaded silicon species, forming a silicon oxide mesoporous structure.

Catalytic application of MSNs

Apart from the extraordinary magnetic nanomaterials properties, MSNs have attractive catalytic activity. They not only have usual catalytic properties but also have magnetic separation capability. They generally boost the chemical reaction and dissociation process in a typical chemical process that also makes the entire process easier. MSNs have a wide application window as will be discussed next (Natter et al. 2000, Preghenella et al. 2005).

Magnetic acid catalysis

Due to the rotation and external magnetic field and magnetic motion of the particles magnetized, the whole process in nanoprobe reaction solid acid catalysis. Changing the size of the magnetic field produced by Lorentz force consequently led in disturbing the reactant molecules. Besides, continuously this makes it facile for the magnetic catalyst to attack at the positive center of the carbonyl ion making it easier for the nucleophilic reagent to attack and thus this esterification reaction is catalyzed (Preghenella et al. 2005, Slowing et al. 2007).

Catalytic hydrogenation

Hydrogenation catalysts are basically derived from ferromagmetic substances. MSNs improve the catalytic hydrogenation and the presence of the noble metal in the catalyst fixed the bed process approach of catalytic hydrogenation. A perquisite of using mesoporous silica nanomaterials in the catalytic hydrogenation is transfer efficiency, fewer side effects and high productivity (Karimi et al. 2015).

Photochemical catalysis

TiO_2 MSNs have unique properties such as high chemical stability, high corrosiveness and compatibility with the human body because of these properties TiO_2 MSNs are used as a photochemical catalyst. Shihong and co-workers prepared isolated TiO_2/SiO_2/$NiFe_2O_4$ (TSN) nanoparticles. This photocatalyst exhibited high super-para-magnetic properties (Shihong et al. 2007).

Merit and demerits of MSNs

- Huge efficiency for adsorption and catalysis.
- Implementation of large molecules that allow them to chemically modify in a number of ways.
- Silanol groups' existence on the surface of the MSNs helps in selectivity and effectivity of functionalizing the MSNs.

- Deactivation of the organic functional groups available on the silica surface either by hydrogen bonding or covalent bonding (Shylesh et al. 2010).
- One of the prime deterrents of the grafting technique for the preparation of MSNs is the reaction at the surface of pores of organosilanes at the initial stage. These further deeply damage molecules at the pore center, creating varying distribution of organic functional groups.
- As per the synthesis methods reported so far for MSNs, two main difficulties occur. First, instant changes in physical properties of the solution is caused by the organosilanes Second, phase separation results due to the poor solubility of the organosilanes in the surfactants or if its condensation or hydrolysis rates differ greatly from that of TEOS.

Applications of nanocatalyst in the various biomedical fields

In the present world of nanotechnology where smaller the size of object the more superior will be its properties, particularly the catalysts of nano-size are applied in different chemical processes and are proved to be favorable to human life. Some of the nanocatalysts applications reported in the last decades are listed below.

In drug delivery

Controlled Drug-Delivery Systems (CDDSs) comprising of different inorganic nanomaterials, such as nano-silica, TiO_2, carbon-based systems and other nanomaterials has gained a lot of attention owing to enhanced therapeutic efficiency and reduced side effects. Basically, the size, shape and morphology of these nanomaterials have a great impact on cell penetration, entry inside the nucleus, cellular uptake (drug release) and biocompatibility. Carbon-based nanomaterials are fascinating systems in biomedical applications. Nano-sized carbon-based materials such as carbon nanotubes have gained significant attention owing to their unique chemical, physical and structural properties. CNTS are used in different fields like field emission sources, electric non-conductors, Li-ion batteries, fuel cells and in capacitors. Apart from these, recently CNTS are used in hydrogen adsorption as well which is attributed to their porous nature, lightweight, stability and economic availability. The tubular structure of CNTs allows for efficient hydrogen uptake. Hydrogen possessing high energy and being environment friendly is assumed to be the energy bearer in the coming time.

Specifically, CNTs have a tubular structure with 2–10 nm in diameter where small molecules can easily reside or can be easily encapsulated. Facile surface modification of CNTs can be done leading to a heterogeneous structure, which can be further stained with non-quenching and non-photobleaching agents. CNTs have also been used as a smart drug carrier for pH triggered release systems. Modification of the surface of CNTs with some pH sensing agents is well utilized and applied in drug delivery. In a report, CNTs are functionalized with polysaccharides sodium alginate (ALG) and chitosan (CHI) and conjugated with folic acid for targeted therapy. These nanocomposites when loaded with DOX showed slower release under physiological condition, i.e., pH of 7.4, while the faster and appreciable amount of DOX was released at a pH of 5.5. Interestingly by tuning the ratio of CHI/ALG on SWCNTs in CHI/ALG-SWCNTs nanocomposite higher drug loading with enhanced therapeutic efficiency can be attained (Shi Kam et al. 2004, Zhang et al. 2009).

In another study, SWCNTs modified with hydrazine have been used as pH-sensitive drug delivery vehicles. DOX was conjugated to the modified SWCNTs by hydrazone bonds which are easily cleaved under acidic pH in TME. Interestingly the drug loading efficiency was greatly enhanced on hydrazine modified SWCNTs as compared to pure SWCNTs. Moreover, CNTs with a large inner diameter are helpful in loading anticancer drugs. For example, cisplatin and DOX were loaded onto mildly oxidized MWCNTs with a large inner diameter. PEG and folic acid were also incorporated onto the nanocomposites to block the release of cisplatin from the inner part of the nanotubes. As a result, the nanocomposite caused higher cytotoxicity to cancer cells than unmodified nanotubes (Gu et al. 2011).

Maja et al. studied the colloidal TiO_2 nanoparticles as a drug delivery carrier of the ruthenium complex to melanoma cells. The developed carrier exhibited slower release of the complex under visible light, while enhanced release was observed under UV irradiation. However, this light-dependent toxicity on melanoma cells was inferred as both components of the system act as photosensitizers, generate reactive oxygen species, causing cell death (Nešić et al. 2017).

In a current study, the deposition of ZnS quantum dots on TiO_2 nanotube surface was done, loaded with 5-fluorouracil (5-FU) and its release pattern studied. These surface-functionalized TiO_2 nanotubes are promising materials to attain targeted drug release specifically to cancer cells and thereby minimizing the toxic effects of the anticancerous drug (Jafari et al. 2020).

Applications of TiO_2 nanostructures: recent advances

Wang et al. developed, a multifaceted approach for functionalization of mesoporous silica for the fabrication of Fe-based Fenton-like nanocatalyst with an accelerated biodegradable property and catalytic-enabled therapeutic functionality. The novel constructed PEG/rFeOx-HMSN nanocatalyst possessed not only coordination-based biodegradation, but was also biocompatible as demonstrated from both *in vitro* and *in vivo* biodegradability and pharmacokinetic studies. From *in vivo* animal studies, a higher rate of tumor-suppression was attained as 85.6 and 91.9% towards 4T1 breast tumor by intravenous Fig. 6.10a and intratumoral Fig. 6.10b administrations of nanocatalysts, which can be further visualized from photographic images of the dissected tumor for both intratumoral Fig. 6.10c and intravenous Fig. 6.10d treatment after completion of an experiment. The authors have presented a schematic presentation Fig. 6.10e how this iron-engineered PEG/rFeOx-HMSN nanocatalyst catalyzes the Fenton-like reactions by utilizing abundant H_2O_2 under acidic conditions and generating toxic hydroxyl radicals, which causes apoptosis of tumor cells and finally killing them (Wang et al. 2018).

Conventional silica nanoparticles possess such control, while Dendritic Fibrous Nano-Silica (DFNS) have them due to its fibrous morphology, diverse surface functionalization probabilities and a large surface area, make them promising

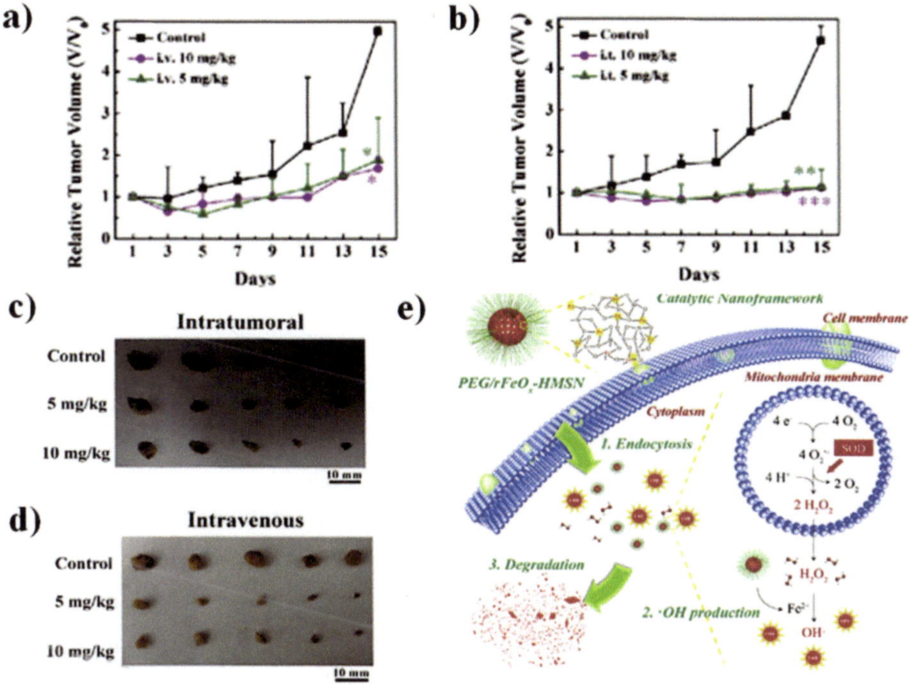

Figure 6.10: Relative tumor volumes of 4T1-bearing mice after administration with different doses of PEG/rFeOx-HMSN nanocatalyst either (a) intravenously or (b) intratumorally during 15 days of treatment. Digital photographs of dissected tumors after the therapeutic process from (c) intratumoral and (d) intravenous groups. (e) Scheme showing the therapeutic mechanism of PEG/rFeOx-HMSN nanocatalyst and hydroxyl radical-mediated tumor cell death. Reprinted with permission from (Wang et al. 2018).

nanocarriers. From this point of view, both pure and magnetic DFNS has been used for the delivery of the potent anticancerous drug DOX. Magnetic DFNS due to their fibrous character had a larger surface area as compared to iron oxide nanoparticles, thereby showing higher drug loading efficacy. From the *in vitro* drug release profile of DOXFe_3O_4/FMSMs it was observed that there is a fast release for the first initial hours due to the physically adsorbed drug at the surface, while slow and sustained release over 20 hours was noticed due to the entrapped drug in the fibrous channels of DFNS (FMSMs) as well as their strong interaction with silica surface silanols. In another report, Du et al. synthesized uniform dendrimeric amino-functionalized silica (HPSNs-NH_2) as nanocarriers similar to DFNS. These nanocarriers were functionalized with low-molecular-weight PEI using succinic anhydride as a linker to enhance biocompatibility and interactions with anticancerous drugs (topotecan (TPT), and nucleic acids. HPSNs-PEI exhibited higher co-loading of the drug and nucleic acid, that showed a better biocompatible character along with improved p-DNA delivery and high transfection efficiencies. In particular, it showed higher killing efficiency to cancer cells through synergetic effects from the co-delivery of the anticancer drug and nucleic acid.

Future perspective and summary

Nanostructured materials and nanotechnology are considered as a key feature of future technology and science progress, due to the great prospective for manipulation in this ultra-small nanoscale system.

Sustainable preparation and catalytic application

Catalytic nanomaterials research area is very centralized on green-inspired, i.e., environment-friendly perspective to synthesize the nanomaterials. Bale and his co-researchers explored protein, e.g., soybean peroxidase, poly-L-lysine, and bovine serum albumin. These can be used to produce Ag nanoparticles on CNTs (Bale et al. 2007, Murphy 2008).

The future role of catalytic nanomaterials

Besides gold and CNTs, a series of other materials have been explored as a catalyst. Many other materials besides gold and carbon nanotubes have been investigated as catalysts. Size tunability and the dispersion ability in the solvent are the key factors to control the selectivity. Due to these properties, a potential application of these solids is associated with its specific area. To decrease the release of pollutants the catalytic system is consumed in automotive devices such as aluminium oxide, impregnated with platinum and palladium nanoparticles with a high degree of dispersion. Materials used as electrodes of fuel cell components are doped cerium nanoparticles (Persson et al. 2006, Rabbani and El-Kaderi 2011).

Colloidal platinum–ruthenium nanoparticle can be achieved by the reduction of platinum and ruthenium ions with citric acid and $NaBH_4$. These are active in the catalytic oxidation of methanol, the reaction of industrial interest for the anode catalyst preparation cells of direct methanol fuel (DMFC, Direct Methanol Fuel Cell) (Liotta et al. 2005).

This chapter discussed the basics of catalysts and the application of various fields as well as in plasmonic nanoparticles and mesoporous silica nanoparticles. Many chemical reactions can be catalyzed using the plasmonic nanoparticle in the presence of light. Photocatalysis problems such as catalytic efficiencies and visible light response have been solved by plasmonic nanocatalysis, which are already producing amazing results. Besides this over the past one decade significant growth has been noticed with advanced synthesis of nanomaterials and more valuable results are anticipated in the coming years.

It is believed that many researchers will elaborate nanocatalysts and more effective endeavors will be dedicated to the new synthesis of novel catalysts with a higher performance and be more cost effective. For practical applications, the catalyst cost can be a restricting factor to overcome this issue, metals like aluminium and copper could be useful. As discussed above, the size tenability, shape and concentration of metal in nanomaterials play a crucial role to determine the intermetallic effect as well as electronic band structure, which can result in a catalytic behavior. Currently, on the basis of various studies by different groups and the application of several fields, one can say that the nanocatalyst will solve various problems. At the same time, it would be logical to envisage that they will perform a very positive role in the field of nanotechnology, catalyst science for the environmental remediation and most important energy issues and other major problems that human society currently faces.

Summary

Nanostructured materials and nanosciences open a new window and show great promises for various potential applications in different fields including catalysis with controlled and optimized properties. Nanofibers, CNTs and nanoparticles display potential applications for hydrogen storage because of high surface area. They also recommend a series of benefits for the physicochemical reactions such as surface interactions, adsorption besides this, bulk absorption, surface coverage, molecular analysis, rapid kinetics, by the surface catalyst. The fundamentally huge surface areas and distinctive adsorbing properties of nanophase materials can aid the separation of gaseous hydrogen, and the high surface area of individual nanoparticles can make short diffusion pathways to the materials' interiors. Larger dispersion of the catalytically active species is enabled by the use of nano-sized dopants and thus makes higher mass transfer reactions easier. Polymer nano composites and functionalized carbon nanotubes have a unique microstructure for physisorption of hydrogen atom/ molecule on the surface and inside the bulk. This chapter briefly conferred various nanomaterials for hydrogen storage and also presented in plasmonic application and most important biomedical advantages of nanocatalysts. Additionally MSNs preparation and catalytic application have been explored. Apart from these future aspects of nanocatalysts application, its energy and environmental concerns also received attention.

References

Anastas, P. T. and J. C. Warner. 1998. Principles of green chemistry. Green Chemistry: Theory and Practice 29–56.

Anpo, M., H. Yamashita, K. Ikeue, Y. Fujii, S. G. Zhang, Y. Ichihashi et al. 1998. Photocatalytic reduction of CO_2 with H_2O on Ti-MCM-41 and Ti-MCM-48 mesoporous zeolite catalysts. Catalysis Today 44(1-4): 327–332.

Bale, S. S., P. Asuri, S. S. Karajanagi, J. S. Dordick and R. S. Kane. 2007. Protein-directed formation of silver nanoparticles on carbon nanotubes. Advanced Materials 19(20): 3167–3170.

Bao, J. L. and E. A. Carter. 2019. Surface-plasmon-induced ammonia decomposition on copper: excited-state reaction pathways revealed by embedded correlated wavefunction theory. ACS Nano 13(9): 9944–9957.

Chaturvedi, S., P. N. Dave and N. Shah. 2012. Applications of nano-catalyst in new era. Journal of Saudi Chemical Society 16(3): 307–325.

Chen, M., D. Kumar, C.-W. Yi and D. W. Goodman. 2005. The promotional effect of gold in catalysis by palladium-gold. Science 310(5746): 291–293.

Della Pina, C., E. Falletta and M. Rossi. 2008. Highly selective oxidation of benzyl alcohol to benzaldehyde catalyzed by bimetallic gold–copper catalyst. Journal of Catalysis 260(2): 384–386.

Dhiman, M. 2020. Plasmonic nanocatalysis for solar energy harvesting and sustainable chemistry. Journal of Materials Chemistry A8(20): 10074–10095.

Dunworth, W. P. and F. Nord. 1954. Noble metal—synthetic polymer catalysts and studies on the mechanism of their action. Advances in Catalysis, Elsevier 6: 125–141.

Fichtner, M. 2005. Nanotechnological aspects in materials for hydrogen storage. Advanced Engineering Materials 7(6): 443–455.

Gawande, M. B., P. S. Branco and R. S. Varma. 2013. Nano-magnetite (Fe_3O_4) as a support for recyclable catalysts in the development of sustainable methodologies. Chemical Society Reviews 42(8): 3371–3393.

Goswami, D. Y. and F. Kreith. 2007. Handbook of Energy Efficiency and Renewable Energy, CRC Press.

Gu, Y.-J., J. Cheng, J. Jin, S. H. Cheng and W.-T. Wong. 2011. Development and evaluation of pH-responsive single-walled carbon nanotube-doxorubicin complexes in cancer cells. International Journal of Nanomedicine 6: 2889.

Jafari, S., B. Mahyad, H. Hashemzadeh, S. Janfaza, T. Gholikhani and L. Tayebi. 2020. Biomedical applications of TiO_2 nanostructures: Recent advances. International Journal of Nanomedicine 15: 3447.

Karimi, L., M. E. Yazdanshenas, R. Khajavi, A. Rashidi and M. Mirjalili. 2015. Optimizing the photocatalytic properties and the synergistic effects of graphene and nano titanium dioxide immobilized on cotton fabric. Applied Surface Science 332: 665–673.

Li, Y. and G. A. Somorjai. 2010. Nanoscale advances in catalysis and energy applications. Nano Letters 10(7): 2289–2295.

Liotta, L., G. Di Carlo, G. Pantaleo and G. Deganello. 2005. Co_3O_4/CeO_2 and $Co_3O_4/CeO_2–ZrO_2$ composite catalysts for methane combustion: Correlation between morphology reduction properties and catalytic activity. Catalysis Communications 6(5): 329–336.

Liu, R., Y. Yu, K. Yoshida, G. Li, H. Jiang, M. Zhang et al. 2010. Physically and chemically mixed TiO_2-supported Pd and Au catalysts: unexpected synergistic effects on selective hydrogenation of citral in supercritical CO_2. Journal of Catalysis 269(1): 191–200.

Mori, K., P. Verma, R. Hayashi, K. Fuku and H. Yamashita. 2015. Color-controlled Ag nanoparticles and nanorods within confined mesopores: microwave-assisted rapid synthesis and application in plasmonic catalysis under visible-light irradiation. Chemistry—A European Journal 21(33): 11885–11893.

Murphy, C. J. 2008. Sustainability as an emerging design criterion in nanoparticle synthesis and applications. Journal of Materials Chemistry 18(19): 2173–2176.

Natter, H., M. Schmelzer, M.-S. Löffler, C. Krill, A. Fitch and R. Hempelmann. 2000. Grain-growth kinetics of nanocrystalline iron studied *in situ* by synchrotron real-time X-ray diffraction. The Journal of Physical Chemistry B104(11): 2467–2476.

Nešić, M., J. Žakula, L. Korićanac, M. Stepić, M. Radoičić, I. Popović et al. 2017. Light controlled metallo-drug delivery system based on the TiO_2-nanoparticles and Ru-complex. Journal of Photochemistry and Photobiology A: Chemistry 347: 55–66.

Niemann, M. U., S. S. Srinivasan, A. R. Phani, A. Kumar, D. Y. Goswami and E. K. Stefanakos. 2008. Nanomaterials for hydrogen storage applications: a review. Journal of Nanomaterials.

Persson, K., A. Ersson, S. Colussi, A. Trovarelli and S. G. Järås. 2006. Catalytic combustion of methane over bimetallic Pd–Pt catalysts: The influence of support materials. Applied Catalysis B: Environmental 66(3-4): 175–185.

Preghenella, M., A. Pegoretti and C. Migliaresi. 2005. Thermo-mechanical characterization of fumed silica-epoxy nanocomposites. Polymer 46(26): 12065–12072.

Rabbani, M. G. and H. M. El-Kaderi. 2011. Template-free synthesis of a highly porous benzimidazole-linked polymer for CO_2 capture and H_2 storage. Chemistry of Materials 23(7): 1650–1653.

Rajendran, A., M. Rajendiran, Z. Yang, H. Fan, T. Cui, Y. Zhang et al. 2019. Functionalized silicas for metal-free and metal-based catalytic applications: a review in perspective of green chemistry. The Chemical Record.

Seayad, A. M. and D. M. Antonelli. 2004. Recent advances in hydrogen storage in metal-containing inorganic nanostructures and related materials. Advanced Materials 16(9-10): 765–777.

Sharma, N., H. Ojha, A. Bhardwaj, D. Pathak and R. Sharma. 2015. Preparation and catalytic applications of nanomaterials: a review. Rsc Advances 5(66): 53381–53403.

Shi Kam, N. W., N. Kam, T. Jessop, P. Wender and H. Dai. 2004. Nanotube molecular transporters: internalization of carbon nanotube–protein conjugates into mammalian cells. Journal of the American Chemical Society 126(22): 6850–6851.

Shihong, X., S. Wenfeng, Y. Jian, C. Mingxia and S. Jianwei. 2007. Preparation and photocatalytic properties of magnetically separable TiO_2 supported on nickel ferrite. Chinese Journal of Chemical Engineering 15(2): 190–195.

Shylesh, S., V. Schunemann and W. Thiel. 2010. Magnetically separable nanocatalysts: bridges between homogeneous and heterogeneous catalysis. Angewandte Chemie International Edition 49(20): 3428–3459.

Singh, J., T. Dutta, K. Kim, M. Rawat, P. Samaddar and P. Kumar. 2018. 'Green' synthesis of metals and their oxide nanoparticles: applications for environmental remediation. Journal of Nanobiotechnology 16(1): 84.

Singh, S. B. and P. K. Tandon. 2014. Catalysis: a brief review on nano-catalyst. Journal of Energy and Chemical Engineering 2(3): 106–115.

Slowing, I., B. Trewyn, S. Giri and V. Lin. 2007. Mesoporous silica nanoparticles for drug delivery and biosensing applications. Advanced Functional Materials 17(8): 1225–1236.

Tandon, P. 2019. Catalytic applications of copper species in organic transformations: A review. Journal of Catalyst and Catalysis 1(2): 21–34.

Thomas, J. M. 2010. The advantages of exploring the interface between heterogeneous and homogeneous catalysis. ChemCatChem. 2(2): 127–132.

Vollath, D. 2008. Nanomaterials an introduction to synthesis, properties and application. Environmental Engineering and Management Journal 7(6): 865–870.

Wang, D., A. Villa, F. Porta, L. Prati and D. Su. 2008. Bimetallic gold/palladium catalysts: Correlation between nanostructure and synergistic effects. The Journal of Physical Chemistry C112(23): 8617–8622.

Wang, L., M. Huo, Y. Chen and J. Shi. 2018. Iron-engineered mesoporous silica nanocatalyst with biodegradable and catalytic framework for tumor-specific therapy. Biomaterials 163: 1–13.

Wu, P., J. Zhu and Z. Xu. 2004. Template-assisted synthesis of mesoporous magnetic nanocomposite particles. Advanced Functional Materials 14(4): 345–351.

Zhang, X., L. Meng, Q. Lu, Z. Fei and P. Dyson. 2009. Targeted delivery and controlled release of doxorubicin to cancer cells using modified single wall carbon nanotubes. Biomaterials 30(30): 6041–6047.

Zhou, B., M. Wen and Q. Wu. 2013. C-isolated Ag–C–Co sandwich sphere-nanostructures and their high activity catalysis induced by surface plasmon resonance. Nanoscale 5(18): 8602–8608.

Zhou, L., D. Swearer, C. Zhang, H. Robatjazi, H. Zhao, H. Henderson et al. 2018. Quantifying hot carrier and thermal contributions in plasmonic photocatalysis. Science 362(6410): 69–72.

CHAPTER 7

Recent Advancement of Electrocatalyst System in CO_2 Reduction

Insights of Fe, Co and Ni Metallo-ligand Clusters in Homogeneous Molecular Level

Jeevithra Dewi Subramaniam and *Pei Meng Woi**

Introduction

Fossil fuel sources hold a large fraction in energy supply to the world's needs such as residential and commercial, transportation and industrial uses (Dresselhaus and Thomas 2001). Without any doubt, the rising energy demand will surpass the limited amounts of fossil resources in the near future. It should be noted that fossil fuels have been used as precursors to manufacture various chemical products such as polymers and synthetic waxes. Chemical resources will be affected appropriately with fossil fuels deficiency. On top of that, elevation of atmospheric CO_2 concentration is another serious issue appearing from mass consumption of fossil fuels (Li et al. 2018). Over the years, fossil resources use has led to the greenhouse gas effect that gradually threatens human survival (Boot Handford et al. 2014). These projected impacts extend beyond climate to severe global warming which caused ocean acidification, as the ocean is the major sink for atmospheric CO_2 (Rosas-Hernandez et al. 2017). Granting a future energy supply that is secure and CO_2-neutral will require switching to non-fossil energy sources such as wind, solar, nuclear and geothermal energy (Appel et al. 2013).

Department of Chemistry, Faculty of Science, Universiti Malaya, 50603 Kuala Lumpur, Malaysia.
* Corresponding author: pmwoi@um.edu.my

Along with this, to mitigate atmospheric CO_2, utilizing CO_2 as a carbon energy source via Carbon Capture Utilization (CCU) technologies is one of the primary focuses of many countries (Do et al. 2018, Simmons et al. 2011). Photocatalytic CO_2 reduction (Nov-Institute 2015) and electrocatalytic CO_2 reduction (Vidal Vázquez et al. 2018) are part of the CCU technologies which prominently advance in the molecular level research for catalytic conversion of CO_2 into value-added chemical fuels and organic materials (Sampson et al. 2014). Therefore, transforming CO_2 into a source of fuel or precursor for chemical products by electrochemically means offer an attractive way to decrease atmospheric concentrations. The challenges presented here are great, but the potential rewards are enormous.

CO_2 is an extremely stable molecule generally produced by fossil fuel. Converting CO_2 to a useful state by activation/reduction is a scientifically challenging problem, which requires appropriate catalysts and energy input. This has attributed to several fundamental challenges in the development of different chemical catalysis pathways, electrochemistry, photochemistry, semiconductor physics and engineering technology advances.

Fundamental of CO_2 reduction

Chemical properties of CO_2

Carbon dioxide is an extremely stable linear triatomic molecule. The central carbon atom is bonded to two oxygen atoms by double bonds with the bond distance of 1.1615 Å. The carbon atom possesses *sp* hybridization and the C-O distance of it is shorter than a C-O double bond involving a sp^2 carbon center as indicated in Fig. 7.1 (Schwarz and Dodson 1989).

CO_2 contains polar bonds due to the difference in electronegativity between C and O. Its electronic structure is best represented as $O^{-\delta} - C^{+2\delta} - O^{-\delta}$, highlighting its susceptibility to nucleophilic attack as the carbon and electrophilic attack at oxygen, often described in terms of its quadrupole moment. With an ionization potential of 13.78 eV (vs. 12.6 for water, 10.0 for ammonia), CO_2 is non-basic and interacts weakly only with Brønsted and Lewis acids. With a carbon localized at LUMO, CO_2 is susceptible to attack by nucleophiles and to reduction.

As mentioned early, the ionization potential and the electron affinity of CO_2 is 13.73 eV and 3.8 eV, respectively. Thus, it is a poor electron donor and a good

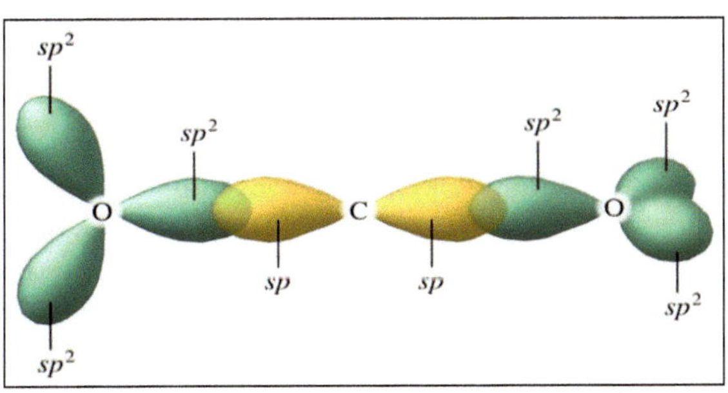

Figure 7.1: The hybridization orbitals of the CO_2 molecule.

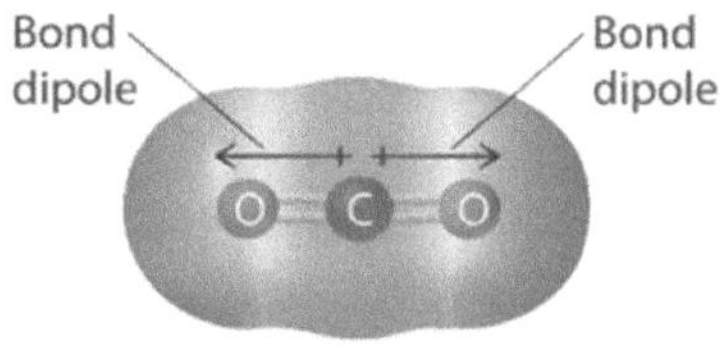

Figure 7.2: Charge distributions of CO_2.

electron acceptor. CO_2 molecule demonstrates numerous definite positions which require specific electronic properties for a potential coordination center (Benson et al. 2009). This is due to the difference in electronegativity of carbon and oxygen atoms as shown in Fig. 7.2.

CO_2 reduction potential

It is now understood that CO_2 are thermodynamically and kinetically stable molecules which require greater amounts of energy to reduce into liquid fuels or fuel precursors such as synthetic gas (CO/H_2). From the indepth discussion by Kleiji et al., Sakkura et al., and Benson et al. the thermodynamic barrier and kinetic barrier are the two challenges in CO_2 reduction (Benson et al. 2009, Kleij 2013, Sakakura et al. 2007).

Thermodynamic barrier

CO_2 is highly unfavorable for one-electron reduction with higher $E^o = -1.90$ V vs. NHE, E6 due to a large reorganizational energy between the linear molecule and bent radical anion (Wang et al. 2015). In order to lower the potential substantially, proton assisted multi-electron transfer reaction is required to produce thermodynamically more stable products. This is summarized in Eqn (1)–(5) (pH 7 in aqueous solution versus NHE 25°C, 1 atm and 1M for the other solutes) (Benson et al. 2009).

$$CO_2 + e^- \rightarrow CO_2^- \qquad E^o = -1.90\text{ V} \qquad \text{(E1)}$$

$$CO_2 + 2H^+ + 2e^- \rightarrow CO + H_2O \qquad E^o = -0.53\text{ V} \qquad \text{(E2)}$$

$$CO_2 + 2H^+ + 2e^- \rightarrow HCOOH \qquad E^o = -0.61\text{ V} \qquad \text{(E3)}$$

$$CO_2 + 4H^+ + 4e^- \rightarrow HCHO + H_2O \qquad E^o = -0.48\text{ V} \qquad \text{(E4)}$$

$$CO_2 + 6H^+ + 6e^- \rightarrow CH_3OH + H_2O \qquad E^o = -0.38\text{ V} \qquad \text{(E5)}$$

$$CO_2 + 8H^{+1} + 8e^- \rightarrow CH_4 + 2H_2O \qquad E^o = -0.24\text{ V} \qquad \text{(E6)}$$

Kinetic barrier

The assembly of the nuclei and formation of chemical bonds to convert the relatively simple CO_2 molecules into more complex and energetic molecules is the significant setback in the CO_2 conversion. According to Kubiak and co-workers there are two strategic ways to accomplish this. The easiest way is the conversion of CO_2 and H_2O into CO and H_2, utilizing Fischer-Töpsch technologies to produce value-added organics such as liquid fuels. The second strategy is known as ‘do it the hard way’ through direct electrocatalytic conversion CO_2 into liquid fuels. The kinetic barrier is superior here. Are there any possibilities to overcome this challenge? One of the methods is to recognize a single catalyst that is capable to perfect the sequence of conversion steps CO_2 to CO, then H_2CO, and then to higher carbon molecules with low kinetic barriers.

Conversion of CO_2 into value-added organics

Several different methods for catalytic approaches for CO_2 reduction are reported (Aresta and Dibenedetto 2007, Mikkelsen et al. 2010, Sakakura and Kohno 2009, Sakakura et al. 2007). The most developed method is using CO_2 as a reagent as shown in Fig. 7.3 (Kleij 2013, Sakakura et al. 2007). Nevertheless, the direct activation of CO_2 for electrophilic or nucleophilic attack by a second substrate through prior complexation to a transition metal catalyst is a highly challenging task. However, the design and implementation of such direct activation strategies may help to define the shape of the future product portfolio based on this renewable carbon feedstock (Kleij 2013, North and Pasquale 2009).

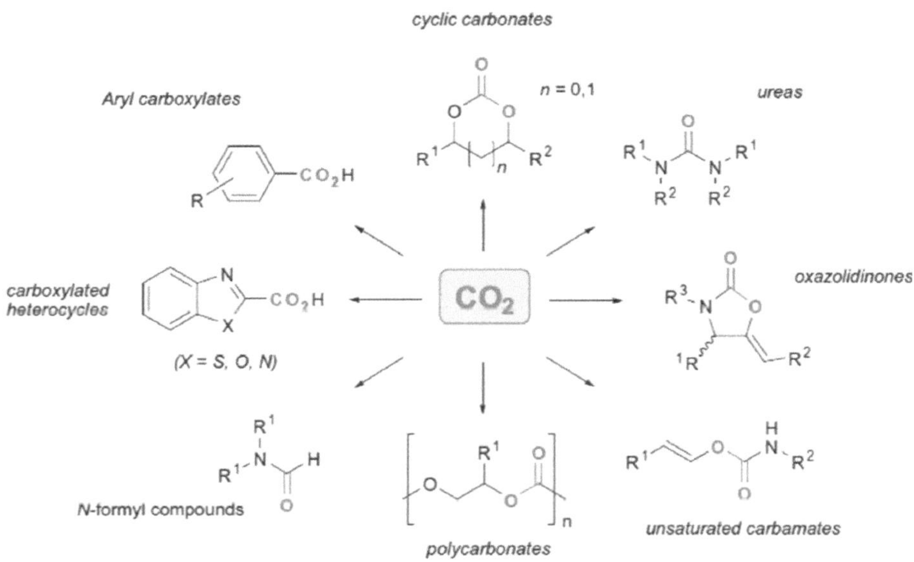

Figure 7.3: CO_2 conversion into value-added organic products using CO_2 as the reagent. Reprinted with the permission (Kleij 2013). Copyright 2013 Elsevier B.V.

Coordination behavior of CO_2 towards transition metal

Coordination mode of CO_2

As mentioned earlier, one possibility to lower the thermodynamic barrier is by proton assisted multi electron transfer process such as those in Eqns. (2)–(6). Transition-metal complexes can have accessible multiple redox states that promote such pathways. Fujita and co-workers analyzed binding of CO_2 to a vacant coordination site at the metal center of a reduced catalyst that was the main step in reducing CO_2 catalytically (Yano and Yachandra 2014). Therefore, to activate the metal center understanding the coordination mode of CO_2 towards the metal center is important.

The coordination of CO_2, results in a net transfer of electron density from the metal to the LUMO of the ligand if the complexation takes place by the double bond or the central carbon atom. The LUMO of CO_2 is an antibonding orbital and therefore the electron transfer should result in a weakened C-O interaction (Okamura et al. 2016, Takeda et al. 2017). As the prime step to activate CO_2, it should accept an electron in its antibonding π orbitals that causes the molecule to bend.

Table 7.1 shows several common coordination modes with transition metal complexes which involve CO_2 (Dobbek et al. 2001, Seefeldt et al. 2009):

a) ***η^1 (C) coordination mode***: via electron donation from metal to the carbon orbital with formation of the metallo-acid derivative (preferred electron-rich metal ion).
b) ***η^1 (O) coordination mode***: via oxygen by donation of the oxygen lone p-electron pair to the vacant orbital of the metal (preferred electron-poor metal ion).
c) ***η^2 (C, O) coordination mode***: via π-complex formation via C=O double bond. The 'back-bonding' from filled metal *d*-orbitals to empty π*-orbitals from CO_2 plays an important role.
d) ***η^2 (O, O) coordination mode***: binding type as 'metal carboxylate'.

Table 7.1: Mode of coordination with metal ion (Kleij 2013, Paparo and Okuda 2017).

Coordination mode	η^1*(C)*	η^1*(O)*	η^2 *(C, O)*	η^2 *(O, O)*
Illustration	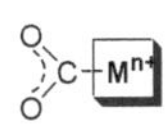	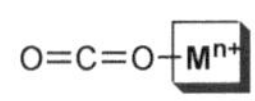	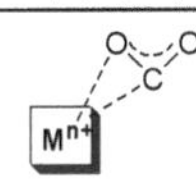	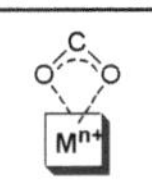

Application of multinuclear metal clusters (Fe, Co, Ni) as a molecular electrocatalyst

Nature's biological systems, especially photosynthesis play a vital role in balancing carbon-neutral cycle. Enzymes have been discovered to catalyze the reversible reduction of CO_2 to CO (CO dehydrogenases) or CO_2 to formate (formate dehydrogenases). Utilization of 3d transition metals (Fe, Ni, Mo, etc.), within the enzyme clusters is a significant design favored by nature.

A popular enzyme known as CODH (CO dehydrogenases) has been identified to catalyze CO_2 to CO. [NiFe]-CO dehydrogenase from Cluster C (Fig. 7.4) with the active site consisting of Ni and Fe centers bridged by an Fe_3S_4 cluster that rigidly positions these two metal centers in close proximity. Scheme 1 describes the mechanism of [NiFe] CODHs in CO_2 reduction into CO (Appel et al. 2013, Froehlich and Kubiak 2012, Jeoung and Dobbek 2007, Volbeda and Fontecilla Camps 2005).

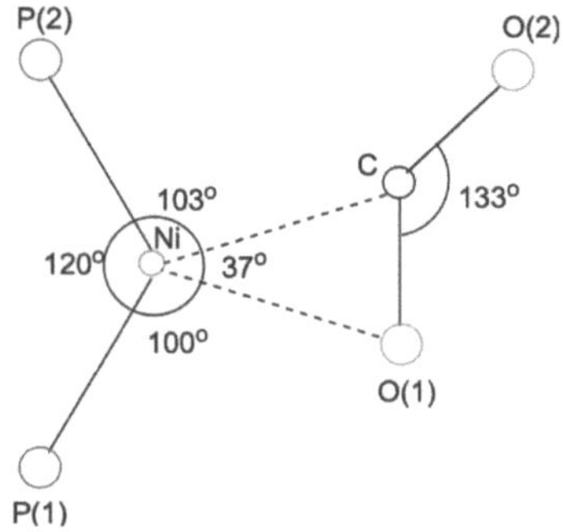

Figure 7.4: Structure of the active site cluster C. (a) Ball-and-stick presentation of [NiFe] CO dehydrogenase (cluster C). (b) Schematic drawing of the inorganic core of cluster C (Dobbek 2019). Copyright 2018 Springer Nature.

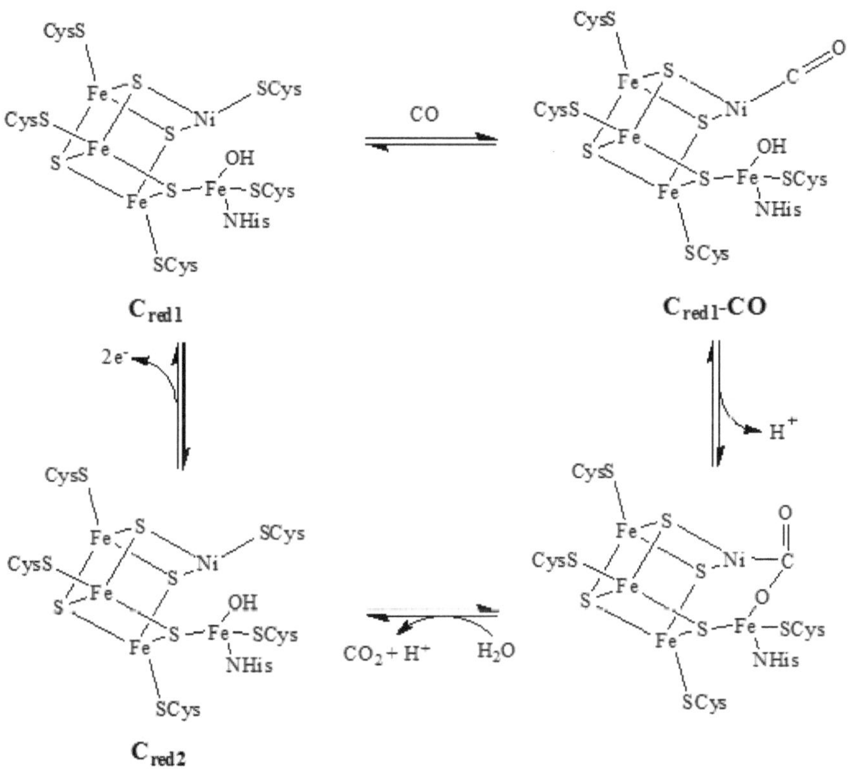

Scheme 1: Proposed mechanism for the reduction of CO_2 to CO by [NiFe] CODH (Gan et al. 2016, Jeoung and Dobbek 2007). Reproduced with permission (Gan et al. 2016). Copyright 2015 Springer Nature.

Electrocatalytic system

An electrocatalyst involves an electron transfer reaction at the electrode and aid acceleration of a chemical reaction. For an efficient and optimal electrocatalyst, both the electron transfer and chemical kinetics are required to be fast. Besides, it should demonstrate an optimum thermodynamic match between the redox potential (E^0) for the electron transfer reaction and the chemical reaction that is being catalyzed, i.e., reduction of CO_2 to HCOOH (Benson et al. 2009).

Generally, electrocatalysts are screened for some of their crucial features as listed below to determine the best overall catalysts (Benson et al. 2009, Schneider et al. 2012a).

1. Redox potentials
2. Current efficiencies
3. Electron transfer rate
4. Chemical kinetics

Electrocatalysts are electron transfer agents which are expected to work ideally near the thermodynamic potential of the reaction to be driven, E^0 (products/substrates). The direct electrochemical reduction of CO_2 on most electrode surfaces demands large overpotential which reduces the efficiency in terms of conversion rates. One method to overcome the above issue is to utilize homogeneous redox catalysts in an indirect electrolysis which lower kinetic barriers. This usually yields an increased and/or entirely different selectivity compared to direct electroreduction (Francke and Little 2014, Francke et al. 2018, Rountree et al. 2014).

The overpotential, η can be the differences between the applied electrode potential, $V_{applied}$, and E^0 (products/substrates), at a given current density as referred to in Eqn (7).

$$\eta = V_{applied} - E^0\ (products/substrates) \tag{E7}$$

Both thermodynamic and kinetic considerations are important here. Clearly, to minimize the overpotential, catalysts need to be developed to have formal potentials, E^0 ($Cat^{n+/0}$) which are well-matched to E^0 (products/substrates), and appreciable rate constants, k_{cat}, for the chemical reduction of substrates to products at this potential. Moreover, the heterogeneous rate constant, k_h, for reduction of the electrocatalyst at the electrode must be high for $V_{applied}$ near E^0 ($Cat^{n+/0}$). Reaction rates for these processes can be estimated from the steady-state limiting current in cyclic voltammetry, or by rotating disk voltammetry studies of the heterogeneous electron transfer kinetics (Benson et al. 2009, Francke et al. 2018, Rountree et al. 2014, Schneider et al. 2012a).

Scheme 2 displays a general approach of an electrocatalytic system which explains the general sense of electrocatalysts involving electron transfer agents that ideally operate near the thermodynamic potential of the reaction to be driven, E^0 (products/substrates) and E^0 ($Cat^{n+/0}$).

Transition metal complexes catalyst

Transition-metal complexes are at the cutting edge of potential catalyst research. These catalysts can have multiple and accessible redox states that have been proved to promote multi-electron transfer reactivity. Moreover, the formal reduction potentials can be systematically tuned through ligand modification to better match the potential required for CO_2 reduction (Morris et al. 2009).

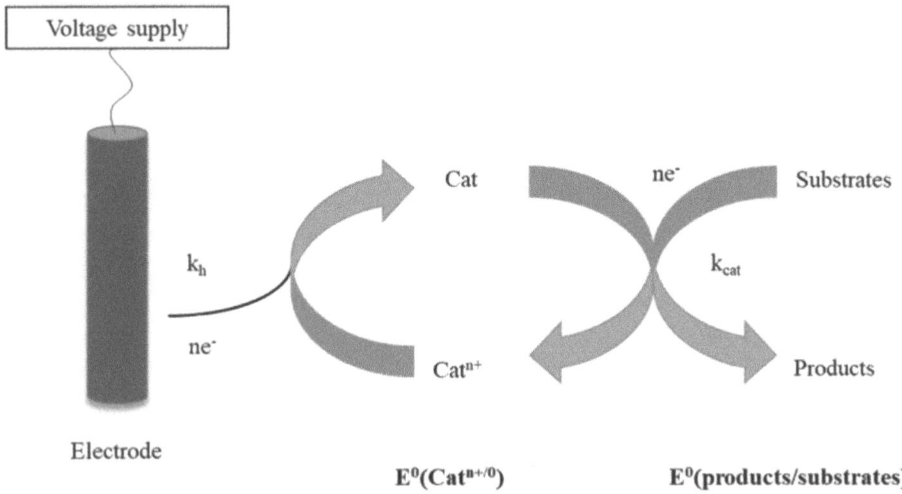

Scheme 2: A general approach of an electrocatalytic system (Benson et al. 2009). Adapted from The Royal Society of Chemistry.

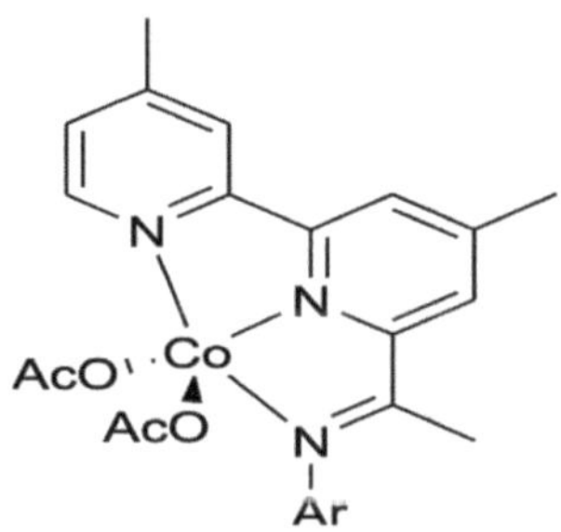

Figure 7.5: Mode of bonding of CO_2 in $[Ni(CO_2)(PCy_3)_2]$,0.75-toluene (Aresta et al. 1975).

Transition-metal complex catalysts of CO_2 electrochemical reduction based on earth-abundant metals have been intensively reviewed (Savéant 2008, Takeda et al. 2017), with the description of the main classes of ligands (polypyridyl, phosphine, cyclam and aza-macrocyclic ligands, porphyrins, phthalocyanines and related macrocycles) and metals (Ni, Co, Mn, Fe), leading to efficient homogeneous electrochemical CO_2 reduction.

The first published crystal structure of CO_2 bound to a transition metal complex in 1975 was reported by Aresta and Nobile. Figure 7.5 illustrates the coordination geometry of η^2-bidentate binding mode involving the carbon atom and one oxygen atom, with significant bending in the CO_2 structure (Aresta et al. 1975). This finding had proved beneficial of transition-metal elements as electrocatalysts to facilitate the CO_2 reduction.

Redox active ligand

Nature's use of redox-active moieties combined with 3d transition metal ions is a powerful strategy to promote multi-electron catalytic reactions. The ability of these moieties to store redox equivalents helps metalloenzymes in promoting multi-electron reactions. In a biomimetic spirit, chemists have recently developed approaches relying on redox-active moieties in the environment of metal centers to catalyze challenging transformations (Praneeth et al. 2012).

Redox-active ligands with larger π-π conjugation often facilitate crucial multi-electron catalytic transformations. Dependence on nature for redox-active moieties to catalyze challenging reactions, either bind as ligands or in the second coordination sphere of a metal cofactor. Nature utilizes redox-active moieties to store and mediate electron-transfer to metal centers thereby communicating noble metals and 3d transitional metals such as Fe, Co onto base metals (Praneeth et al. 2012, Rakowski Dubois and Dubois 2009).

The properties and behaviors of transition metal complexes rely on bridging ligand as it can hold one or more transition metals together. The bridging ligand can act as electronic communicators between the metal centers and depending on their structures, can act as electronic insulators too. In addition, the nature and topology of the ligand can allow the metals to cooperate for a specific reaction (García Antón et al. 2012).

For example, Tanaka and the co-workers demonstrated the electrochemical water oxidation reaction using a ruthenium catalyst containing a *o*-quinone ligand (Wada et al. 2000). Quinone adopts three different oxidation states: (1) quinone, (2) semiquinone, and (3) catecholate. The resulting dinuclear complex displays tremendous water oxidation activity with high turnover numbers (TON > 30000). Compared to other water-oxidation catalysts, which rely on high valent ruthenium oxo species, Tanaka's system shuttles between Ru^{II} and Ru^{III}, coupled with ligand-centered (i.e., semiquinone/quinone) redox couples (Ghosh and Baik 2011, Wada et al. 2000, Wada et al. 2001).

Peng Kang and co-workers developed an imino bipyridine cobalt (II) complex (Fig. 7.6) for electrocatalytic reduction of CO_2 to formate in MeCN which possesses π-conjugation system. The single tridentate ligand in this study allows more open sites, and large steric of the diisopropyl moieties can prevent dimerization of the active hydride intermediate which contribute to increased catalytic reactivity. The redox active imino bipyridial ligand is effective in delocalizing electron density from the Co center to the ligand, thus making the Co center less electron rich (Liu et al. 2018).

In conclusion, employing redox-active ligand with transition metals as the electrocatalysts for CO_2 reduction enhances the efficiency and selectivity besides reducing the overpotential.

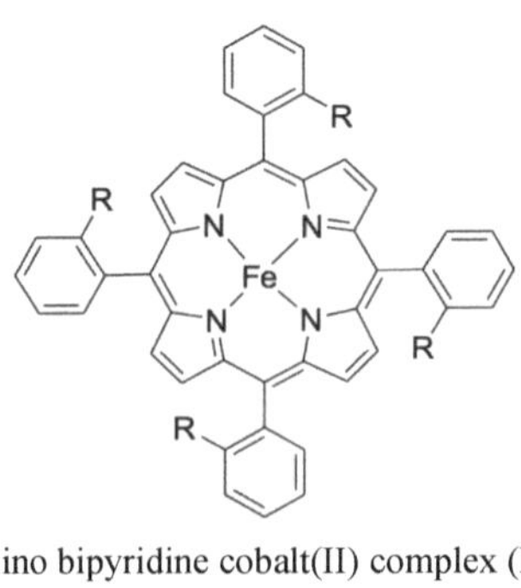

Figure 7.6: Imino bipyridine cobalt(II) complex (Liu et al. 2018).

Iron-based electrocatalyst in CO_2 reduction

Iron complexes are the most frequently studied catalysts for CO_2 electroreduction, even though the reported evidence is confined mostly to porphyrins (Hammouche et al. 1991), cyclopentadienone complexes (Rosas-Hernández et al. 2017) and iron carbonyl clusters (Taheri and Berben 2016a, Taheri and Berben 2016b, Taheri et al. 2015). Iron-based molecular electrocatalysts for the conversion of CO_2 to CO depends mostly on porphyrin-based ligand platforms. Iron porphyrin complexes represent the state of the art in the homogeneous electrocatalytic reduction of CO_2 in terms of catalytic activity, selectivity and overpotential.

In 1991, iron (0) tetraphenylporphyrins (FeTPP) (Fig. 7.7) were reported by the Saveant's group to reduce CO_2 to CO at –1.64 V vs. SCE in DMF. Figure 7.8 shows CV of FeTPP that exhibits clear Fe(III/II), Fe(II/I) and Fe(I/0) couples, the last of which (at –1.64 V *vs.* SCE) developed a catalytic current under CO_2. However, the porphyrins were destroyed by carboxylation and/or hydrogenation of the ring after a few catalytic cycles in DMF, with tetraalkylammonium salts as a supporting electrolyte. In the presence of a hard electrophile such as Mg^{2+} ion improved the rate of the reaction, the production of CO and the stability of the catalyst. The process of breaking C-O bond of CO_2 in the coordination sphere of the complexes into CO was accelerated by Mg^{2+} ions at room temperature. This combination action of FeTPP and Mg^{2+} as a bimetallic catalysis where an electron-rich center initiated the reduction process and an electron-deficient center assisted the transformation of the bond system (Benson et al. 2009, Hammouche et al. 1991).

In 1996, the same research group reported that in the presence of weak Brønsted acids such as 1-propanol, 2-pyrrolidone and CF_3CH_2OH triggers efficiency and

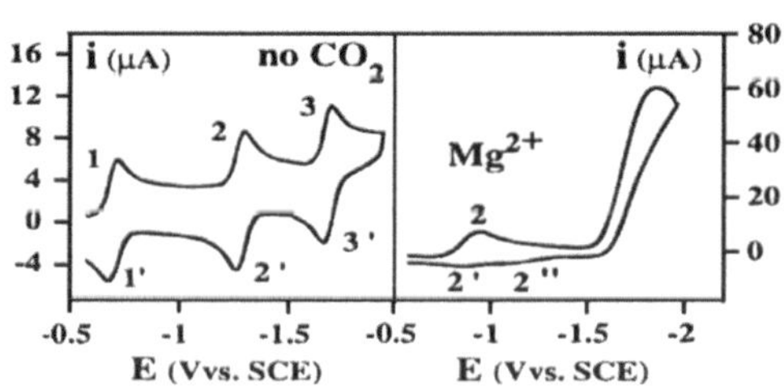

Figure 7.7: Iron tetraphenylporphyrin (FeTPP) catalyst structure shown to reduce CO_2 to CO (Hammouche et al. 1991).

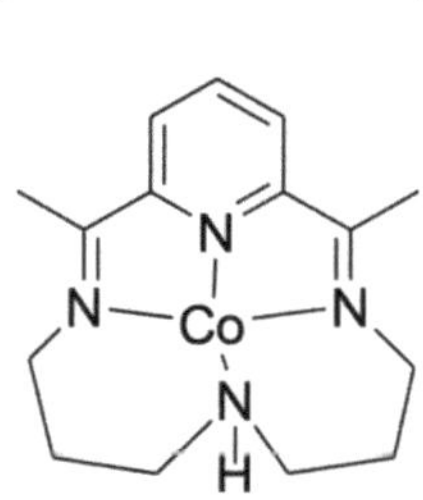

Figure 7.8: Cyclic voltammetry of FeTPP (1 mM) in DMF + 0.1 M NEt_4ClO_4 at a glassy carbon electrode in the absence of CO_2 (left) and in the presence of CO_2 and Mg^{2+} (right). Scan rate = 100 mVs^{-1} (Hammouche et al. 1991). Copyright 1996 American Chemical Society.

Table 7.2: List of iron complexes for electrochemical reduction of CO_2.

Entry	**Iron complexes**	**Conditions (solvent, proton source, electrode)**	**Potential vs SCE (V)**	**Products (FE%)**	**Ref.**
1	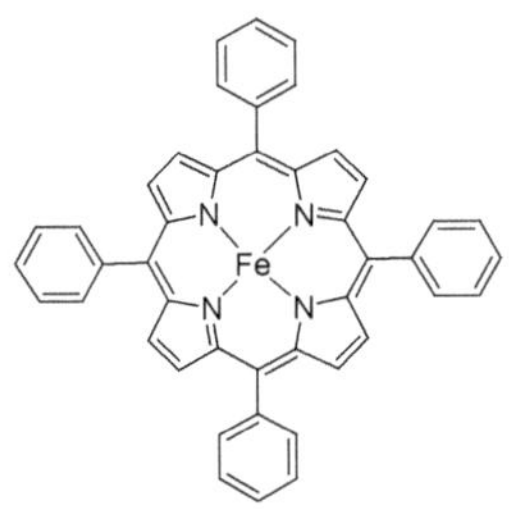	DMF (+ 1 M phenol), Hg pool	−1.70	CO (94%)	(Hammouche et al. 1991)
		DMF (+ 6.7 M 1-propanol), Hg pool	−1.70	CO (60%) $HCOO^-$ (35%)	(Bhugun et al. 1996)
2	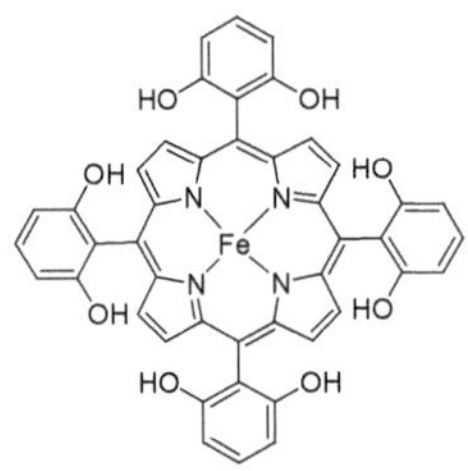	DMF (+ 2M H_2O), glassy carbon	−1.35	CO (94%)	(Costentin et al. 2012)
3	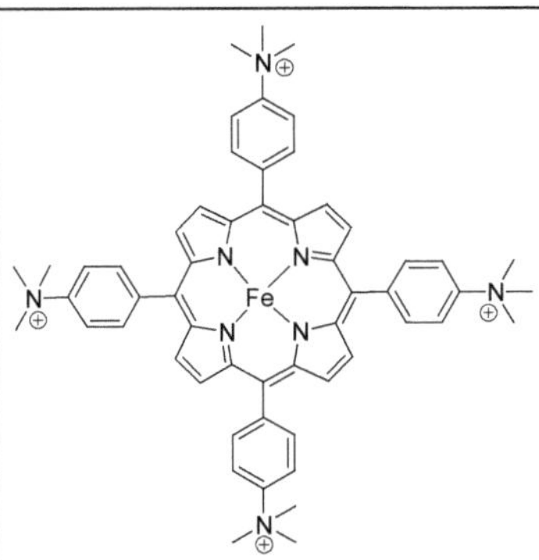	H_2O (pH 7.2), glassy carbon	−1.1	CO (90%)	(Costentin et al. 2015)
4	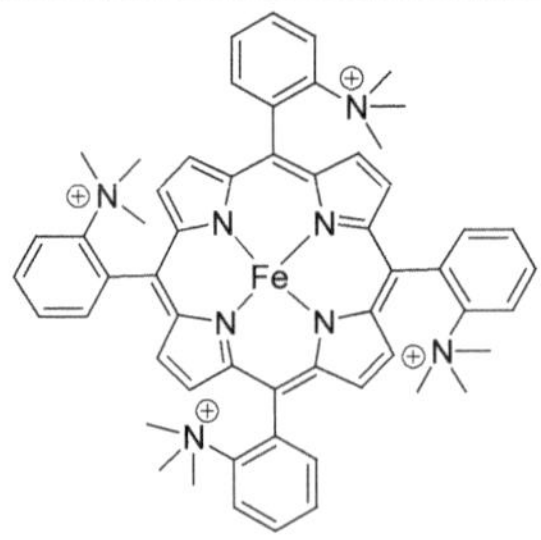	DMF (+ 3 M phenol), glassy carbon	−1.2	CO (100%)	(Azcarate et al. 2016)

Table 7.2 Contd. ...

...Table 7.2 Contd.

Entry	Iron complexes	Conditions (solvent, proton source, electrode)	Potential vs SCE (V)	Products (FE%)	Ref.
5	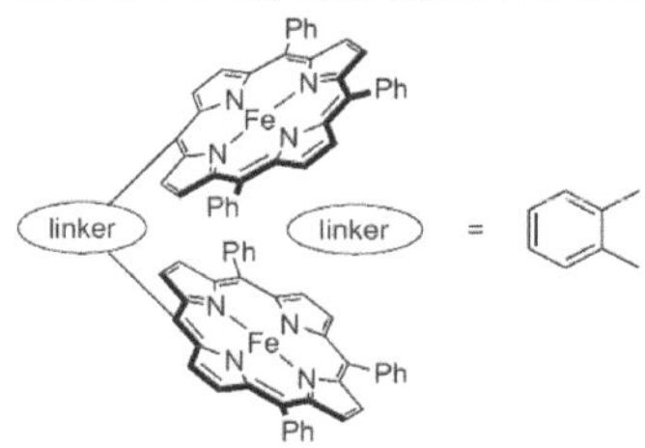	DMF (+10% H_2O)	–1.59	CO (88%) H_2 (12%)	(Mohamed et al. 2015)
6	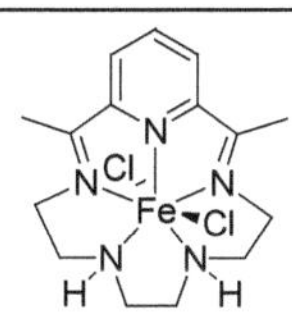	DMF, glassy carbon	–1.25	HCOOH (75%)	(Chen et al. 2015)
7	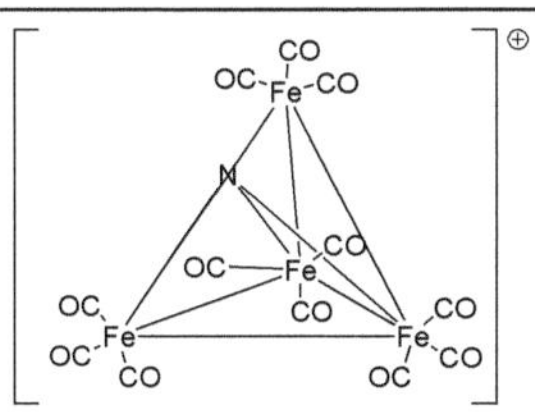	CH_3CN (+ 5% H_2O), glassy carbon	–1.20	$HCOO^-$ (94%)	(Rail and Berben 2011)
		H_2O (pH 7), glassy carbon	–1.20	$HCOO^-$ (96%)	

lifetime of electrocatalysis of CO_2 reduction by FeTPP. The main product was carbon monoxide, while formic acid was formed at a certain limit. The formation of formic acid involved a reaction pathway where the iron–CO_2 interactions are weaker. The effect of the acid synergist is an instance of electrophilic assistance in a two-electron push–pull mechanism where pulling the electron pair out of the substrate by means of the synergist is as important as pushing electrons from the catalyst into the substrate (Bhugun et al. 1996).

In a further development (2012), Saveant's group found that modification of FeTPP through the introduction of phenolic groups in all ortho and ortho′ positions of the phenyl groups (FeTDHPP) (Table 7.2, Entry 2) considerably speeds up catalysis of this reaction by the electrogenerated iron(0) complex. The catalyst demonstrates CO faradaic yield higher than 90% by 50 million turnovers over 4 hours of electrolysis at low overpotential (0.465 V), with no observed degradation. This is due to the high local concentration of protons associated with the phenolic hydroxyl substituents (Costentin et al. 2012).

Iron porphyrin complexes are extensively reported for electrocatalytically reduced CO_2 into CO. However, Berben and the co-workers showed that the iron carbonyl cluster $[Fe_4N(CO)_{12}]^-$ was a selective electrocatalyst for the reduction of CO_2 to formate in aqueous solutions. During controlled potential electrolysis at –1.2 V vs. SCE and pH 7, the electrocatalyst produced a format with a high current density of 4 mA cm^{-2} for over 24 hours, and FE of 96% (Taheri and Berben 2016a, Taheri and Berben 2016b, Taheri et al. 2015). Table 7.2 summarizes the electrochemical reduction of CO_2 using iron complexes.

Cobalt-based electrocatalyst in CO_2 reduction

Several investigations on cobalt complexes as homogeneous catalyst for CO_2 reduction have been reported (Arana et al. 1992, Chen et al. 2015, Tinnemans et al. 1984). The use of tetraazamacrocycle (Fig. 7.9) was first investigated by (Che et al. 1988, Tinnemans et al. 1984).

Figure 7.10 illustrates the CV in MeCN at a carbon electrode produces CO in 20–30% faradaic yield at 940 mV overpotential, with no H_2 as by-product. The active catalytic species is the Co^{I} ligand radical anion species (i.e., the doubly reduced species issued from the Co^{II} neutral) (Takeda et al. 2017).

In recent studies, an efficient Co catalyst for the CO_2-to-CO conversion was discovered with four aminopyridine macrocycle as ligands, linked by pendant amine groups (Fig. 7.11). Highly selective CO formation (98%) could be sustained for 2 hours in a DMF solution saturated with CO_2 and containing 1.2 M of trifluoroethanol

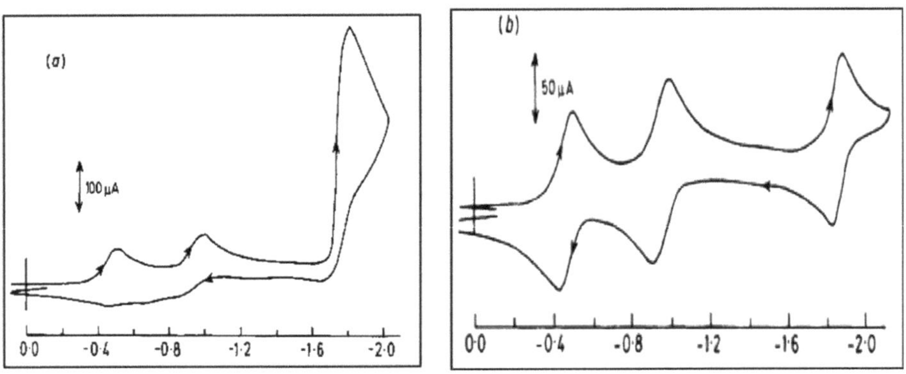

Figure 7.9: Cobalt tetraazamacrocycle (Che et al. 1988, Tinnemans et al. 1984).

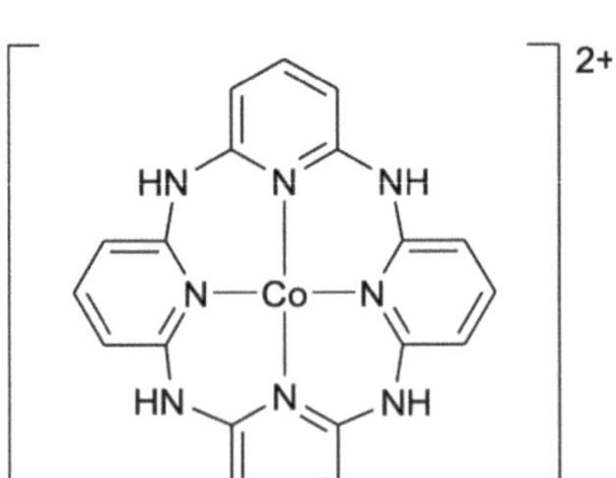

Figure 7.10: Cyclic voltammogram of (ca. 1 mmol) in 0.1 mol dm^{-3} $[NBu_4]BF_4$ in acetonitrile solution in (a) the presence and (b) the absence of CO_2. Working electrode, pyrolytic graphite; scan rate, 100 mVs^{-1}. Reprinted with the permission from (Che et al. 1988). Copyright 1988 Dalton Transaction.

Table 7.3: List of cobalt complexes for electrochemical reduction of CO_2.

Entry	Cobalt complexes	Conditions (solvent, proton source, electrode)	Potential vs. SCE (V)	Products (FE%)	Ref.
1	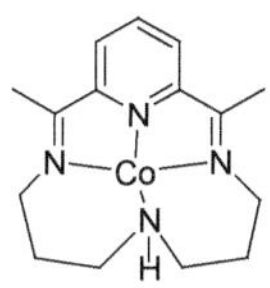	MeCN, pyrolytic graphite	–1.54	CO (20–30%)	(Che et al. 1988)
		MeCN (+ 10 M H_2O), glassy carbon	–1.76	CO (45%) H_2 (30%)	(Zhang et al. 2015)
2	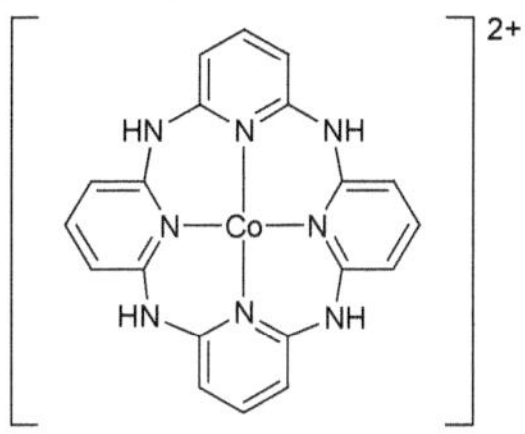	DMF (+ 1.2 M TFE), glassy carbon	–2.35	CO (98%)	(Chapovetsky et al. 2016)
3	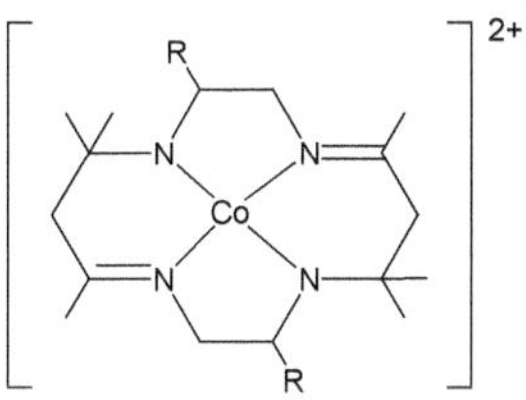 R= H, Me	$MeCN/H_2O$ (2:1), Hg electrode	–1.6 (R=H)	CO (93%)	(Fisher and Eisenberg 1980)
			–1.5 (R=Me)	CO (90%)	
4	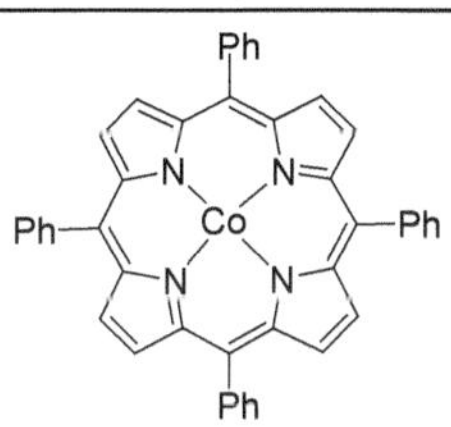	95% MeCN + 5% DMF (+ 0.28 M TFE), GC electrode	–1.89	CO (72%)	(Alenezi 2016)
5	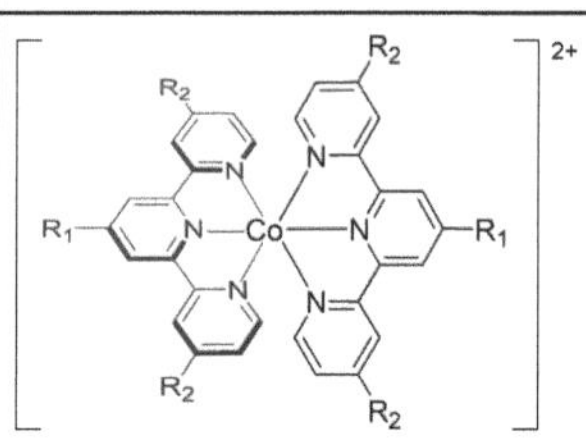 R^1, R^2 = H	DMF (+5% H_2O), Hg electrode	–1.87	CO (12%) H_2 (5%)	(Elgrishi et al. 2014)
6	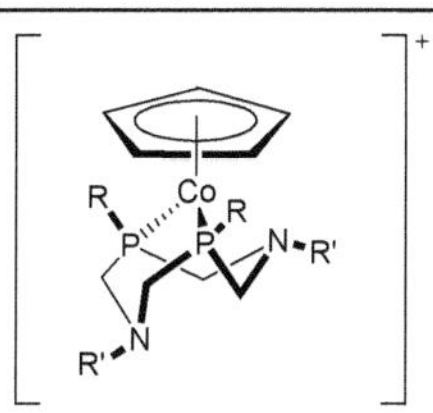 R^1,R^2 = Ph	DMF (+5.6M H_2O), GC electrode	–1.94	HCOOH (38%) CO (1%) H_2 (67%)	(Roy et al. 2017)

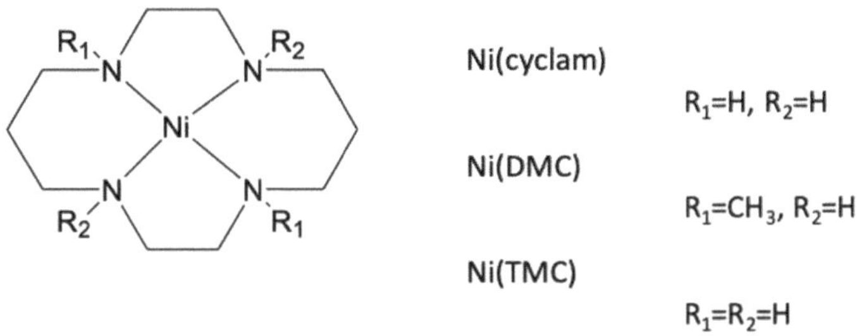

Figure 7.11: Cobalt aminopyridine macrocycle as ligands, linked by pendant amine groups (Chapovetsky et al. 2016).

as a weak acid to help the catalysis with 680 mV overpotential. In brief, 6.2 TON of CO was formed, and it was suggested that the pendant NH groups helps stabilize the Co^{I}-CO_2 adduct (Chapovetsky et al. 2016, Takeda et al. 2017). Table 7.3 summarizes electrochemical reduction of CO_2 using various cobalt complexes.

Nickel-based electrocatalyst in CO_2 reduction

The d^8 electronic state in the valence orbital of nickel(II) favors a planar four coordinate geometry when it coordinates with organic ligands. Reported structures of tetraaza-macrocyclic planar ligands (Table 7.4, Entry 1) possessing tetradentate atoms, results in stable nickel complexes (Beley et al. 1986).

Initially, Sauvage and co-workers discovered that nickel cyclam exhibits high catalytic activity and selectivity towards CO_2-to-CO conversion in aqueous system at a mercury working electrode (Beley et al. 1986). Observations have speculated that a strong chemical absorption exists between the surface of the mercury working electrode and the cyclam complex. Several researches have been conducted to reveal the high catalytic performance and have summarized that this performance may be correlated with the employed working electrodes and the conformations of the cyclam ligand as shown in Fig. 7.12 (Froehlich and Kubiak 2012, Wu et al. 2017). A control experiment using a glassy carbon electrode instead of a mercury working electrode was performed and it found that CO_2 reduction by Ni (cyclam) is a homogeneous reaction. However, a lower efficiency was obtained with the glassy carbon electrode than with the mercury electrode. This may due to the fact that the mercury induces favorable electronic effects to facilitate CO_2 binding to Ni^{I} (Froehlich and Kubiak 2012, Wu et al. 2017).

Research on cyclam complexes has been reported extensively since the 1980s. Ligands such as macrocycle (Meshitsuka et al. 1974), polypyridine (Lam et al. 1995) and N-heterocyclic carbene (Thoi et al. 2013) which are based on nickel as the center metal are emerging in electrocatalytic CO_2 reduction as well (Fig. 7.13). A list of nickel complexes is summarized in the Table 7.4.

Table 7.4: List of nickel complexes for electrochemical reduction of CO_2.

Entry	Nickel complexes	Conditions (solvent, proton source, electrode)	Potential vs. SCE (V)	Products (FE%)	Ref.
1	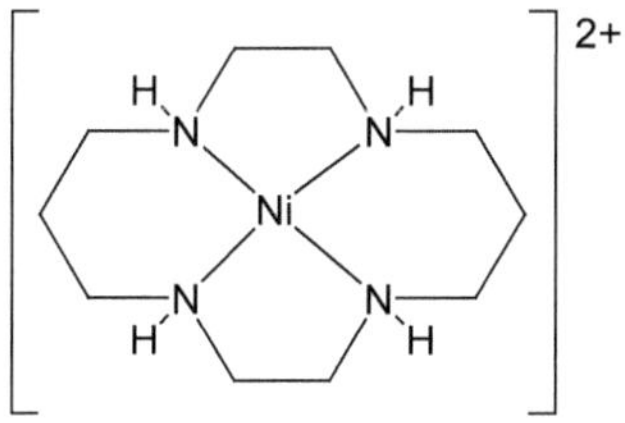	H_2O at pH 4.1, Hg	−1.24	CO (99%)	(Beley et al. 1986)
		DMF, Hg	−1.4	CO (24%) $HCOO^-$ (75%)	(Collin et al. 1988)
		H_2O, GC	−1.54	CO (90%)	(Froehlich and Kubiak 2012)
		H_2O at pH 5, Hg	−1.20	CO (84%)	(Schneider et al. 2012b)
2	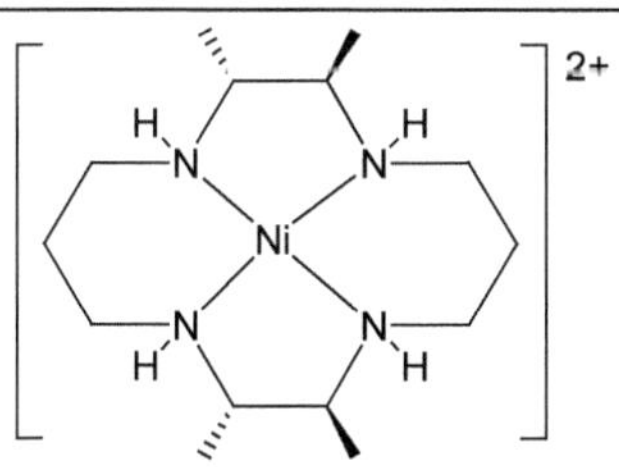	H_2O at pH 5, Hg	−1.20	CO (88%)	(Schneider et al. 2012b)
3	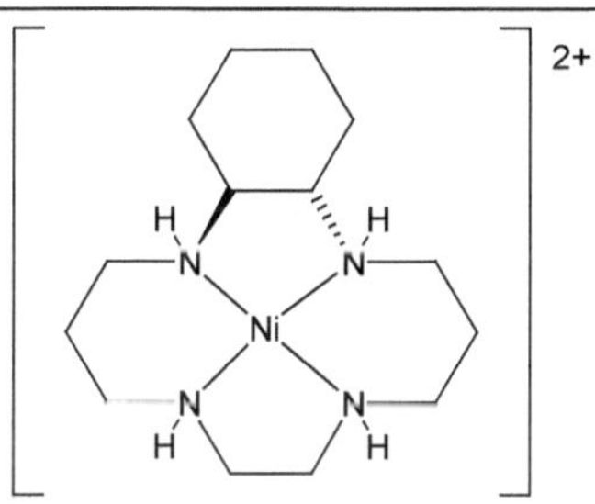	H_2O at pH 5, Hg	−1.20	CO (88%)	(Schneider et al. 2012b)
4	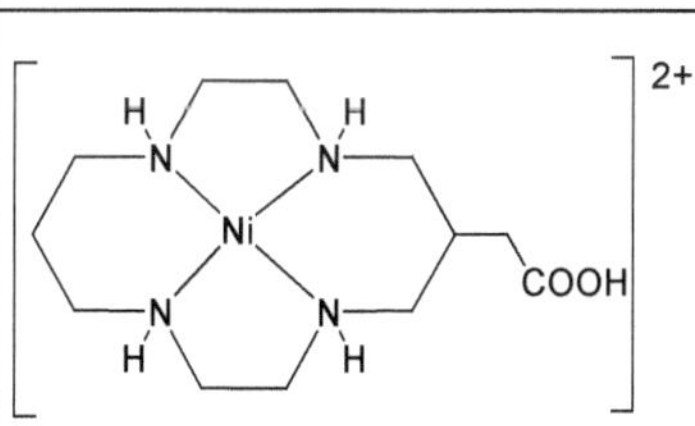	H_2O at pH 2, Hg	−1.23	CO (66%) H_2 (15%)	(Schneider et al. 2012b)
5	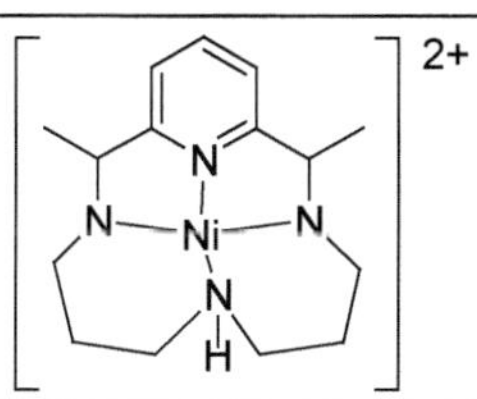	$MeCN/H_2O$ (v/v – 1.2), Hg	−1.30	CO (44%)	(Fisher and Eisenberg 1980)

Table 7.4 Contd. ...

...Table 7.4 Contd.

Entry	Nickel complexes	Conditions (solvent, proton source, electrode)	Potential vs. SCE (V)	Products (FE%)	Ref.
6	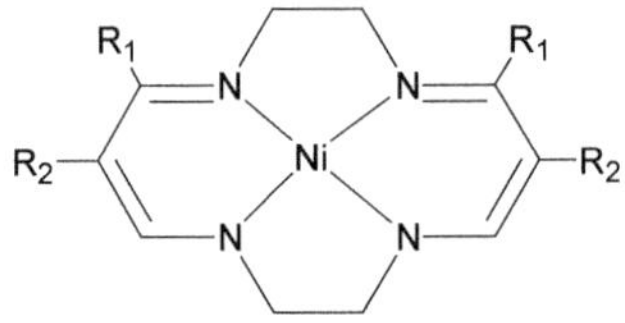 R_1 = Me; R_2 = COOEt	MeCN, Hg	−1.58	$C_2O_4^{2-}$ (90%)	(Rudolph et al. 2000)
7	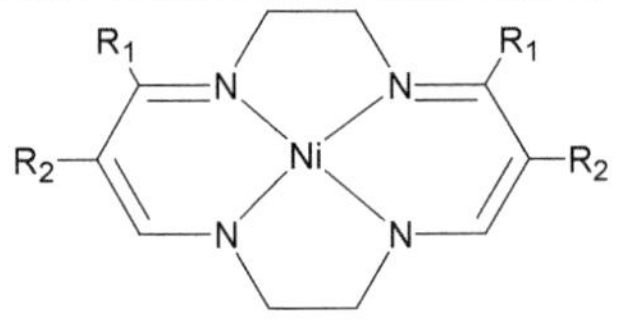 R_1 = Me; R_2 = $COCH_3$	MeCN, Hg	−1.51	$C_2O_4^{2-}$ (78%)	(Rudolph et al. 2000)
8		MeCN, GC	−1.70	CO (trace amount)	(Lam et al. 1995)
9	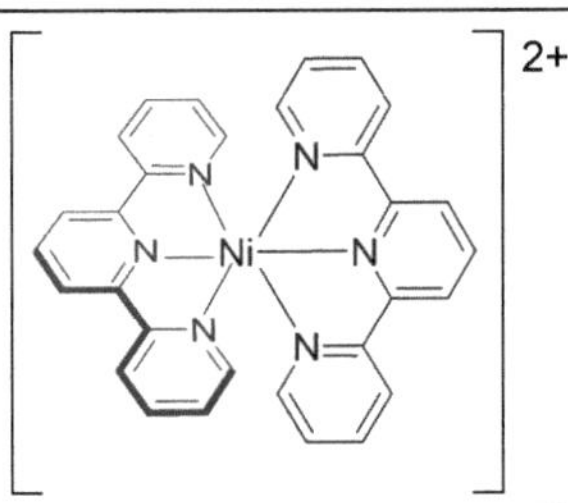	DMF (5% H_2O), Hg	−1.56	CO (20%)	(Elgrishi et al. 2014)
10	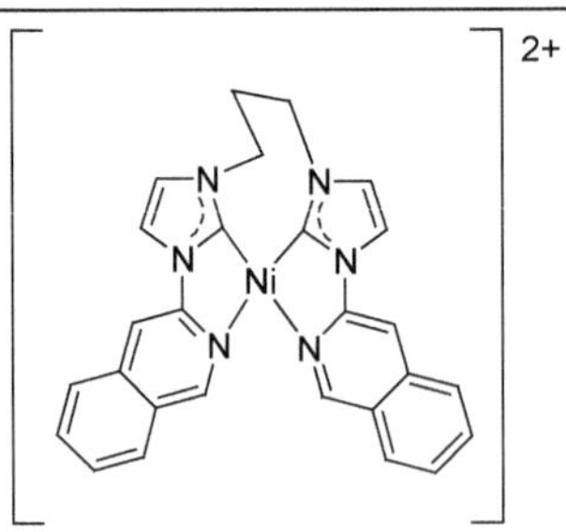	MeCN, GC	−1.80	(22%)	(Thoi et al. 2013)

Table 7.4 Contd. ...

...Table 7.4 Contd.

Entry	Nickel complexes	Conditions (solvent, proton source, electrode)	Potential vs. SCE (V)	Products (FE%)	Ref.
11	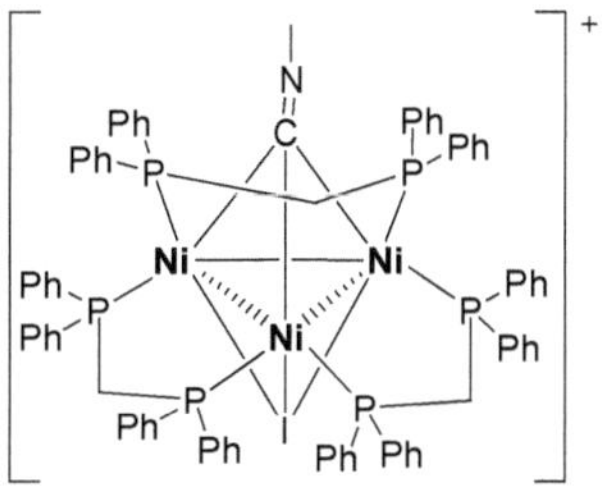	MeCN, Pt	−1.14	CO (unknown) $HCOO^-$ (unknown)	(Ratliff et al. 1992)
12	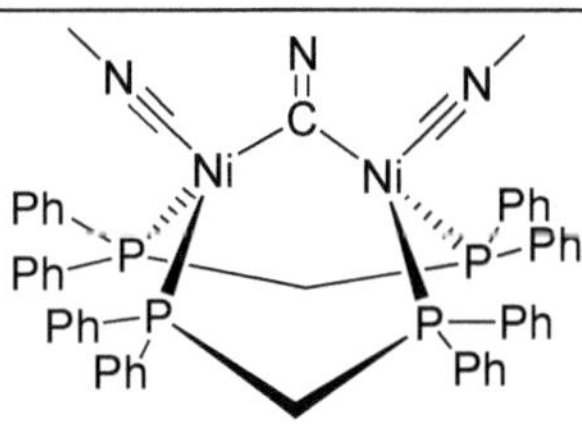	MeCN, GC	−0.87	CO (unknown)	(DeLaet et al. 1987)
13	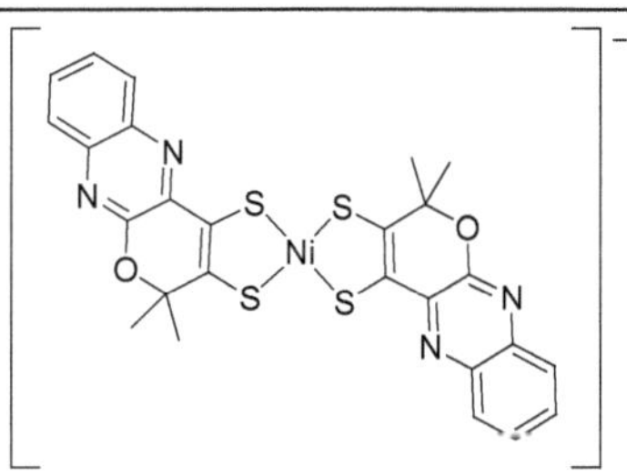	MeCN (2M TFEH), Hg/Au amalgam	−1.95	$HCOO^-$ (60%) CO (15%)	(Fogeron et al. 2018)

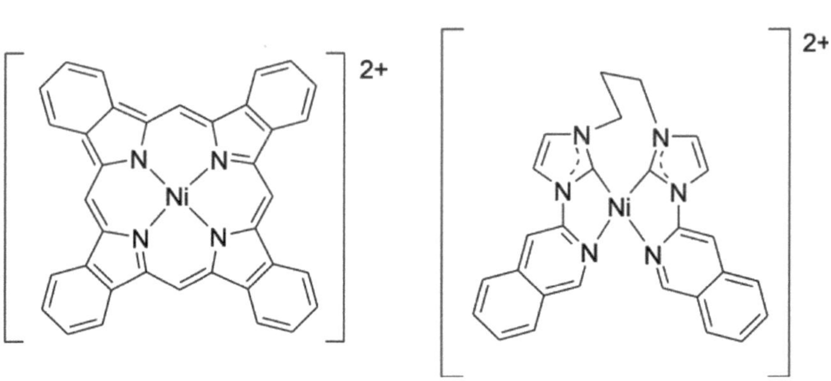

Figure 7.12: Structures of Ni(cyclam), Ni(DMC) (DMC = 1,8 Dimethyl-1,4,8,11-tetraazacyclotetradecane), and Ni(TMC) (TMC = 1,4,8,11-Tetramethyl-1,4,8,11-tetraazacyclotetradecane) (Beley et al. 1986, Froehlich and Kubiak 2012).

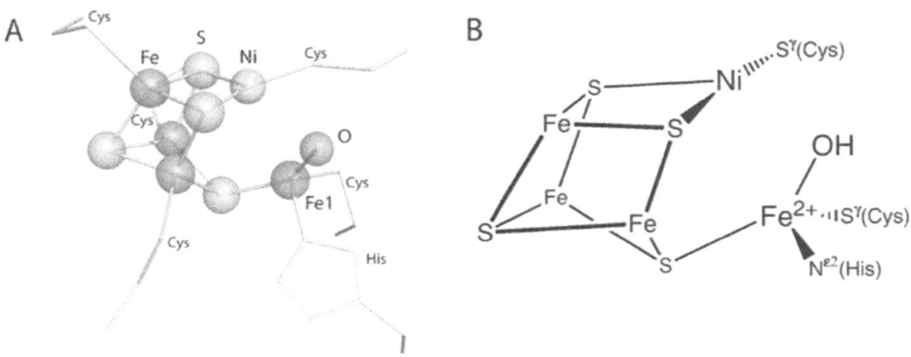

Figure 7.13: (Left) nickel phthalocyanine; (right) nickel N-heterocyclic carbene.

Summary

In this perspective, the Fe, Co and Ni electrocatalysts and homogeneous approaches for CO_2 reduction were summarized. The conversion of molecules that are precursors or directly utilized as fuels are mentioned here. At this stage, it is useful to draw conclusions from work in the past, one that the fundamental reorganization energies of both the CO_2 molecule and the catalysts that reduce CO_2 are extremely important considerations. The second is the delibration on the multi electron transfer and Proton-Coupled Electro Transfer (PCET) processes. Transition metal complexes hold accessible multiple redox states that can facilitate the electron or/and proton transfer. Electron shuttling by the electron reservoir behavior of redox-active ligands enhance the selectivity and efficiency, besides reducing the overpotential. Thus, the cooperation of redox-active ligand with Fe, Co and Ni would be an effective metal complex system for CO_2 reduction.

References

Alenezi, K. 2016. Electrocatalytic study of carbon dioxide reduction by Co(TPP)Cl complex. Journal of Chemistry, 7.

Appel, A. M., J. E. Bercaw, A. B. Bocarsly, H. Dobbek, D. L. DuBois, M. Dupuis et al. 2013. Frontiers, opportunities, and challenges in biochemical and chemical catalysis of CO_2 fixation. Chemical Reviews 113: 6621–58.

Arana, C., S. Yan, K. M. Keshavarz, K. T. Potts and H. D. Abruna. 1992. Electrocatalytic reduction of carbon dioxide with iron, cobalt, and nickel complexes of terdentate ligands. Inorganic Chemistry 31: 3680–3682.

Aresta, M., C. F. Nobile, V. G. Albano, E. Forni and M. Manassero. 1975. New nickel–carbon dioxide complex: synthesis, properties, and crystallographic characterization of (carbon dioxide)-bis(tricyclohexylphosphine)nickel. Journal of the Chemical Society, Chemical Communications 636–637.

Aresta, M. and A. Dibenedetto. 2007. Utilisation of CO_2 as a chemical feedstock: opportunities and challenges. Dalton Transactions 2975–2992.

Azcarate, I., C. Costentin, M. Robert and J. -M. Savéant. 2016. Through-space charge interaction substituent effects in molecular catalysis leading to the design of the most efficient catalyst of CO_2-to-CO electrochemical conversion. Journal of the American Chemical Society 138: 16639–16644.

Beley, M., J. P. Collin, R. Ruppert and J. P. Sauvage. 1986. Electrocatalytic reduction of carbon dioxide by nickel cyclam^{2+} in water: study of the factors affecting the efficiency and the selectivity of the process. Journal of the American Chemical Society 108: 7461–7467.

Benson, E. E., C. P. Kubiak, A. J. Sathrum and J. M. Smieja. 2009. Electrocatalytic and homogeneous approaches to conversion of CO_2 to liquid fuels. Chemical Society Reviews 38: 89–99.

Bhugun, I., D. Lexa and J. -M. Savéant. 1996. Catalysis of the electrochemical reduction of carbon dioxide by Iron(0) porphyrins: Synergystic effect of weak Brönsted acids. Journal of the American Chemical Society 118.

Boot-Handford, M. E., J. C. Abanades, E. J. Anthony, M. J. Blunt, S. Brandani, N. Mac Dowell et al. 2014. Carbon capture and storage update. Energy & Environmental Science 7: 130–189.

Chapovetsky, A., T. H. Do, R. Haiges, M. K. Takase and S. C. Marinescu. 2016. Proton-assisted reduction of CO_2 by cobalt aminopyridine macrocycles. Journal of the American Chemical Society 138: 5765–5768.

Che, C. M., S. T. Mak, W. O. Lee, K. W. Fung and T. C. W. Mak. 1988. Electrochemical studies of nickel(II) and cobalt(II) complexes of tetra-azamacrocycles bearing a pyridine functional group and X-ray structures of $[Ni(L_3)Cl]ClO_4$ and $[Ni(L_3)][ClO_4]_2 \cdot H_2O$ {L_3 = meso-2,3,7,11,12-pentamethyl-3,7,11,17-tetra-azabicyclo[11.3.1]heptadeca-1(17),13,15-triene}. Journal of the Chemical Society, Dalton Transactions 2153–2159.

Chen, L., Z. Guo, X. G. Wei, C. Gallenkamp, J. Bonin, E. Anxolabéhère-Mallart et al. 2015. Molecular catalysis of the electrochemical and photochemical reduction of CO_2 with earth-abundant metal complexes. Selective production of CO vs HCOOH by switching of the metal center. Journal of the American Chemical Society 137: 10918–10921.

Collin, J. P., A. Jouaiti and J. P. Sauvage. 1988. Electrocatalytic properties of (tetraazacyclotetradecane) nickel(2+) and $Ni_2(biscyclam)^{4+}$ with respect to carbon dioxide and water reduction. Inorganic Chemistry 27: 1986–1990.

Costentin, C., S. Drouet, M. Robert and J. -M. Savéant. 2012. A local proton source enhances CO_2 electroreduction to CO by a molecular Fe catalyst. Science 338: 90.

Costentin, C., M. Robert, J. M. Savéant and A. Tatin. 2015. Efficient and selective molecular catalyst for the CO_2 to CO electrochemical conversion in water. Proceedings of the National Academy of Sciences 112: 6882.

DeLaet, D. L., R. Del Rosario, P. E. Fanwick and C. P. Kubiak. 1987. Carbon dioxide chemistry and electrochemistry of a binuclear cradle complex of nickel(0), $Ni_2(\mu\text{-CNMe})(CNMe)_2(PPh_2CH_2PPh_2)_2$. Journal of the American Chemical Society 109: 754–758.

Do, J. Y., R. K. Chava, K. K. Mandari, N. K. Park, H. J. Ryu, M. W. Seo et al. 2018. Selective methane production from visible-light-driven photocatalytic carbon dioxide reduction using the surface plasmon resonance effect of superfine silver nanoparticles anchored on lithium titanium dioxide nanocubes (Ag@LixTiO_2). Applied Catalysis B: Environmental 237: 895–910.

Dobbek, H. Svetlitchnyi, V. Gremer, L. Huber, R. Meyer and Ortwin. 2001. Crystal structure of a carbon monoxide dehydrogenase reveals a [Ni-4Fe-5S] cluster. Science 293: 1281.

Dobbek, H. 2019. Mechanism of Ni,Fe-containing carbon monoxide dehydrogenases. pp. 153–166. *In*: Ribbe, M. W., (ed.). Metallocofactors that Activate Small Molecules: With Focus on Bioinorganic Chemistry. Springer International Publishing, Cham.

Dresselhaus, M. S. and I. L. Thomas. 2001. Alternative energy technologies. Nature 414: 332.

Elgrishi, N., M. B. Chambers, V. Artero and M. Fontecave. 2014. Terpyridine complexes of first row transition metals and electrochemical reduction of CO_2 to CO. Physical Chemistry Chemical Physics 16: 13635–13644.

Fisher, B. J. and R. Eisenberg. 1980. Electrocatalytic reduction of carbon dioxide by using macrocycles of nickel and cobalt. Journal of the American Chemical Society 102: 7361–7363.

Fogeron, T., T. K. Todorova, J. -P. Porcher, M. Gomez-Mingot, L. -M. Chamoreau, C. Mellot-Draznieks et al. 2018. A bioinspired Nickel(bis-dithiolene) complex as a homogeneous catalyst for carbon dioxide electroreduction. ACS Catalysis 8: 2030–2038.

Francke, R. and R. D. Little. 2014. Redox catalysis in organic electrosynthesis: basic principles and recent developments. Chemical Society Reviews 43: 2492–2521.

Francke, R., B. Schille and M. Roemelt. 2018. Homogeneously catalyzed electroreduction of carbon dioxide—methods, mechanisms, and catalysts. Chemical Reviews 118: 4631–4701.

Froehlich, J. D. and C. P. Kubiak. 2012. Homogeneous CO_2 reduction by Ni(cyclam) at a glassy carbon electrode. Inorganic Chemistry 51: 3932–3934.

Gan, L., D. Jennings, J. Laureanti and A. K. Jones. 2016. Biomimetic complexes for production of dihydrogen and reduction of CO_2. pp. 233–272. *In*: Kalck, P. (ed.). Homo- and Heterobimetallic Complexes in Catalysis: Cooperative Catalysis. Springer International Publishing, Cham.

García Antón, J., R. Bofill, L. Escriche, A. Llobet and X. Sala. 2012. Transition-metal complexes containing the dinucleating tetra-N-dentate 3,5-bis(2-pyridyl)pyrazole (Hbpp) ligand—a robust scaffold for multiple applications including the catalytic oxidation of water to molecular oxygen. European Journal of Inorganic Chemistry 2012: 4775–4789.

Ghosh, S. and M. -H. Baik. 2011. The mechanism of O-O bond formation in Tanaka's water oxidation catalyst. Angewandte Chemie International Edition 51: 1221–1224.

Hammouche, M., D. Lexa, M. Momenteau and J. M. Saveant. 1991. Chemical catalysis of electrochemical reactions. Homogeneous catalysis of the electrochemical reduction of carbon dioxide by iron("0") porphyrins. Role of the addition of magnesium cations. Journal of the American Chemical Society 113: 8455–8466.

Jeoung, J. H. and H. Dobbek. 2007. Carbon dioxide activation at the Ni,Fe-cluster of anaerobic carbon monoxide dehydrogenase. Science 318: 1461.

Kleij, A. W. 2013. Carbon dioxide activation and conversion. pp. 559–587. *In*: Suib, S. L. (ed.). New and Future Developments in Catalysis. Elsevier, Amsterdam.

Lam, K. M., K. Y. Wong, S. M. Yang and C. M. Che. 1995. Cobalt and nickel complexes of 2,2′ : 6′,2″: 6″,2‴-quaterpyridine as catalysts for electrochemical reduction of carbon dioxide. Journal of the Chemical Society, Dalton Transactions 1103–1107.

Li, Y., Q. Liu, W. Huang and J. Yang. 2018. Solubilities of CO_2 capture absorbents methyl benzoate, ethyl hexanoate and methyl heptanoate. The Journal of Chemical Thermodynamics 127: 25–32.

Liu, F. -W., J. Bi, Y. Sun, S. Luo and P. Kang. 2018. Cobalt complex with redox-active imino bipyridyl ligand for electrocatalytic reduction of carbon Dioxide to formate. ChemSusChem. 11: 1656–1663.

Meshitsuka, S., M. Ichikawa and K. Tamaru. 1974. Electrocatalysis by metal phthalocyanines in the reduction of carbon dioxide. Journal of the Chemical Society, Chemical Communications 158–159.

Mikkelsen, M., M. Jørgensen and F. C. Krebs. 2010. The teraton challenge. A review of fixation and transformation of carbon dioxide. Energy & Environmental Science 3: 43–81.

Mohamed, E. A., Z. N. Zahran and Y. Naruta. 2015. Efficient electrocatalytic CO_2 reduction with a molecular cofacial iron porphyrin dimer. Chemical Communications 51: 16900–16903.

Morris, A. J., G. J. Meyer and E. Fujita. 2009. Molecular approaches to the photocatalytic reduction of carbon dioxide for solar fuels. Accounts of Chemical Research 42: 1983–1994.

North, M. and R. Pasquale. 2009. Mechanism of cyclic carbonate synthesis from epoxides and CO_2. Angewandte Chemie International Edition 48: 2946–2948.

Nov-Institute. 2015. Sustainable fuels, chemicals and polymers from sun and CO_2 big visions—but also big potential. *In*: nova-institute (ed.). Bio-based News.

Okamura, M., M. Kondo, R. Kuga, Y. Kurashige, T. Yanai, S. Hayami et al. 2016. A pentanuclear iron catalyst designed for water oxidation. Nature 530: 465.

Paparo, A. and J. Okuda. 2017. Carbon dioxide complexes: Bonding modes and synthetic methods. Coordination Chemistry Reviews 334: 136–149.

Praneeth, V. K. K., M. R. Ringenberg and T. R. Ward. 2012. Redox-active ligands in catalysis. Angewandte Chemie International Edition 51: 10228–10234.

Rail, M. D. and L. A. Berben. 2011. Directing the reactivity of $[HFe_4N(CO)_{12}]^-$ toward H^+ or CO_2 reduction by understanding the electrocatalytic mechanism. Journal of the American Chemical Society 133: 18577–18579.

Rakowski Dubois, M. and D. L. Dubois. 2009. Development of molecular electrocatalysts for CO_2 reduction and H_2 production/oxidation. Accounts of Chemical Research 42: 1974–1982.

Ratliff, K. S., R. E. Lentz and C. P. Kubiak. 1992. Carbon dioxide chemistry of the trinuclear complex $[Ni_3(\mu_3\text{-CNMe})(\mu_3\text{-I})(dppm)_3][PF_6]$. Electrocatalytic reduction of carbon dioxide. Organometallics 11: 1986–1988.

Rosas-Hernandez, A., C. Steinlechner, H. Junge and M. Beller. 2017. Photo- and electrochemical valorization of carbon dioxide using earth-abundant molecular catalysts. Topics in Current Chemistry 376: 1.

Rosas-Hernández, A., H. Junge, M. Beller, M. Roemelt and R. Francke. 2017. Cyclopentadienone iron complexes as efficient and selective catalysts for the electroreduction of CO_2 to CO. Catalysis Science & Technology 7: 459–465.

Rountree, E. S., B. D. McCarthy, T. T. Eisenhart and J. L. Dempsey. 2014. Evaluation of homogeneous electrocatalysts by cyclic voltammetry. Inorganic Chemistry 53: 9983–10002.

Roy, S., B. Sharma, J. Pécaut, P. Simon, M. Fontecave, P. D. Tran et al. 2017. Molecular cobalt complexes with pendant amines for selective electrocatalytic reduction of carbon dioxide to formic acid. Journal of the American Chemical Society 139: 3685–3696.

Rudolph, M., S. Dautz and E. -G. Jäger. 2000. Macrocyclic [N_4^{2-}] coordinated nickel complexes as catalysts for the formation of oxalate by electrochemical reduction of carbon dioxide. Journal of the American Chemical Society 122: 10821–10830.

Sakakura, T., J. C. Choi and H. Yasuda. 2007. Transformation of carbon dioxide. Chemical Reviews 107: 2365–2387.

Sakakura, T. and K. Kohno. 2009. The synthesis of organic carbonates from carbon dioxide. Chemical Communications 1312–1330.

Sampson, M. D., A. D. Nguyen, K. A. Grice, C. E. Moore, A. L. Rheingold and C. P. Kubiak 2014. Manganese catalysts with bulky bipyridine ligands for the electrocatalytic reduction of carbon dioxide: eliminating dimerization and altering catalysis. Journal of the American Chemical Society 136: 5460–71.

Savéant, J. M. 2008. Molecular catalysis of electrochemical reactions: mechanistic aspects. Chemical Reviews 108: 2348–2378.

Schneider, J., H. Jia, J. T. Muckerman and E. Fujita. 2012a. Thermodynamics and kinetics of CO_2, CO, and H^+ binding to the metal centre of CO_2 reduction catalysts. Chemical Society Reviews 41: 2036–2051.

Schneider, J., H. Jia, K. Kobiro, D. E. Cabelli, J. T. Muckerman and E. Fujita. 2012b. Nickel(ii) macrocycles: highly efficient electrocatalysts for the selective reduction of CO_2 to CO. Energy & Environmental Science 5: 9502–9510.

Schwarz, H. A. and R. W. Dodson. 1989. Reduction potentials of CO_2^- and the alcohol radicals. The Journal of Physical Chemistry 93: 409–414.

Seefeldt, L. C., B. M. Hoffman and D. R. Dean. 2009. Mechanism of Mo-dependent nitrogenase. Annual Review of Biochemistry 78: 701.

Simmons, J. M., H. Wu, W. Zhou and T. Yildirim. 2011. Carbon capture in metal–organic frameworks—a comparative study. Energy & Environmental Science 4: 2177–2185.

Taheri, A., E. J. Thompson, J. C. Fettinger and L. A. Berben. 2015. An iron electrocatalyst for selective reduction of CO_2 to formate in water: Including thermochemical insights. ACS Catalysis 5: 7140–7151.

Taheri, A. and L. A. Berben. 2016a. Tailoring electrocatalysts for selective CO_2 or H^+ reduction: Iron carbonyl clusters as a case study. Inorganic Chemistry 55: 378–385.

Taheri, A. and L. A. Berben. 2016b. Making C–H bonds with CO_2: production of formate by molecular electrocatalysts. Chemical Communications 52: 1768–1777.

Takeda, H., C. Cometto, O. Ishitani and M. Robert. 2017. Electrons, photons, protons and earth-abundant metal complexes for molecular catalysis of CO_2 reduction. ACS Catalysis 7: 70–88.

Thoi, V. S., N. Kornienko, C. G. Margarit, P. Yang and C. J. Chang. 2013. Visible-light photoredox catalysis: selective reduction of carbon dioxide to carbon monoxide by a nickel n-heterocyclic carbene–isoquinoline complex. Journal of the American Chemical Society 135: 14413–14424.

Tinnemans, A. H. A., T. P. M. Koster, D. H. M. W. Thewissen and A. Mackor. 1984. Tetraaza-macrocyclic cobalt(II) and nickel(II) complexes as electron-transfer agents in the photo(electro)chemical and electrochemical reduction of carbon dioxide. Recueil des Travaux Chimiques des Pays-Bas 103: 288–295.

Vidal Vázquez, F., J. Kihlman, A. Mylvaganam, P. Simell, M. L. Koskinen Soivi and V. Alopaeus. 2018. Modeling of nickel-based hydrotalcite catalyst coated on heat exchanger reactors for CO_2 methanation. Chemical Engineering Journal 349: 694–707.

Volbeda, A. and J. C. Fontecilla Camps. 2005. Structural bases for the catalytic mechanism of Ni-containing carbon monoxide dehydrogenases. Dalton Transactions 3443–3450.

Wada, T., K. Tsuge and K. Tanaka. 2000. Electrochemical oxidation of water to dioxygen catalyzed by the oxidized form of the bis(ruthenium–hydroxo) complex in H_2O. Angewandte Chemie International Edition 39: 1479–1482.

Wada, T., K. Tsuge and K. Tanaka. 2001. Syntheses and redox properties of bis(hydroxoruthenium) complexes with quinone and bipyridine ligands. Water-oxidation catalysis. Inorganic Chemistry 40: 329–337.

Wang, W. H., Y. Himeda, J. T. Muckerman, G. F. Manbeck and E. Fujita. 2015. CO_2 hydrogenation to formate and methanol as an alternative to photo- and electrochemical CO_2 reduction. Chemical Reviews 115: 12936–12973.

Wu, Y., B. Rudshteyn, A. Zhanaidarova, J. D. Froehlich, W. Ding, C. P. Kubiak et al. 2017. Electrode-ligand interactions dramatically enhance CO_2 conversion to CO by the $[Ni(cyclam)](PF_6)_2$ catalyst. ACS Catalysis 7: 5282–5288.

Yano, J. and V. Yachandra. 2014. Mn_4Ca cluster in photosynthesis: Where and how water is oxidized to dioxygen. Chemical Reviews 114: 4175–4205.

Zhang, M., M. El Roz, H. Frei, J. L. Mendoza Cortes, M. Head-Gordon, D. C. Lacy et al. 2015. Visible light sensitized CO_2 activation by the tetraaza $[Co^{II}N_4H(MeCN)]^{2+}$ complex investigated by FT-IR spectroscopy and DFT calculations. The Journal of Physical Chemistry C 119: 4645–4654.

CHAPTER 8

Design of Alloy Electrocatalysts for Oxygen Reduction Reaction from First-Principles Viewpoint

Do Ngoc Son[1,2,*] and *Rohit Srivastava*[3]

Introduction

The development of human civilization relates closely to the understanding of the universe and applies these insights into practice, and at the same time, has to minimize negative impacts on the earth and environment. Theoretical and experimental methods have been often used to uncover insights. The experiment is indispensable to prove practically that a certain system or property can be realized. However, due to the trial and error nature, the experiment could cause undesirable impacts on the environment and are also costly. At the beginning of the history of scientific advancement, the experiment contributes to the discovery and adjustment of theory. However, with the increasing support of information and computer technology, theoretical methods have accurately described and predicted many properties of systems and new materials that were later realized and proved by experiments. Theoretical methods (Ramasubramaniam and Carter 2007) including modeling, computation, simulation and machine learning (Takahashi and Tanaka 2016) will gradually replace the experimental method in the early and important, but costly stage of material development for applications that is the design, property discovery

[1] Computational Physics Laboratory Ho Chi Minh City University of Technology, 268 Ly Thuong Kiet Street, Ward 14, District 10, Ho Chi Minh City, Vietnam.

[2] Vietnam National University Ho Chi Minh City, Quarter 6, Linh Trung Ward, Thu Duc District, Ho Chi Minh City, Vietnam.

[3] Catalysis Research Lab, School of Petroleum Technology, Pandit Deendayal Petroleum University Gandhinagar-382007, Gujarat, India.

* Corresponding author: dnson@hcmut.edu.vn

and optimization of materials, before applying the experimental method in testing and fabrication. Thus making the world's development more sustainable and economical. The application of computer and information technology has become the top priority to advance the fourth technological revolution, and the field of materials science and technology is not out of this scope (Schleder et al. 2019).

Materials design plays a key role in a wide range of industrial applications. However, the development is still slow because it relies heavily on previously known materials. Material design and development need to be accelerated, which includes (1) improving known properties, (2) finding new materials, and (3) exploring new properties. Recently, computational material design has been applied in many different fields and become an important contributor to the development of materials science and technology. It is a combination of many different theoretical methods from first-principles calculations to continuum mechanics (Saito 1999, Schleder et al. 2019). Recently, NASA established the NASA 2040 vision, which aims at making computational material design become a consistently applied approach in their technological development (Liu et al. 2018). In this chapter, the design methodology of alloy materials for chemical reactions based on the density functional theory calculations combined with the thermodynamics model and then apply to PdCo alloy for the oxygen reduction reaction are presented. The results contribute to the field of proton exchange membrane fuel cells. The design process is rather complicated and difficult to fully explore the relationship of material structure versus properties for the matter of interest. Therefore, this chapter may not be a complete presentation. However, it has been attempted to systematically present the design steps in a comprehensive manner to progress the realization of computational materials design.

Proton exchange membrane fuel cells

The alternative fuel industry is gaining great importance in today's world of depleting oil, coal and natural gas reserves. There are many new technologies emerging in this area and various types of fuel cells have been developed. A fuel cell is an electrochemical device that directly converts the chemical energy of supplied fuel and oxidant to the electrical energy. Each fuel cell consists of three main components, which are two electrodes (anode, cathode) separated by the electrolyte. Fuel cells are commonly classified according to the electrolyte used (Mench 2008), in which, the Proton Exchange Membrane (PEM) fuel cell has attracted much attention among the others because of its high efficiency and advancement. This fuel cell has been considered as a key factor for the medium and long term future diversification of clean energy supply with applications in the portable devices, stationary power generation and automotive sectors. A schematic description of a PEM fuel cell is shown in Fig. 8.1. A membrane electrode assembly, which is the key part of a PEM fuel cell, consists of a polymer electrolyte in contact with an anode and a cathode on each side. The membrane plays a dual role of protons conduction and gas separation to the other side of the cell.

As described in Fig. 8.1, fuel such as hydrogen gas or alcohol is supplied to the anode, while oxygen gas is delivered to the cathode. Hydrogen gas and alcohols are oxidized on the anode to generate protons and electrons. The protons transfer

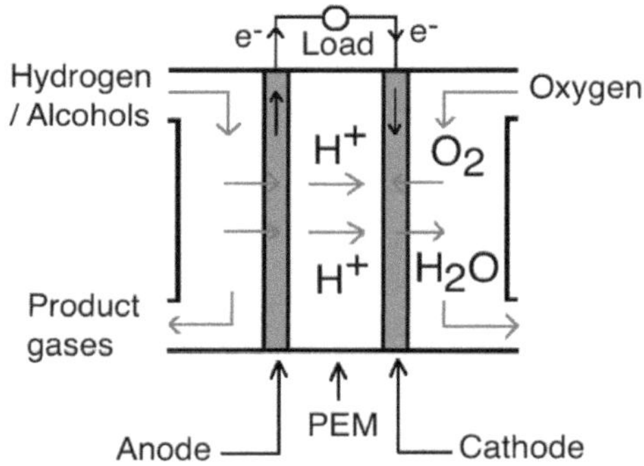

Figure 8.1: Diagram of PEM fuel cells.

to the cathode through the humidified polymer electrolyte membrane, whereas the electrons travel through an external circuit to the cathode to create an electrical current. At the cathode, the protons recombine with the electrons and then react with the oxygen gas to form water that is then depleted out of the cell. The oxidation of the hydrogen and alcohol molecules is relatively easy using a catalyst. However, the reduction of the oxygen molecules on the cathode is more difficult because of their double bond nature, which causes significant activation loss. Therefore, the main challenge of fuel cell development is the design of effective catalysts for the oxygen reduction reaction.

Oxygen reduction reaction

Oxygen Reduction Reaction (ORR) is presented by the equation:

$$O_2 + 4\,(H^+ + e^-) \rightleftarrows 2\,H_2O. \qquad (1)$$

This reaction has slow kinetics and therefore affects the performance of the PEM fuel cell. Therefore, catalysts have been used to boost the reaction. So far platinum is still the best catalyst option for the oxygen reduction reaction because of its excellent activity. However, it is expensive, rare and unstable in the acidic medium of the ORR. Therefore, many different materials for instance non-precious metal catalysts, metal-free catalysts and single-atom catalysts have been developed to replace and reduce the dosage of the platinum (Wang et al. 2019, Li et al. 2019). At present, the most successfully commercialized catalyst is the carbon-supported Pt, often described as Pt/C. It has also been proved that alloying of Pt with less expensive metals and constructing the core-shell nanoparticles could improve the stability of the catalyst, while the ORR activity has a comparable magnitude to that of pure Pt. Simultaneously, many Pt-based alloys such as Pt_3Ni were found to exhibit the surpassing performance to that of commercial Pt/C catalyst. The Pt-skin structures are the best ones among Pt-based alloys. Although the Pt-based catalysts have high ORR activity, a further reduction of Pt loading is required. Complete elimination of the Pt usage is possible by considering the alloys of other transition metals. The Pd-based binary alloys with inexpensive metals such as Co, Fe, Ni, Cu have appeared as good replacements for Pt-based catalysts. Among them, PdCo was found to be the most promising. Experiments showed that acid treatment and annealing resulted in the Pd-rich skin that can significantly improve the ORR activity and the catalyst stability, which were found to be better than those of pure Pt. Besides the transition metal oxides including spinel and perovskite-type oxides, nitrides, phosphides, which contain earth-abundant and inexpensive metal elements such as Mn, Co, Fe, Ni, Al, La, Cr, Ti, have been investigated. Metal-free catalysts such as macrocycle complexes and heteroatom-doped carbon materials are also considered (Li et al. 2019). The most popular metal-free catalysts are porphyrin and phthalocyanine (macrocycle molecules) and the nitrogen, phosphorous, boron and sulfur-doped graphene (heteroatom-doped carbon materials). The macrocycle complexes and the graphene have low intrinsic ORR activity; however, many researchers have shown that doping can significantly enhance the ORR activity (Wang et al. 2019). Doping with nitrogen is especially a great strategy to modify the metal-free catalysts. Furthermore, doping the metal-free catalysts with transition metals can further improve the ORR activity. This scheme has advantages compared to metal catalysts as it has low metal loading and maximizes the utilization of the doped metals. Recently, transition metal dichalcogenides have also been proposed for the ORR catalyst (Na et al. 2020). From the review of literature, one finds that the PdCo alloy is an excellent candidate that compromises the ORR activity and their stability with their cost compared to pure Pt. However, the details of the microscopic structure and composition–activity and stability relationships are still unclear. Therefore, we worked out this topic in our earlier articles (Son and Takahashi 2012, Son et al. 2015). Next the design process of the alloys for chemical reactions based on the density functional theory calculations and the thermodynamics model have been systematically generalized.

First-principles design methodology of alloy materials for chemical reactions

There are many criteria that are needed to verify catalysts. However, the most important ones should be reaction activity and catalyst stability. Generally, the theoretical material-design study for reaction on the surface is performed in the following steps:

1) Review the literature to find a good candidate for electro-catalysts: this step is the same as the standard procedure for any research. However, this step can be reinforced by materials screening, data mining and machine learning to gain more insightful predictions (Schleder et al. 2019). One can get some information on the cost, abundance, environmentally benign, reactivity, stability and maybe the structure and composition—reactivity and stability relationship of candidates.
2) Model the cathode, usually by a supercell for thin films, with an appropriate unit cell size. The dimensions of unit cell proportionally increase with the size of the adsorbate. The adsorbate has to be well-installed inside the unit cell boundary. This step requires information such as the phase, composition, surface plane and the symmetry, etc., of the cathode materials.

3) Design substrates with different arrangements of doping atoms for the specific content of doping, for instance, Co atoms arranged in different sets of positions in the unit cell of Pd. This step requires simplifying the symmetry-equivalent structures to obtain the symmetry-independent ones. Solid-state physics expresses that the symmetry-independent structures exhibit distinguishable energetics.
4) Select appropriate theoretical approaches with appropriate parameters for specific materials of the substrates: one often uses quantum approaches such as density functional theory and quantum molecular dynamics to study the active sites of the adsorbate, and thermodynamics model for the understanding of the energetics of the intermediate reaction steps and the propensity of forming the products under influences of temperature, pressure, electrode potential and pH level. The density functional theory calculations are used to understand the energetics of the substrates at the atomic scale and 0K. Cluster expansion and CALPHAD can be applied to explore in a broader context of space and conditions (Cao et al. 2018, Ohtani 1999). For example, CALPHAD shows the phase boundary of the alloy materials concerning the influences of temperature, pressure and chemical potential.
5) Study the energetics of the substrates and select the samples of the substrates for the next steps.
6) Search for the reaction intermediates and their active sites on the samples by calculating their binding energy.
7) Propose the reaction mechanisms based on the information of the obtained reaction intermediates and their binding energy.
8) Use the thermodynamics model to understand the energetics of intermediate reaction steps, the rate-limiting step and the activation barrier of the reaction for each mechanism.
9) Estimate the reaction rate using the Arrhenius equation or approximately compare the activation barrier of the reaction on the samples with that on the references. The reaction activity is usually measured in the kinetics model through the exchange current–bias voltage relationship in the cyclic voltammetry experiments. However, in the thermodynamics model, there is no direct comparison to the experiment, but the comparison could be made with the references. It means that one has to calculate the reaction rate for the selected substrates and the references in the same footing and computational methods.
10) Study the dissolution potential and segregation energy of the samples and compare them with the references to understand the propensity of the stability of the samples.
11) Correlate the reaction activity with the electronic structure properties. This step perhaps facilitates the prediction of new material candidates for the reaction based on the understanding of the electronic properties.
12) Validate the obtained results with experiments and further optimize the structure of the substrates for the reaction. Suggest new catalyst materials. Depending on

the complexity of the obtained results and also the experimental and theoretical data collected from the literature, one can use machine learning (Schleder et al. 2019) to optimize and suggest new candidates of electro-catalysts for the reaction.

This procedure can apply to other catalysts and chemical reactions as well. The demonstration of this methodology to PdCo alloy toward oxygen reduction reaction is presented next.

Application of design methodology to PdCo alloy for oxygen reduction reaction

Our earlier work on the PdCo alloy for the oxygen reduction reaction (Son and Takahashi 2012, Son et al. 2015) with the point-by-point correspondence are summarized here.

Step 1: By reviewing the literature, it was found that Pd and Co are less expensive and more abundant than Pt. Therefore, it is worth paying attention to the alloy of Pd and Co for the ORR. Besides, the PdCo alloy has higher ORR activity and better stability than pure Pt and commercial Pt/C (Savadogo et al. 2004). Annealing the PdCo alloy can further improve the activity and stability. The annealed PdCo alloy mainly exhibits the face-centered cubic (111) framework (Wang et al. 2007). However, the geometry structure and composition—activity and stability relationship is unknown.

Step 2: With the information found in Step 1, the cathode of PdCo was modeled by a slab with five atomic layers and a vacuum space of 13 Å. The vacuum space is large enough to avoid the interaction of the slab with its image along the normal of the surface. The (2×2) unit cell in the face-centered cubic (111) framework was chosen with the lattice constant that was optimized for each Co percentage. The outermost layers of the slab are Pd skin because the experiment showed that the surface of the PdCo alloy is enriched with Pd due to the annealing and acid treatment.

Step 3: For each Co content of 10, 20, 30, 40, 45, 60% within the slab model in Step 2, it was possible to create all possible sets of Co arrangement in the Pd host and then simplify for symmetry-independent structures of the PdCo unit cell. It was found the number of symmetry-independent Co arrangements is 36, 60, 110, 63, 25, and 1 for 10, 20, 30, 40, 45, and 60% Co, respectively.

Step 4: Density Functional Theory (DFT) calculations are appropriate for the study of chemical reactions on catalyst surfaces because of its high performance and reasonable accuracy. The plane-wave basis set, the right cutoff energy of 400 eV and the k-point mesh of $13 \times 13 \times 1$ are used for the slab model. The convergence of the electronic density and the total energy has been examined. The Perdew-Burke-Ernzerhof derivative of the spin-polarized generalized gradient approximation for the exchange-correlation energy and the projector augmented wave method for the valent electron-ion interaction should be used. The spin-polarized calculations should be performed for the transition metals. The dipole corrections should be included to exclude any extra interaction between the slab with its periodic images. In this chapter, only the density functional theory calculations for the mentioned unit cell were focused. For considering large substrates, one has to use the cluster expansion and the CALPHAD method; however, they require much more effort. Also, it is time-consuming to study the ORR on a larger surface of the catalyst.

Step 5: To correctly describe the energetics of the PdCo structures, one has to optimize the lattice constant using the DFT calculations. The lattice constant is 3.88, 3.85, 3.82, 3.80, 3.79, and 3.72 Å for 10, 20, 30, 40, 45, and 60% Co, respectively. For each Co content, the energetics of the PdCo structures was understood by considering the energy landscapes of all structures. For example, for 30% Co, it was shown in Fig. 8.2 that the Octahedral-Like (OL) structure is the most stable. By calculating the enthalpy of mixing for the most stable structure of each Co content, one can figure out the tendency of forming the PdCo alloy toward the variation of Co content, as seen in Fig. 8.3 (Son and Takahashi 2012). It is not possible to study the ORR on all the generated substrates and also the stability of all substrates. Therefore, one has to select only a few of them for the investigation of the ORR. It may be reasonable to choose the lowest energy structure and several special structures in each energy landscape, see our previous publication (Son et al. 2015). The most stable, i.e., lowest energy structures are also shown in Fig. 8.3. The best enthalpy of mixing was found at about 30% Co. It means that the PdCo alloy most favorably forms at this Co percentage. To clarify the composition–activity relationship in the next steps, one can choose a set of the structures shown in Fig. 8.4.

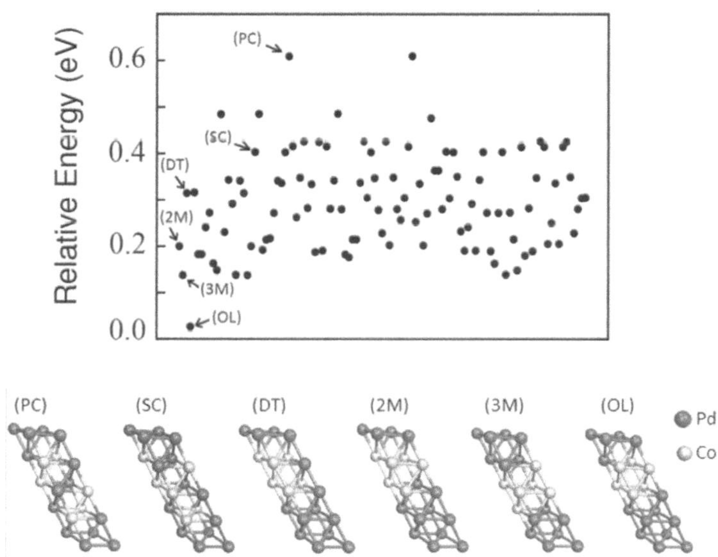

Figure 8.2: Energy landscape for 30% Co. The total energy of each structure is presented relative to that of the lowest energy structure (OL). According to the Co arrangement: Partially Clustering (PC), Scattered (SC), Double Triangle (DT), monolayer at the second (2M) and third (3M) atomic layer, and Octahedral-Like (OL). Reprinted with permission from J. Phys. Chem. C 116(2012): 6200–6207. Copyright (2012) American Chemical Society.

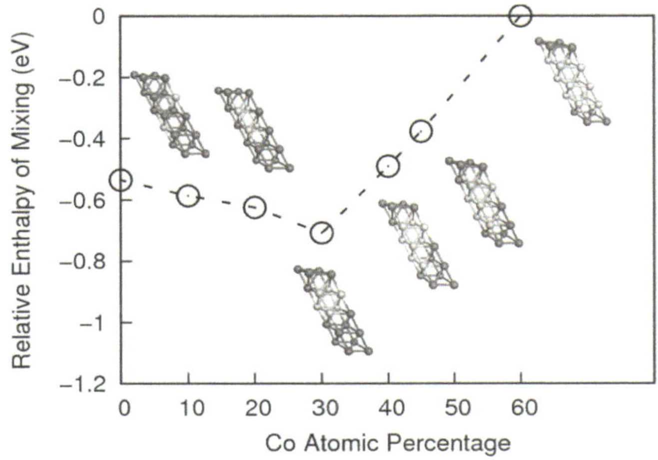

Figure 8.3: Enthalpy of Co mixing in the Pd host relative to the phase-separated structure of 60% Co. The lowest energetics geometry structure is also presented for each Co percentage. Reprinted with permission from J. Phys. Chem. C 116(2012): 6200–6207. Copyright (2012) American Chemical Society.

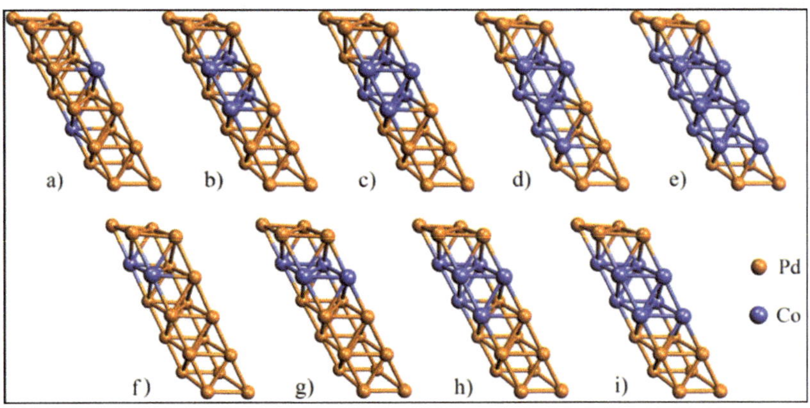

Figure 8.4: From (a) to (e) are the most stable substrates of PdCo for Co content of 10, 20, 30, 40, and 60%, respectively. From (f) to (i) denote the substrates with the maximum number of Co atoms underneath the surface layer for 10, 20, 30, and 40% Co. Reprinted with permission from J. Phys. Chem. C 119(2015): 24364–24372. Copyright (2015) American Chemical Society.

Step 6: The ORR is one of the most complicated reactions consisting of many intermediate steps. The left side of the equation (1) involves an oxygen molecule, four protons and four electrons. Therefore, one can assume different scenarios that may occur. Some physical instinctive should be used to build the structures: First, in the supercell periodic model, it is rather difficult to describe point charges such as the protons and electrons separately. Therefore, one has to assume that the protons and electrons combine in the manner: $\frac{1}{2} H_2 \rightarrow H^+ + e^-$. This equation is also the oxidation of hydrogen gas on the anode of hydrogen fuel cells. Simultaneously, if one applies this equation in simulation in the context of the thermodynamics model, it is considered as a reference to the hydrogen standard electrode. Therefore, in this chapter, hydrogen atom and hydrogen molecule to load into the simulation cell instead was used for loading proton and electron separately because it is simply not possible or very expensive to precisely describe the point charges in the supercell periodic model. Second, the oxygen gas is more sensitive than hydrogen gas on transition metals (Karlberg 2006). In reality, the transition metals are easily oxidized at ambient conditions such as ambient humidity, temperature and pressure. The conditions are somewhat similar to those in the ORR environment. Therefore, the adsorption of the oxygen molecule is largely stronger than that of the hydrogen molecule, and hence, one can assume that the oxygen molecule approaches the catalyst surface before hydrogen atoms do. Third, the oxygen gas may dissociate or remain in a molecular form during the approach to the alloy surfaces. Thus, one could start the search for ORR intermediates by assuming the associative and dissociative scenarios of the oxygen molecule, which form O_2* and O* on the surface of the substrates. In the DFT calculations, O_2 and O are loaded onto the surface of each substrate. The DFT geometry optimization results in the local stable adsorption configurations of O_2* and O* on the surface. With continuous loading of the hydrogen atom onto each stable adsorption configuration, it is possible to find intermediates that are *OOH, HOOH*, HO* + HO*, HO* + O*, *OH + H_2O, O* + H_2O, *OH, H_2O on the surface of the substrates. It needs to be noted that for different substrates or different materials, one could perhaps find different intermediates. Thus, for new catalysts, one has to examine the possibilities of forming new intermediates. Details about the type, geometry structure, adsorption energy, adsorption site of the ORR intermediates on the PdCo alloy were given in the earlier publications (Son and Takahashi 2012, Son et al. 2015). Here, the most important information for proposing the reaction mechanisms should be the adsorption energy of the intermediates. In this work, the effects of reaction intermediate coverage and water medium were ignored.

Step 7: The proposal of reaction mechanisms requires experience. Literature often starts with the mechanisms that are available elsewhere, and then go to test the existence of the ORR intermediates. This procedure limits the ability of the DFT method to discover new intermediates of the reaction. Therefore, our procedure begins with searching for intermediates before proposing mechanisms. This procedure can be applied to any chemical reaction on electro-catalysts. The reaction mechanisms are proposed based on the information about the sequence of forming the intermediates. For each step of loading H on the previously formed intermediates, one has to compare the adsorption energy of the intermediates with the same species and number of atoms but with different structures and adsorption sites. The intermediate with a lower adsorption energy tends to form a more stable state therefore located in the reaction step is likely to occur later.
The dissociative mechanism:

$$\tfrac{1}{2}\, O_2 + * \rightarrow O* \tag{2}$$

$$O* + (H^+ + e^-) \rightarrow HO* \tag{3}$$

$$HO* + (H^+ + e^-) \rightarrow H_2O + * \tag{4}$$

The associative mechanism:

$$O_2 + * \rightarrow O_2* \tag{5}$$

$$O_2* + (H^+ + e^-) \rightarrow HOO* \tag{6}$$

$$HOO* \rightarrow HO* + O* \tag{7}$$

$$(HO* + O*) + (H^+ + e^-) \rightarrow HO* + HO* \tag{8}$$

$$(HO* + O*) + (H^+ + e^-) \rightarrow O* + H_2O \tag{9}$$

$$O* + (H^+ + e^-) \rightarrow HO* \tag{10}$$

$$HO* + (H^+ + e^-) \rightarrow H_2O + * \tag{11}$$

Step 8: The Gibbs free energy differences were formulated as follows (Son et al. 2015, Nørskov et al. 2004). For the dissociative mechanism:

$$\Delta G_0(U) = G_{H_2O}(U) - G_{O*+H_2}(U) = \Delta G_0(0) + eU \tag{12}$$

$$\Delta G_1(U) = G_{HO*+(1/2)H_2}(U) - G_{O*+H_2}(U) = \Delta G_1(0) + eU \tag{13}$$

$$\Delta G_2(U) = G_{H_2O}(U) - G_{HO*+(1/2)H_2}(U) = \Delta G_2(0) + eU \tag{14}$$

For the associative mechanism:

$$\Delta G_0(U) = G_{2H_2O}(U) - G_{O_2^*+2H_2}(U) = \Delta G_0(0) + eU \tag{15}$$

$$\Delta G_1(U) = G_{O_2^*+2H_2}(U) - G_{HOO*+(3/2)H_2}(U) = \Delta G_1(0) + eU \tag{16}$$

$$\Delta G_2(U) = G_{2H_2O}(U) - G_{HOO*+(3/2)H_2}(U) = \Delta G_2(0) + eU \tag{17}$$

$$\Delta G_3(U) = G_{2H_2O}(U) - G_{HO*+O*+(3/2)H_2}(U) = \Delta G_3(0) + eU \tag{18}$$

$$\Delta G_4(U) = G_{2H_2O}(U) - G_{O*+H_2O+H_2}(U) = \Delta G_4(0) + eU \tag{19}$$

$$\Delta G_5(U) = G_{2H_2O}(U) - G_{HO*+HO*+H_2}(U) = \Delta G_5(0) + eU \tag{20}$$

where, the Gibbs free energy denoted by G, and $\Delta G_i(0) = \Delta E + \Delta ZPE - T\Delta S$ at the electrode potential of 0 V with the reaction energy (ΔE), the change of zero-point energies (ΔZPE), and the changes of entropies (ΔS). In this work, one can assume that the pH level equals zero, which is equivalent to the condition ½ $H_2 = H^+ + e^-$ or the reference to the hydrogen standard electrode, and the pressure of 1 bar. One can obtain ΔE and ΔZPE from the DFT calculations, ΔS from the earlier work (Nørskov et al. 2004). With the obtained results for Gibbs free energy differences for the given temperature of 300 K and the electrode potential of 1.23 V, one can build the thermodynamics diagrams shown in Figs. 8.5 and 8.6 for the dissociative and associative mechanisms, respectively. While analyzing these diagrams, it was found that the rate-limiting step is the first hydrogenation step with the highest thermodynamics barrier, $\Delta G_1(U)$.

Step 9: The thermodynamics barrier is identified by $\Delta G_1(U)$, which is found to be in the order: Pd-Co 10% > Pd-Co 20% > Pd-Co 30% > Pd-Co 40% > Pd-Co 60% as shown in Fig. 8.7. It is not necessary to perform detailed calculations of the reaction

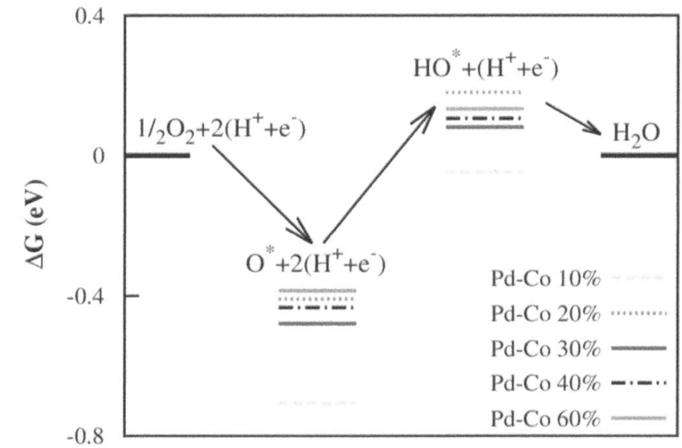

Figure 8.5: The Gibbs free energy diagram for the dissociative mechanism. Reprinted with permission from J. Phys. Chem. C 119(2015): 24364–24372. Copyright (2015) American Chemical Society.

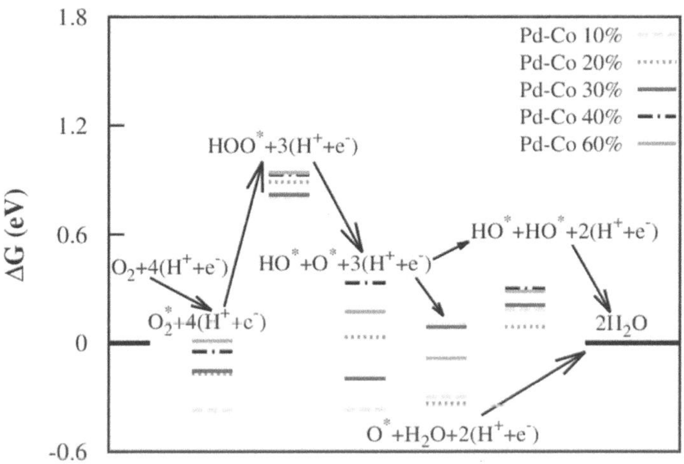

Figure 8.6: The Gibbs free energy diagram for the associative mechanism. Reprinted with permission from J. Phys. Chem. C 119(2015): 24364–24372. Copyright (2015) American Chemical Society.

rate but it can be compared directly based on the thermodynamics barrier. The lower barrier gives rise to higher activity that is approximately exp $(-\Delta G_1(U)/k_BT)$. One can see that the ORR activity is not the highest, even though the best mixing was found at 30% Co. The thermodynamics barrier for the rate-limiting step of the ORR was also obtained on pure Pt that is of about 0.8 and 1.1 eV for the dissociative and associative mechanisms, respectively. The barrier for the ORR on the PdCo alloy is significantly lower than that on Pt. Therefore, the PdCo alloy can be an excellent candidate to replace Pt. Furthermore, maximizing the number of Co atoms below the Pd skin maximizes the ORR activity (Son and Takahashi 2012).

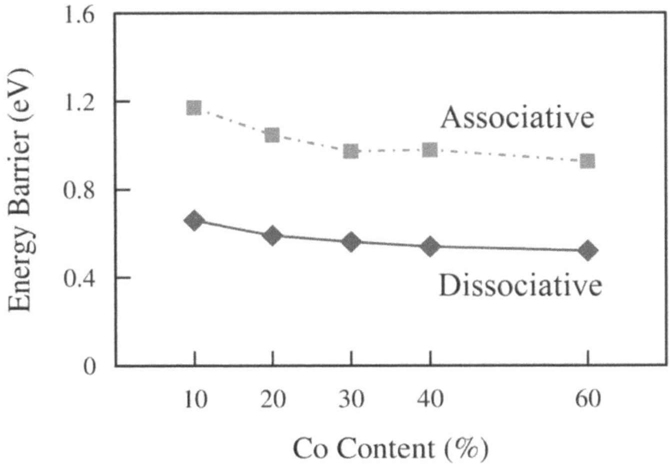

Figure 8.7: The energy barrier $\Delta G_1(U)$ of the rate-limiting step in the dissociative and associative mechanisms for each Co percentage. Reprinted with permission from J. Phys. Chem. C 119(2015): 24364–24372. Copyright (2015) American Chemical Society.

Step 10: The stability of the PdCo alloy by calculating the dissolution potential *D* of the PdCo slab by using the method of Greeley and Norskov was also studied (Greeley and Norskov 2007):

$$(D - D_{Pd}^{0})ne = E_{Pd}^{Bulk} - \frac{E_{PdCo}^{Slab} - E_{PdCo}^{Slab\,w/o\,Pd\,skin}}{N_{Surface}}. \tag{21}$$

The dissolution potential of bulk Pd, $D_{Pd}^{0} = 0.95$ V. The oxidation state of Pd is $n = 2$. $N_{Surface} = 8$. It was found that the PdCo alloy is less sensitive to the dissolution than Pd with the obtained dissolution potential ~ 1.1 V (Son and Takahashi 2012), which is higher than that of Pd, 0.95 V. It means that the PdCo alloy is more stable than Pd under working conditions of PEM fuel cells.

Step 11: Correlating the ORR activity with the electronic properties may facilitate the development of electro-catalysts. The electronic density of states, charge density difference, electronic band structure, etc., can be considered. However, the quantitative parameters are preferred because it is easier to make comparisons and descriptions. Therefore, one often considers the d-band center. Figure 8.8 shows the correlation of the d-band center, the ORR energy barrier and the Co content of the PdCo alloy. In the PdCo alloy with Pd skin, the increase of the Co content shifts the d-band center to a more negative value and decreases the energy barrier; and thus, increases the ORR activity.

Step 12: Step 9 shows that the PdCo alloy has better ORR activity compared to Pt. This result is in good agreement with the experiment, done for ~ 30% Co (Savadogo et al. 2004). Furthermore, Fe and Ni have similar electronic properties with Co. Therefore, they can be promising candidates. With this suggestion in mind, research methodology for Pd-skin/$Pd_3Fe(111)$ was performed and it was found that Pd-skin/

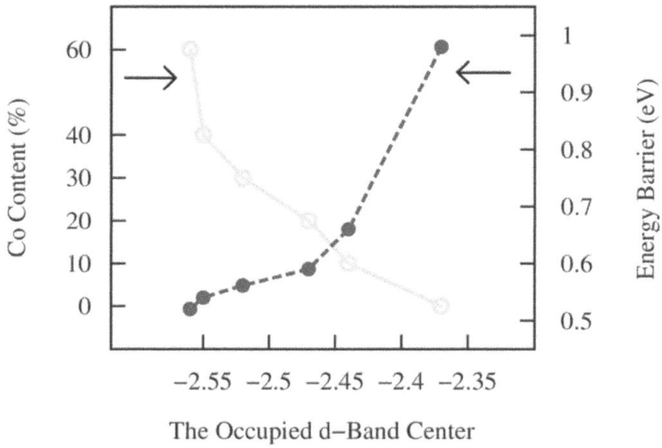

Figure 8.8: The occupied d-band center of PdCo alloy versus Co percentage and the thermodynamics barrier for the ORR. Reprinted with permission from J. Phys. Chem. C 119(2015): 24364–24372. Copyright (2015) American Chemical Society.

$Pd_3Fe(111)$ is also an excellent catalyst for the ORR, while Fe is less expensive and more environmentally benign than Co (Son et al. 2017, Son et al. 2018).

Conclusion

In this chapter, the methodology for computational materials designs of alloy catalysts for chemical reactions were introduced. Although the PdCo alloy toward oxygen reduction reaction was chosen as a demonstration for this methodology, one can definitely apply it to any catalyst and any chemical reaction with a little modification. It was also shown that the PdCo alloy is an excellent candidate for the ORR catalyst. It would be more complete to take into account machine learning into consideration; however, it will be presented in the future work of our group.

Acknowledgement

This research work was funded by SERB-DST, New Delhi, under the scheme of "ASEAN-India Research & Development" (Grant No: IMRC/AISTDF/CRD/2018/000048).

References

Cao, L., C. Li and T. Mueller. 2018. The use of cluster expansions to predict the structures and properties of surfaces and nanostructured materials. Journal of Chemical Information and Modeling 58: 2401–2413.

Greeley, J. and J. K. Norskov. 2007. Electrochemical dissolution of surface alloys in acids: Thermodynamic trends from first-principles calculations. Electrochimica Acta 52: 5829–5836.

Karlberg, G. S. 2006. Adsorption trends for water, hydroxyl, oxygen, and hydrogen on transition-metal and platinum-skin surfaces. Physical Review B 74: 153414

Li, Y., Q. Li, H. Wang, L. Zhang, D. P. Wilkinson and J. Zhang. 2019. Recent progresses in oxygen reduction reaction electrocatalysts for electrochemical energy applications. Electrochemical Energy Reviews 2: 518–538.

Liu, X., D. Furrer, J. Kosters and J. Holmes. 2018. Vision 2040: A roadmap for integrated, multiscale modeling and simulation of materials and systems. http://www.sti.nasa.gov.

Mench, M. M. 2008. Fuel Cell Engines. John Wiley & Sons, Inc., Hoboken, New Jersey.

Na, O. M., N. T. X. Huynh, P. T. Thi, V. Chihaia and D. N. Son. 2020. Mechanism and activity of the oxygen reduction reaction on WTe2 transition metal dichalcogenide with Te vacancy. RSC Advances 10: 8460.

Nørskov, J. K., J. Rossmeisl, A. Logadottir, L. Lindqvist, J. R. Kitchin, T. Bligaard et al. 2004. Origin of the overpotential for oxygen reduction at a fuel-cell cathode. Journal of Physical Chemistry B 108: 17886–17892.

Ohtani, H. 1999. CALPHAD approach to materials design. pp. 105–134. *In*: Saito, T. (ed.). Computational Materials Design. Springer Series in Materials Science, vol 34. Springer-Verlag, Berlin, Heidelberg. https://doi.org/10.1007/978-3-662-03923-6_4.

Ramasubramaniam, A. and E. A. Carter. 2007. Coupled quantum-atomistic and quantum-continuum mechanics methods in materials research. MRS Bulletin 32: 913–918.

Saito, T. 1999. Computational Materials Design. Springer-Verlag, Berlin, Heidelberg.

Savadogo, O., K. Lee, K. Oishi, S. Mitsushima, N. Kamiya and K. -I. Ota. 2004. New palladium alloys catalyst for the oxygen reduction reaction in an acid medium. Electrochemistry Communications 6: 105–109.

Schleder, G. R., A. C. M. Padilha, C. M. Acosta, M. Costa and A. Fazzio. 2019. From DFT to machine learning: recent approaches to materials science—a review. Journal of Physics: Materials 2: 032001.

Son, D. N. and K. Takahashi. 2012. Selectivity of palladium−cobalt surface alloy toward oxygen reduction reaction. Journal of Physical Chemistry C 116: 6200–6207.

Son, D. N., O. K. Le, V. Chihaia and K. Takahashi. 2015. Effects of Co content in Pd-Skin/PdCo alloys for oxygen reduction reaction: density functional theory predictions. Journal of Physical Chemistry C 119: 24364–24372.

Son, D. N., P. N. Thanh, N. D. Quang, K. Takahashi and M. P. Pham-Ho. 2017. First-principles study of Pd-skin/Pd_3Fe(111) electrocatalyst for oxygen reduction reaction. Journal of Applied Electrochemistry 47: 747–754.

Son, D. N., P. V. Cao, T. T. T. Hanh, V. Chihaia and M. P. Pham-Ho. 2018. Influences of electrode potential on mechanism of oxygen reduction reaction on Pd-Skin/Pd_3Fe(111) electrocatalyst: insights from DFT-based calculations. Electrocatalysis 9: 10–21.

Takahashi, K. and Y. Tanaka. 2016. Material synthesis and design from first principle calculations and machine learning. Computational Materials Science 112: 364–367.

Wang, W., D. Zheng, C. Du, Z. Zou, X. Zhang, B. Xia, H. Yang and D. L. Akins. 2007. Carbon-supported Pd-Co bimetallic nanoparticles as electrocatalysts for the oxygen reduction reaction. Journal of Power Sources 167: 243–249.

Wang, X., Z. Li, Y. Qu, T. Yuan, W. Wang, Y. Wu et al. 2019. Review of metal catalysts for oxygen reduction reaction: from nanoscale engineering to atomic design. Chem 5: 1486–1511.

Chapter 9

Metal-Organic Frameworks Catalyst for Energy Applications

Deepa Rajwar,[1,*,#] *Gayatri Chauhan*[2,#] and *Alok Kumar*[3]

Introduction

Remarkable achievements in the progress of Metal-Organic Frameworks (MOFs) over the last two decades have led the way for their applications in diverse areas of research, such as heterogeneous catalysis, gas storage, luminescent materials, adsorption, drug delivery, ferroelectric materials, biomedical imaging, optoelectronic devices, gas separation, chemical sensing, nonlinear optics, chirality and many more (Zhang et al. 2009, Farha et al. 2010, Wee et al. 2012, Linares et al. 2014, Choi et al. 2015, Medishetty et al. 2017, Chen et al. 2018, Zhu et al. 2018, Cadiau et al. 2020).

Yaghi's group started the pioneering work in the field of MOFs and reported the first stable, highly porous MOF (MOF-5) with the high surface area (4,500 m^2g^{-1}) (Li et al. 1999). A review by Wang et al. reported the recent progress in the use of MOFs and their derivatives for energy applications. It emphasized the challenges in achieving current energy targets by performance in energy storage and conversion (Wang et al. 2017). Apart from non-renewable sources of energy, such as fossil fuels, which have dominated society for a long time, other energy sources such as batteries, fuel cells and photovoltaic devices for energy harvesting are extremely relevant in the modern world to fulfill the current need. A recent review highlights important applications of MOFs with different structures and functionalities for renewable energy and environmental applications (Zhang et al. 2017). This chapter covers the

[1] School of Materials Science and Engineering, NTU, Singapore.
[2] Department of Physics, Banasthali University Banasthali, India.
[3] Department of Biochemistry and Molecular Biology, University of Maryland, Baltimore, USA.
* Corresponding author: rajwardeepa@gmail.com
# Author is an independent researcher and not currently associated with the institute, but part of the work is carried out by the author at this institute.

present and the future of research at the interface of catalysis, new porous materials for energy, based on metal-ion/metal-ion clusters and organic ligands. It also covers the progress made in the field of MOFs over the last two decades.

The basic design principle of metal-organic frameworks

MOFs are a subclass of Metal-Organic Materials (MOMs), first introduced by Cook et al. (Cook et al. 2013). Another class of important materials in this category are self-assembled coordination complexes (SSCs) or Metal-Organic Networks (MONs). Both MOFs and SSCs are porous and, therefore, are important for catalysis and gas storage. These highly ordered structures can combine the excellent catalytic properties of inorganic materials with the processability of organic materials. Both covalent and non-covalent interactions can form the macromolecular structures. Although covalently bonded porous structures such as Covalent Organic Frameworks (COFs) were extensively studied in the past but their solubility and the processability are of significant concern (Ma and Scott 2018, Cui et al. 2020). Non-covalent self-assembly such as metal coordination is an efficient way to prepare processable functional organic macromolecular materials. Reversibility is the most important property of this kind of self-assembly, which gives the ability for self-correction of defects formed during the assembly process. Extended pi conjugation in conjugated SSCs offers unique electronic and electroluminescent properties for functional materials. The two and three-dimensional (2D and 3D) representative networks of self-assembled terpyridines organic linker through zinc metal-ion coordination are shown in Fig. 9.1a (Gao et al. 2014). Bridging the gap between MOFs and SSCs is crucial to the development of new functional materials for future energy needs (Caskey et al. 2008, Zheng et al. 2009).

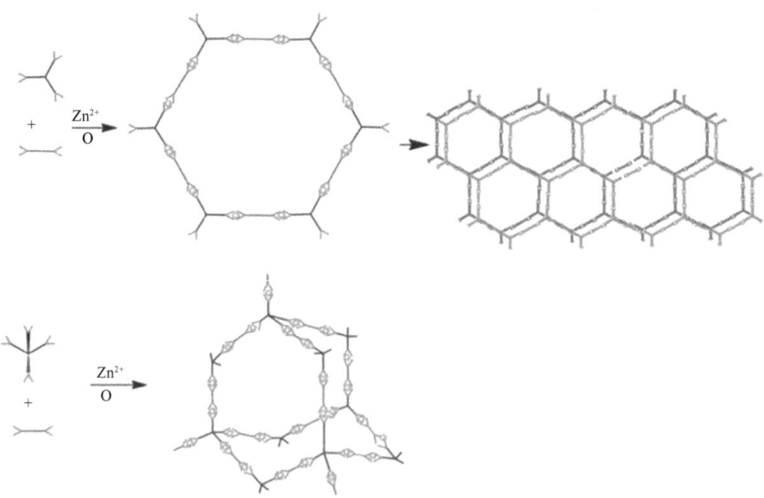

Figure 9.1(a): Two-component (a) 2D super graphite-like structure and representation of π-π stacking interaction between monolayers to form a 3D porous network (b) 3D diamond-like self-assembled coordination network of tetragonal organic units (adopted from Gao et al. 2014).

Organic synthesis of MOFs allows the formation of a variety of ligands structures with different shapes, sizes and functionalities. In MOFs, metal-ligand interaction takes place by the contribution of a lone pair of electrons from the ligand, which acts as a Lewis base to the metal ion, a Lewis acid. This gives rise to an electrostatic interaction between the newly formed positively charged metal ion and negatively charged/polarized donor atom of the ligand. Post-modification of MOFs is possible through 'isorecticular chemistry principles' which allows changing the functionality and metrics of a porous structure without altering its underlying topology (Doonan et al. 2009). This opens the way for a variety of MOF structures which can be used as catalysts. For instance, post-synthetic modification of MOF, namely UiO-66, was carried out to achieve catalytic performance by incorporation of weak and strong functional groups together with platinum nanoparticles (Choi et al. 2007). Imidazole and zinc metal-ion based MOF, ZIF-8 was reported to encapsulate nanoparticles of different sizes and shapes (Lu et al. 2012). In a new approach for achieving an extended multi-layer of MOF nanoparticles, poly(methyl methacrylate) (PMMA) supported self-assembled MOF monolayers (SAMMs) at the liquid-solid interface was reported using a histamine anchor (Katayama et al. 2019). In a different approach, semiconducting nanostructures as self-sacrificing templates were used to synthesize three dimensional (3D) MOF (ZIF-67) with excellent electrocatalytic performance (Cai et al. 2017).

Selection of organic ligands for MOFs

The organic ligands are basic building blocks, adding flexibility and diversity to the MOF structures. These MOF structures can be formed by designing a ligand, which adopts different symmetry conformations through the free internal bond rotation. Related to this, the reticular chemistry concept 'stitching molecular building blocks into external structures by strong bonds' was introduced by Yaghi's group (Kalmutzki et al. 2018). The organic building blocks of these materials mainly consist of carboxylic compounds based on phenyl and polyphenyl molecules, imidazole and others. Figure 9.1b shows the structure of some representative organic ligands.

Porosity is one of the most remarkable features of MOFs, especially for their use in catalytic applications. Depending on the pore size of nanostructure materials, they can be categorized as microporous (< 2 nm), mesoporous (2–50 nm) and macroporous (< 50 nm) MOFs. The materials that have metal/metal-ion centers linked to each other by organic ligands, such as 1,4-benzenedicarboxylic acid (BDC), benzenetricarboxylic acid (BTC), methylimidazole (MeIM), and terpyridine, are porous coordination compounds with open network structures. Unlike previously known porous zeolites materials, these materials can be formed into many structures because of a large possibility for the selection of the organic linkers and metal ions/ metal-ion clusters.

Microporous MOF-508 based on BDC, 4,4'-bipyridine, and Zn^{2+} were reported to be the first microporous MOF for the separation of alkanes (Chen et al. 2006). Earlier, zeolite-based materials were used for this purpose. Similarly, the solvothermal reactions of H_2BDC or azobenzene-4,4′-dicarboxylic acid (H_2ABD) with zinc ions/clusters led to the formation of four crystalline materials (Nguyen et al.

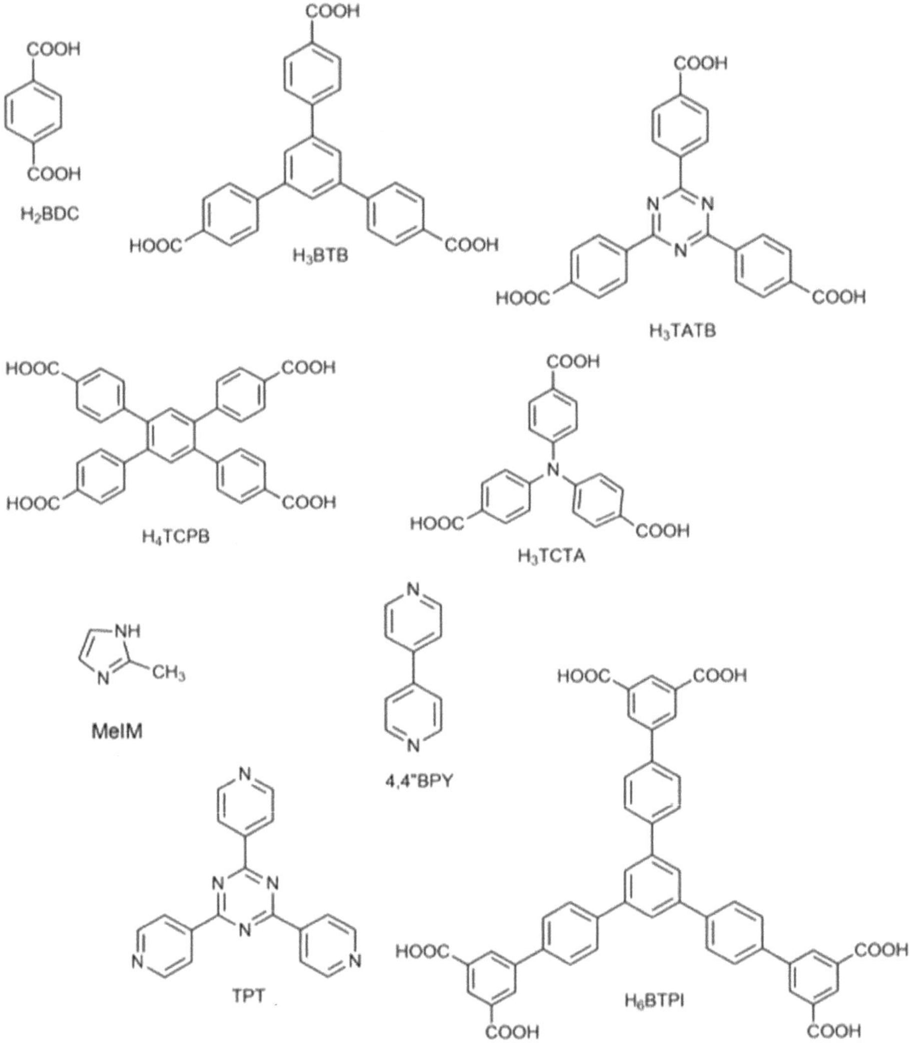

Figure 9.1(b): Structure of some representative organic ligands as building units of MOFs.

2011). A mesoporous framework, NOTT-119, from C_3-symmetric hexacarboxylate and $Cu(NO_3)$ metal cluster was reported (Yan et al. 2011). It showed argon sorption behavior with large desorption hysteresis and high thermal stabilities. NOTT-116 constructed from the elongated nanosized hexacarboxylate linker $(L_2)^{6-}$, showed a very high BET surface area of 4664 m^2g^{-1}, and total H_2 adsorption capacity of 9.2 wt% at 77 K and 50 bar. NOT-112, based on C_3-symmetric tricarboxylate ligand with smaller length scales has a lower BET surface area of 3800 m^2g^{-1} but a higher maximum excess H_2 uptake compared to NOTT-116 (Yan et al. 2010). Other than carboxylate containing MOFs, a multi-component MOF with tetrahedral cages with a large cavity and small pores, which consists of C_3-symmetric triamines,

2-formylpyridine as organic linkers with Fe^{2+} as metal ions were reported (Bilbeisi et al. 2012). It was shown that Fe_4L_4 face-capped tetrahedral cages and Fe_2L_3 helicates can be combined due to identical Fe-Fe distances in both the networks, with the potential to incorporate several guest species in a size-selective fashion useful for heterogeneous catalysis (Chen et al. 2018). Bipyridines and terpyridines were proposed as building blocks for SSCs decades ago (Schubert et al. 2001, Wild et al. 2011, Sun et al. 2019). There is a lot of perspective for the development of terpyridines to produce 2D and 3D organic frameworks for many applications. For example, transition metal-complexes of π-conjugated bis- and tris (terpyridine) possess a great point of view to be used as new functional materials in electronic and optoelectronic devices.

Selection of metal-ions for MOFs

Metal-ions belonging to p-block, d-block transition metal ions, lanthanides and actinides of f-block are used widely in developing MOMs. Metals like Zn, Cu, Ni, Mn, Al, Cr, Mg, Zr, Fe, In, Ga, Mo, Cr, Co, Ru, Cd, Ag, Na, and La are used in inorganic units. Extended structures of MOFs can be built through Secondary Building Units (SBUs), which is a combination of organic or inorganic entities, such as metal oxide clusters and aromatic polycarboxylates as shown in Fig. 9.2

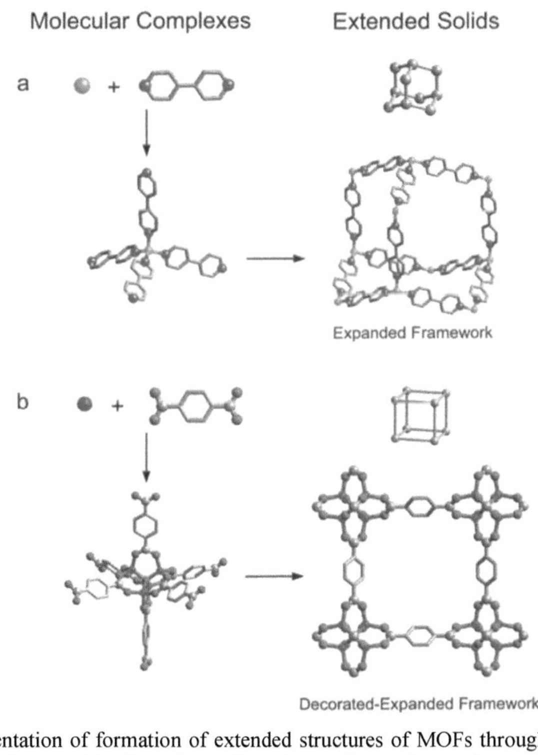

Figure 9.2: Representation of formation of extended structures of MOFs through Secondary Building Units (SBUs), which is a combination of organic or inorganic entities, such as metal oxide clusters and aromatic polycarboxylates (Reprinted with permission from Eddaoudi et al. 2001).

(Eddaoudi et al. 2001, Tranchemontagne et al. 2009, Kalmutzki et al. 2018). Some of the inorganic building units are $Zn_4O(COO)_6$, $M3O(COO)_6$ [M = Zn, Cr, In and Ga], $Zr_6O_4(OH)_4(COO)_{12}$, $Cu_4Cl(C_2H_2N)_4$, $Fe(C_2O_2)_2(H_2O)_2$, $Al(OH)(CO_2)_2VO(CO_2)_2$, $Mn_3(COO)_8$, and CuS_3.

The first mesoporous 3D MOF“IRMOF-16” was reported in 2002 by Li et al., consisting of ZnO_4 tetrahedra as a coordinating site with an organic linker TPDC. This belongs to the family of isoreticular MOF-5 (Li et al. 1999). A 3D MOF consisting of two ligands BIB and 4,4’-ADB (BIB=1,4-bis(2-methyl-imidazol-1-yl) butane, 4,4’-ADB=azobenzene-4,4’-dicarboxylica acid) with trinuclear Zn^{2+} cluster were reported. Despite being an interpenetrated network, it turned out to be robust because of the zinc-ion cluster as SUBs (Liu et al. 2012). This MOF is a 3D structure with trinuclear Zn^{2+} clusters $Zn_4O(COO)_6$ as SBU. Similarly, a mesoporous network with unique tetranucleariron–oxo clusters, which acts as 8-connectors, are linked by 3-connecting BTBs was reported (Choi et al. 2007). The role of heavy transition metal ions, such as Pt^{2+}, Ru^{2+}, Os^{2+}, and Ir^{2+} in the formation of SSCs was researched thoroughly in the past (Wild et al. 2010, Chakrabarty et al. 2011). In particular, the chemistry of electron-poor square-planar divalent Pt^{2+} and Pd^{2+} ions and electron-rich nitrogen-containing moieties is well reported in the self-assembly process. Another well-explored metal ion with low cost and non-toxicity is (d10) Zn^{2+} ions. The advantage of using Zn^{2+} ions in SSCs is the non-existence of Metal to Ligand Charge Transfer (MLCT) processes, and there is only the possibility of intraligand (IL) charge transfer. Hence MONs with Zn^{2+} ions are excellent candidates in optoelectronics (Gao et al. 2014, Rajwar et al. 2014).

MOFs as catalyst materials

Catalysts materials play a pivotal role in achieving environmental sustainability. Heterogeneous catalysts are an important class of materials to convert energy efficiently for the viability of green technologies, as well as to reduce carbon footprints (Huang et al. 2020, Wee et al. 2012).

Electrocatalysis by which electricity can be converted into hydrogen offers a clean and renewable energy solution to the present energy needs. To address the challenges in the designing of MOF-based electrocatalysts, experimental and theoretical studies on electrochemical Oxygen Evolution Reaction (OER) and its enhancement as well as actual sites of MOFs responsible for this were studied in detail (Shi et al. 2019). In this context, oxygen reduction reaction (ORR) catalysts materials with MOFs are the promising candidates (Huang et al. 2020).

The current challenge in developing efficient MOFs for energy devices is their limited conductivity and processability (So et al. 2015). The future of the catalytic application of MOMs largely depends on the solubility of these materials. Therefore, one should always keep in mind the solubility (in polar as well as in nonpolar solvents) while designing the network structures. Recent progress in this regard is a highly crystalline NMOF for energy harvesting applications. This MOF was made up of p-conjugated dicarboxylate ligands and [$Ln(OAc)_3$] (Ln = Gd, Eu, Yb)] metal oxides (Zhang et al. 2011). With further progress, in recent reports, titanium-based robust MOF ACM-1 was synthesized by a one-step synthesis for its photocatalytic application. High mobility of photogenerated electrons and hole localization in organic linker resulting in a long exciton lifetime makes this MOF an excellent candidate for future applications (Cadiau et al. 2020).

MOFs are being discovered for many useful applications, interpenetrating the boundaries of science, engineering and technology. In this direction, MOFs were reviewed for their Non Linear Optical properties (NLOs), and some reports suggest their better NLO properties in comparison to other reported materials (Medishetty et al. 2017). A two-component 3D SSC was reported with basic building units of terpyridine based linear and tetrahedral organic ligand by coordination with zinc metal-ions. As such, these networks interpenetrate with very little porosity but turn out to be exceptional materials for light-harvesting. The energy transfer was reported from tetragonal donors to linear acceptor molecules under two-photon excitation (Lim et al. 2012, He et al. 2013). Many structures already exist and more are to be discovered, but their catalytic application is yet to be explored. This can be done with a combined effort of research communities with diverse backgrounds with industrial collaborations as well. A detailed discussion on the use of MOF based catalyst materials for energy application will be described next.

Metal-organic frameworks as catalyst for energy applications

The nano MOF terminology includes nanosized MOF's pore, MOFs with metal nanoparticles, and the formation of their composites with quantum dots, graphene, CNTs and semiconductor nanoparticles (Kaur et al. 2016). These are distinguished by their high surface area, porosity, reactivity and enhanced stability. These promising features of nano-MOFs act as a catalyst in enhancing the performance of energy devices. Here the focus will be on the application of nano-MOF on energy generation and storage devices.

Light-harvesting devices

Light-harvesting technology has received considerable attraction as it enables the utilization of the most ambient light source, i.e., the sun. Conversion of solar light into electrical energy (photovoltaic) or chemical energy (photocatalytic) is the two-broad aspects of light-harvesting.

Photovoltaic applications

Dye sensitize solar cell (DSSC)

MOF offers a major opportunity in light-harvesting and energy transfer in DSSC. The nanoscale films of MOFs are ideal for sensitization because charge migration distance remains low at the nanoscale. As the energy is usually harvested far from the MOF/semiconductor interface can be lost thermally (Hasselman et al. 2006, Maza et al. 2016, Kuyuldar et al. 2019). The building block of DSSC and working principle are given in Fig. 9.3. A dye molecule acts as a photoactive material and therefore, is also called a sensitizer. For instance, the light-harvesting properties of porphyrin-based nano MOFs drew considerable attention (Li et al. 2017, Hosoyamada et al. 2018). The coupling of porphyrin-based MOFs with CdSe/ZnS core/shell quantum

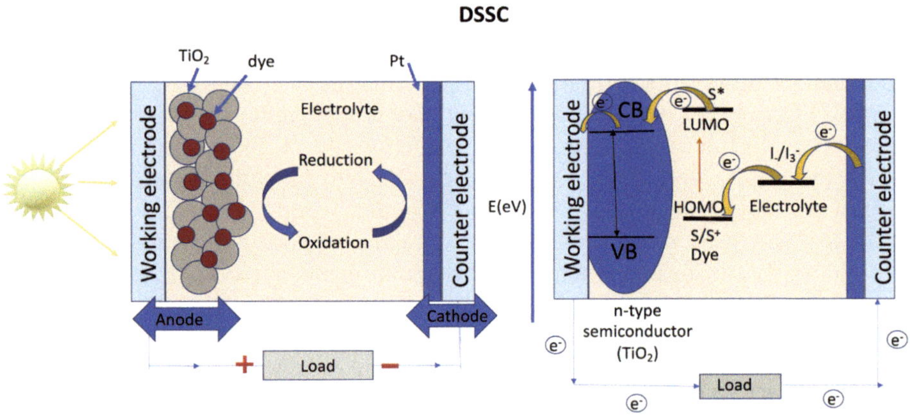

Figure 9.3: The (a) schematic diagram and (b) working principle of DSSC.

dots for enhanced light harvesting was used. More than 50% increase is observed in the photon harvested by the QD–MOF hybrid monolayer (Jin et al. 2013). CdS nanoparticle coupled with porphyrin MOF, layer by layer (lbl) deposition technique of functionalized MOF on the surface of TiO_2 thin film, also improves photocurrent density (Lee et al. 2014, Liu et al. 2015, Lu et al. 2016). Tang et al. reported TiO_2 derived from MIL-125 with Cu_2ZnSnS_4 and tin oxide/MoS_2 as a photoanode and achieved cell efficiencies of 8.10 and 8.96%, respectively (Tang et al. 2016, 2017).

As a major application, MOF is used as an alternative to electrocatalyst material (usually expensive platinum) to design the cathode. Use of Cobalt Sulfide (CoS) nanoparticles of a MOF (ZIF-67) with variable particle sizes of 50 to 320 nm resulted in the improvement of the efficiency by 8.1% due to the high external surface area and high roughness factors which enhanced interaction with dye molecules (Hsu et al. 2015).

Organic solar cell (OSC)

The nanosheets of tellurophene-based 2D MOF as an electron extraction layer in organic solar cells reduces charge recombination, alters work function and improves conductivity, which is attributed to the Power Conversion Efficiency (PCE) of 10.39% (Xing et al. 2018). Nanosheets of zinc-porphyrin based MOF in the active layer (P3HT:PCBM) of organic solar cells also increase the PCE almost twice to the reference device. The crystallization of P3HT causes an increase in hole mobility, reduced grain size and high absorption (Sasitharan et al. 2020).

Perovskite solar cell (PVSC)

The most challenging issue in PVSC is the lifetime and presence of toxic lead. A stable and porous indium based MOF was synthesized by Zhou et al. In-BTC nanocrystal used as an additive in perovskite precursor was shown in a schematic and energy diagram in Fig. 9.4. This improves the morphology and crystallinity and simultaneously reduces the grain boundaries and defects of perovskite films leading to PCE 19.63%, higher than the pristine device (Zhou et al. 2020). PVSCs achieve enhanced PCE 21.44%, and improved lifetime due to POM@Cu-BTC nanoparticles

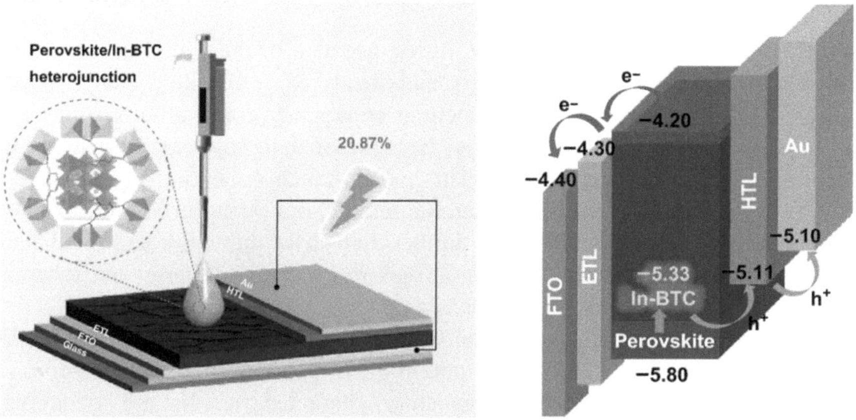

Figure 9.4: Schematic and energy level diagram of the device (Reprinted with permission from Zhou et al. 2020).

doped in HTM layer (Dong et al. 2019) and also confining perovskite QDs in MOF enhances stability and efficiency (Zhang et al. 2018, Zhang et al. 2019).

Photocatalytic application

H_2 production and CO_2 reduction processes

The common factor of photocatalytic processes for poor efficiency is insufficient light-harvesting, larger charge carrier transport, low porosity and poor stability. In the Hydrogen Evolution Reaction (HER), the photosensitizer is used to produce the electron, and then a catalyst reduces the proton source into H_2 using that electron. A photosensitizer was developed by using Ir/Ru metal-ions and porphyrin units as the ligands in the MOFs (Zhang et al. 2015, Kim et al. 2016). These approaches promote light-harvesting and facilitate charge separation. Noble-metal NPs based MOF enhances the efficiency of hydrogen generation (Fateeva et al. 2012, Wang 2012). A UiO-67 derivative nano-MOF shows a self-repairing property to tackle the stability of the molecular catalyst (Kim et al. 2016). The core-shell nanostructures with iron oxide (a-Fe_2O_3) core and a TiO_2 (anatase phase) shell drastically increase the amount of H_2 generated (deKrafft et al. 2012). Nano-flakes of NiCo-MOF were also grown on Ni mesh and Co@MOF composite (Nasalevich et al. 2015, Khalid et al. 2018). Incorporation of polyoxymethalaye (POM) cluster shows a significantly improved catalytic HER performance (Zhang et al. 2015).

Using MOF based catalysts for the reduction of CO_2 into fuel (CH_4 or CO) by using solar energy is a promising application. A porphyrin-based MOF catalyst, PCN-222 (Xu et al. 2015), and zirconium MOF based on $\{[RuII(H_2bpydc)(terpy)(CO)](PF_6)_2\}_n$ show excellent activity in CO_2 reduction under visible-light irradiation (Kajiwara et al. 2016). The integration of semiconductors and MOFs is another promising solution to enhance the efficiency of exciton generation and charge separation (Li et al. 2014).

Super-capacitors

Super-Capacitors (SCs) are largely preferred because of their high-power density, rapid charge and discharge, high safety and steady cycle life. SCs can be used to assemble the batteries or fuel cells, opening emergency doors in airbus and brake assistance, etc. (Wang et al. 2020). But because of their low energy density, it is still a step away from practical use. This performance issue can be addressed by (1) modifying the porous electrode material and (2) optimization of the electrolyte. SCs possess higher surface areas and thinner dielectrics that decreases the distance between the electrodes and hence increases the capacitance and energy. The schematic diagram shown in Fig. 9.5 will give a basic understanding of the device.

Nano MOFs are of great potential in enhancing the capacitive capabilities of SCs. Nanoscale MOFs as a metal compound of the precursor involves the pore size gradation and surface hydrophilicity structural characteristics that are conducive to facilitate fast ionic transport and diffusion and therefore show higher electrochemical activity (Chen et al. 2019, Yang et al. 2019).

In a recent report, a 3D nano-electrode material Co/Zn–S@rGO was prepared by combining Zn0.76Co0.24S nanoparticles with rGO film. Co/Zn–S@rGO-7 samples showed high capacitance of 1640 F g^{-1} at the current density of 1 A g^{-1} in 6 M KOH alkaline aqueous electrolyte, and the high energy density of 91.8 W h kg^{-1} and the power density of 800 W h kg^{-1}. The Co/Zn–S polyhedron dispersed in the rGO interlamellar to form porous structures, which not only can provide abundant paths for electrolyte-ion diffusion, but also prevented the agglomeration of rGO and offered more reaction sites (Xin et al. 2020). Qu and co-workers also synthesized hollow Ni-Co-Se nano-polyhedron by a selenide reaction on ZIF-67 as an electrode material. It exhibits high specific capacitance of 1668 F g^{-1} at 1 A g^{-1}, high energy density 38.5 W h kg^{-1}, power density 32.0 W kg^{-1} and excellent cyclic stability (82.3% retention after 5000 cycles) (Qu et al. 2020).

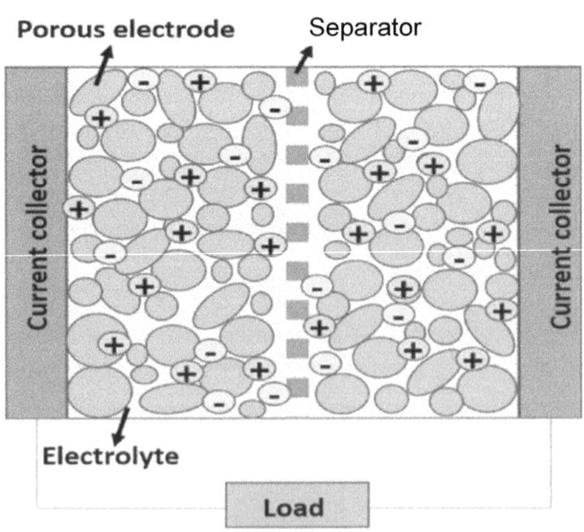

Figure 9.5: Schematic diagram of super-capacitor.

Ni-Co@Co-MOF/NF nanomaterial electrode improves the conductivity of electrons and accelerates the transfer of ions results in specific capacitance of 2697 F g^{-1} at 1 A g^{-1}, the power density is 853 W kg^{-1} and the energy density is 61.4 W h kg^{-1} (Jiang et al. 2020).

The ZIF-67 (Wang et al. 2015) nanocrystals, NENU-5 nanocrystal with pyrrole, i.e., copper MOF HKUST-1 with PM0.2 immobilized in its pores (Wang et al. 2018) and MOF/polymer hybrid material on carbon fibres using UiO-66 nanocrystals (Qi et al. 2018) coated electrodes showed very high areal capacitance because of large electron transfer capabilities, active sites and high conductivity of nanosize MOF electrolyte accessibility.

The high pseudocapacitance of SCs can most importantly be achieved by the nanosheets array arranged at a nano size distance. This nanosheet structure of MOFs provides high contact area to electrolyte and better accessibility (Lee et al. 2013, Yang et al. 2017).

Batteries

Rechargeable batteries are potential candidates for the growing market of portable electronics, electric vehicles and storage. Modulation of the irregular output power produced by renewable energies with limitations of low energy density and high cost triggered the extensive exploration of the new material for next-generation energy storage devices.

Nano MOF in metal-ion battery

The redox-active metal nodes, which are present in MOFs and lithium-stabilizing ligand moieties can boost the amount of stored Li-ions and can increase the theoretical capacity of metal-ion batteries (Wei et al. 2017). Nano MOFs are attractive because they provide close contact with the electrolyte for faster ion transfer (Wu and Guo 2019). Recently for lithium-ion batteries a novel core-shell structure with MoO_2 nanobelts as the core and a MOF-derived carbon layer as the shell, exhibited the specific capacity of 1049 mA h g^{-1} at 0.1 A g^{-1} for the 50th cycle, rate capability of 710 mA h g^{-1} at 5 A g^{-1} and cycling stability of 787.7 mA h g^{-1} at 1 A g^{-1} after 300 cycles (Zhang et al. 2019). A hierarchical mesoporous flower-like Ni-BTC MOF (Ni-BTCEtOH) assembled from 2D nanosheets also shows the specific capacity of 1085 mA h g^{-1} at 100 mA g^{-1} and cycling stability of 1000 mA g^{-1} for 1000 cycles with and greater rate performance, which makes it superior to other reported MOF-based anode materials (Gan et al. 2018).

For a Na-ion battery, metal selenide nanotubes (Co/(NiCo)Se_2), by ion-exchange and subsequent salinization of ZIF-67 was used as an anode which favors high electrical conductivity (Park et al. 2017). Using Nano MOF as an additive in the electrolyte with enhanced charge transportation and reduced impurity by absorption was reported (Zhu et al. 2014). The cycling life and rate performance of the cell can also be improved by balancing ion transfer through a separator. NH_2-functionalized titanium-based MIL-125 (Liu et al. 2017), UiO-66-NH_2 nanocrystals on glass fibers (Shen et al. 2019) are examples of nano MOF based separator contributing to the improved areal capacity.

Nano-MOF in Li-S battery

The Li metal batteries (Li-S and Li-Se) are of greater importance due to their high theoretical energy density ~ 2500 Wh kg^{-1}, low cost, nontoxicity and ease in availability. However their practical use suffers by the insulating behavior of sulfur and intermediate lithium polysulfides (Li_2S_2 and Li_2S). These discharged materials are sluggish in nature and soluble in an electrolyte, which causes the shuttle effect. This effect limits the charging efficiency and fast capacity. MOFs provide a platform for the trapping of sulfur as a porous nanostructured host (Baumann et al. 2019) reducing the polysulfide movement and accommodating volume expansion. The entrapment of the polysulfide in MOFs can be enhanced by introducing the coordinated metal site and Lewis acid sites at the linker (Zhou et al. 2015, Wang et al. 2015, Hong et al. 2018). It is found that the smaller particle size of MOF leads to better and enlarged contact with an electrolyte which reduces the shuttle effect (Zhou et al. 2015) and cage-like pore structure of MOF are more efficient for polysulfide immobilization (Jiang et al. 2018).

Another aspect of MOF in LiS battery is as a polysulfide sieve at the separator. It selectively sieves Li^+ to weaken the shuttle effect. This is achieved by pore size selection of nano MOF. In a study $Cu_3(BTC)_2$ (HKUST-1) was used to construct MOF@GO separator with micropores of 9 Å which is significantly smaller than the diameters of lithium polysulfide, thus well suited for blocking polysulfides (Bai et al. 2016).

Nano-MOF in Li-O_2 battery

In this category, the oxygen is reduced (ORR) at the porous cathode to produce electricity with a high theoretical specific capacity of 3842 mA h g^{-1} as shown in Fig. 9.6. Using the ambient air as a source of O_2, causes the contamination in the electrolyte and redox reactions at the cathode. There is poor reversibility and solubility of reduction products, i.e., Li_2O_2 and Li_2O. These factors dramatically limit the efficiency and cyclability of the Li-O_2 battery.

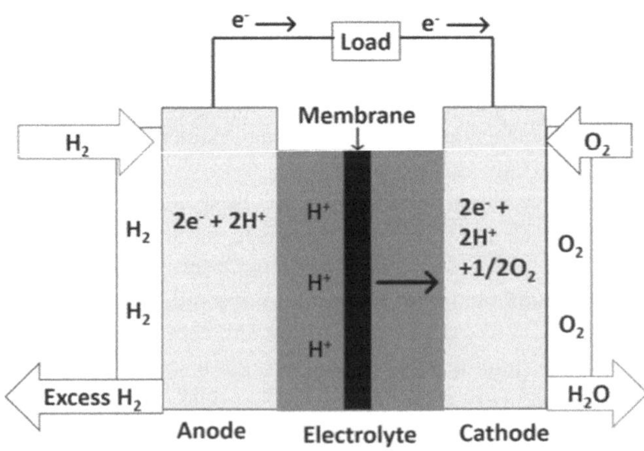

Figure 9.6: A scheme for a Li-O_2 battery.

Using nano MOF as a catalyst to ease the ORR at the cathode can be an effective approach to solve the above issues. 2D nano MOF sheets create a more active site, which enhances the catalytic effect. The Dy-BTC nanospheres based electrode demonstrated a significantly higher initial discharge specific capacity (7618 mA h g^{-1}) (Liu et al. 2020). 2D Mn-MOF, Co-MOF-74 also improve the cell capacity by enhancing the catalytic performance (Yan et al. 2017, Yuan et al. 2019).

Hydrogen and other fuel cells

Fuel cells offer a carbon-free method of converting chemical energy directly into electrical energy. Fuel cells can be categorized as Direct Methanol Fuel Cells (DMFCs), Polymer Electrolyte Membrane Fuel Cells (PEMFCs), alkaline fuel cells, phosphoric acid fuel cells, solid oxide fuel cells and molten carbonate fuel cells (Steele and Heinzel 2010). Among these PEMFC and DMFC merits with low working temperature, high power density and quick start-up (Song et al. 2016).

Polymer electrolyte membrane fuel cells

It is the most promising type of fuel cell for both research as well as commercial purposes. Figure 9.7 shows the working mechanism of the PEMFC. The fuel cell is a combination of an anode where H_2 is oxidized, an electron blocking and proton exchange membrane/electrolyte, and a cathode where oxygen captures electron through outer circuit and proton through the membrane to reduce it to water. The traditional material used for proton exchange is Nafion, which leads to reduced efficiency because of its low temperature and humid atmosphere limitations. There are two types of proton conduction of MOFs (a) below 100°C and under aqueous electrolyte conditions and (b) above 100°C and under non-aqueous electrolyte conditions.

At low-temperature conditions, there is a fine balance of water and heat. Excessive loss of water due to high temperature can lead to electrolyte loss, and overload of water can lead to catalyst poisoning. A novel MOF-anchored-nanofiber framework (UiO-66-NH_2@NFs) was designed by Wang et al. The homogeneous UiO-66-NH_2@NFs/Nafion membrane with a moderate UiO-66-NH_2 filler amount of 8 wt% presents excellent performance on elevating proton conductivity of 0.27 S cm^{-1} (80°C, 100% RH). The interconnected UiO-66-NH_2@NFs framework

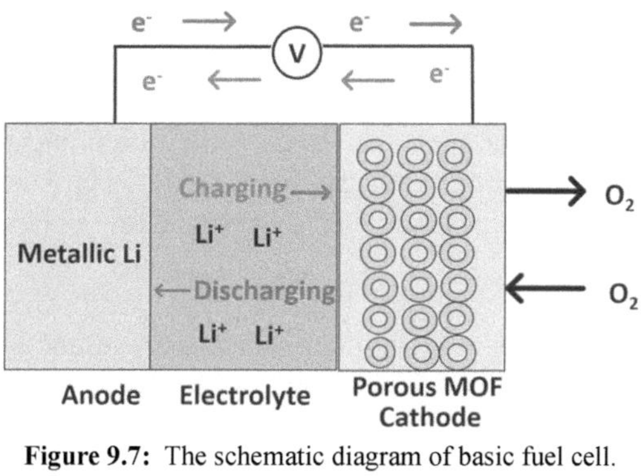

Figure 9.7: The schematic diagram of basic fuel cell.

with hydrophilic SPES nanofibers offers consecutive proton transfer channels, making protons easily transfer by the hydration channels (Wang et al. 2020). In another work, Im@MOF-801 and Im-MOF-801 exhibited excellent stabilities in both water and acid. The proton conductivities of these hybrid membranes were much higher (0.023–0.066 S cm^{-1} at 30°C, 0.050–0.128 S cm^{-1} at 90°C) than that of pure C-SPAEKS (0.010S cm^{-1} at 30°C, 0.024 S cm^{-1} at 90°C). Moreover, C-SPAEKS/Im-MOF-801-4% could maintain a stable proton conductivity for more than 100 hours (Zhang et al. 2020).

At high temperature, a proton becomes less efficient, and only the hopping mechanism of conduction can be used. Yang et al. showed in their work that ZIF-8@GO/Nafion hybrid membrane provides good water retention capability. The unique monolayer structure of ZIF-8@GO also facilitates a rapid transfer of water or proton and promotes it significantly. Even at 120°C and 40% RH, ZIF-8@GO/Nafion-1 displays a proton conductivity as high as 0.28 ± 0.058 S cm^{-1}, almost 55 times higher than that of recast Nafion (0.005 S cm^{-1}). Meanwhile, the membrane's methanol permeation and activation energy are decreased (Yang et al. 2015). Nano-MOFs inserted into the polymeric matrix (Nafion and SPEEK) impart a strong influence in the ionomeric chains by favoring the formation of H^+ pathways along with the hydrophilic channels give a conductivity as high as 0.306 S cm^{-1} with MIL-101-SO3H (7.5 wt%) in SPEEK (Li et al. 2014, Escorihuela et al. 2018).

At the cathode, the costly platinum-based catalyst for Oxygen Reduction Reaction (ORR) limits its commercial application. ORR at the cathode can occur in two ways. (a) in acidic medium reduced by $4e^-$ to form water using the proton that was transferred from the anode. This reaction is sluggish and causes instability. (b) in a basic medium reduced by $2e^-$ to form the hydroxide, which is then transferred to the anode. In this reaction formation of hydrogen peroxide limits its efficiency. Chen et al. reported Fe-Co-N-C electrocatalysts fabricated from a cationic Co (II)-based MOF precursor that dopes the Fe element by anion exchange. Favorable microstructure (e.g., high surface area, dominant mesopores) and multiple active sites are responsible for improved ORR performance (Chen et al. 2019).

Direct alcohol fuel cells (DAFC)

The challenges of the conventional fuel cell in terms of purity, storage and distribution of hydrogen shifts the attention to DAFC. Methanol is a more efficient fuel as it oxidizes easily compared to other alcohols, and ethanol is preferred because it can be produced in large quantities from biomass. The most challenging issue in DAFC is catalytic oxidation without compromising structure stability and efficient electron transfer. In a recent work Noor et al. studied the oxidation of methanol on NiO based MOFs and its composites with reduced graphene oxides. The 5 wt% rGO/NiO-MOF composite demonstrated exceptional results having a low potential value of 0.8 V and low impedance with the highest peak current density value of 275.85 mA cm^{-2} (Noor et al. 2019). In another report, Tripathy et al. synthesized a Co containing Metal-Organic Framework (Co-MOF) that serves as a bifunctional electrocatalyst for the ORR and OER in an alkaline medium. The Co-MOFs catalyze the ORR with lower onset potential by a four-electron reduction path (Tripathy et al. 2019).

Summary and future outlook

The properties of MOFs to maximize their potential as active catalyst materials in energy application devices can be controlled by synthetic design. Different geometries of MOFs are possible through metal-ligand coordination as well as different shapes of organic linkers. These can be linear, trigonal, tetragonal and in many more forms. Catalytic activity not only depends on the metal center of the networks, but end functional groups such as carboxylates, amines, imidazole of the linker molecules also provide catalytic activities. By careful selection of substituents and metal ions, for example, by changing the number and type of end groups of the bridging ligands, it is possible to develop large macromolecular structures by non-covalent interactions which offer vast opportunities for the development of energy-efficient devices and catalysts.

References

Bai, S., X. Liu, K. Zhu, S. Wu and H. Zhou. 2016. Metal–organic framework-based separator for lithium–sulfur batteries. Nature Energy 1: 16094.

Baumann, A. E., D. A. Burns, B. Liu and V. S. Thoi. 2019. Metal-organic framework functionalization and design strategies for advanced electrochemical energy storage devices. Communications Chemistry 2: 86.

Bilbeisi, R. A., J. K. Clegg, N. Elgrishi, X. de Hatten, M. Devillard, B. Breiner et al. 2012. Subcomponent self-assembly and guest-binding properties of face-capped $Fe_4L_4^{8+}$ capsules. Journal of the American Chemical Society 134: 5110–5119.

Cadiau, A., N. Kolobov, S. Srinivasan, M. Goesten, H. Haspel, A. Bavykina et al. 2020. A new titanium metal organic framework with visible-light responsive photocatalytic activity. Angewandte Chemie International Edition 59: 13468–1347.

Cai, G., W. Zhang, L. Jiao, S. -H. Yu and H. -L. Jiang. 2017. Template-directed growth of well-aligned MOF arrays and derived self-supporting electrodes for water splitting. Chem. 2: 791–802.

Caskey, D. C., T. Yamamoto, C. Addicott, R. K. Shoemaker, J. Vacek, A. M. Hawkridge et al. 2008. Coordination-driven face-directed self-assembly of trigonal prisms. Face-based conformational chirality. Journal of the American Chemical Society 130: 7620–7628.

Chakrabarty, R., P. S. Mukherjee and P. J. Stang. 2011. Supramolecular coordination: self-assembly of finite two- and three-dimensional ensembles. Chemical Reviews 111(2009): 6810–6918.

Chen, B., C. Liang, J. Yang, D. S. Contreras, Y. L. Clancy, E. B. Lobkovsky et al. 2006. A microporous metal–organic framework for gas-chromatographic separation of alkanes. Angewandte Chemie International Edition 45: 1390–1393.

Chen, C., J. -J. Zhou, Y. -L. Li, Q. Li, H. -M. Chen, K. Tao et al. 2019. $NiCo_2S_4$@Ni_3S_2 hybrid nanoarray on Ni foam for high-performance supercapacitors. New Journal of Chemistry 43: 7344–7349.

Chen, X., J. Huang, Y. Huang, J. Du, Y. Jiang, Y. Zhao et al. 2019. Efficient Fe-Co-N-C electrocatalyst towards oxygen reduction derived from a cationic coII-based metal–organic framework modified by anion-exchange with potassium ferricyanide. Chemistry—An Asian Journal 14: 995–1003.

Chen, Y. -Z., R. Zhang, L. Jiao and H. -L. Jiang. 2018. Metal–organic framework-derived porous materials for catalysis. Coordination Chemistry Reviews 362: 1–23.

Choi, K. M., K. Na, G. A. Somorjai and O. M. Yaghi. 2015. Chemical environment control and enhanced catalytic performance of platinum nanoparticles embedded in nanocrystalline metal–organic frameworks. Journal of the American Chemical Society 137: 7810–7816.

Choi, S. B., M. J. Seo, M. Cho, Y. Kim, M. K. Jin, D. -Y. Jung et al. 2007. A porous and interpenetrated metal–organic framework comprising tetranuclear ironIII–oxo clusters and tripodal organic carboxylates and its implications for (3,8)-coordinated networks. Crystal Growth & Design 7: 2290–2293.

Cook, T. R., Y. -R. Zheng and P. J. Stang. 2013. Metal–organic frameworks and self-assembled supramolecular coordination complexes: comparing and contrasting the design, synthesis, and functionality of metal–organic materials. Chemical Reviews 113: 734–777.

Cui, D., D. F. Perepichka, J. M. MacLeod and F. Rosei. 2020. Surface-confined single-layer covalent organic frameworks: design, synthesis and application. Chemical Society Reviews 49: 2020–2038.

deKrafft, K. E., C. Wang and W. Lin. 2012. Metal-organic framework templated synthesis of Fe_2O_3/TiO_2 nanocomposite for hydrogen production. Advanced Materials 24: 2014–2018.

Dong, Y., J. Zhang, Y. Yang, L. Qiu, D. Xia, K. Lin et al. 2019. Self-assembly of hybrid oxidant POM@ Cu-BTC for enhanced efficiency and long-term stability of perovskite solar cells. Angewandte Chemie International Edition 58: 17610–17615.

Doonan, C. J., W. Morris, H. Furukawa and O. M. Yaghi. 2009. Isoreticular metalation of metal–organic frameworks. Journal of the American Chemical Society 131: 9492–9493.

Eddaoudi, M., D. B. Moler, H. Li, B. Chen, T. M. Reineke, M. O'Keeffe et al. 2001. Modular chemistry: secondary building units as a basis for the design of highly porous and robust metal–organic carboxylate frameworks. Accounts of Chemical Research 34: 319–330.

Escorihuela, J., R. Narducci, V. Compañ and F. Costantino. 2018. Proton conductivity of composite polyelectrolyte membranes with metal-organic frameworks for fuel cell applications. Advanced Materials Interfaces 6: 1801146.

Farha, O. K., A. Özgür Yazaydın, I. Eryazici, C. D. Malliakas, B. G. Hauser, M. G. Kanatzidis et al. 2010. *De novo* synthesis of a metal–organic framework material featuring ultrahigh surface area and gas storage capacities. Nature Chemistry 2: 944–948.

Fateeva, A., P. A. Chater, C. P. Ireland, A. A. Tahir, Y. Z. Khimyak, P. V. Wiper et al. 2012. A water-stable porphyrin-based metal-organic framework active for visible-light photocatalysis. Angewandte Chemie International Edition 51: 7440–7444.

Gan, Q., H. He, K. Zhao, Z. He and S. Liu. 2018. Morphology-dependent electrochemical performance of Ni-1,3,5-benzenetricarboxylate metal-organic frameworks as an anode material for Li-ion batteries. Journal of Colloid and Interface Science 530: 127–136.

Gao, Y., D. Rajwar and A. C. Grimsdale. 2014. Self-assembly of conjugated units using metal-terpyridine coordination. Macromolecular Rapid Communications 35: 1727–1740.

Hasselman, G. M., D. F. Watson, J. R. Stromberg, D. F. Bocian, D. Holten, J. S. Lindsey et al. 2006. Theoretical solar-to-electrical energy-conversion efficiencies of perylene–porphyrin light-harvesting arrays. The Journal of Physical Chemistry B 110: 25430–25440.

He, T., R. Chen, Z. B. Lim, D. Rajwar, L. Ma, Y. Wang et al. 2013. Efficient energy transfer under two-photon excitation in a 3D, supramolecular, Zn(II)-coordinated, self-assembled organic network. Advanced Optical Materials 2: 40–47.

Hong, X. -J., T. -X. Tan, Y. -K. Guo, X. -Y. Tang, J. -Y. Wang, W. Qin et al. 2018. Confinement of polysulfides within bi-functional metal–organic frameworks for high performance lithium–sulfur batteries. Nanoscale 10: 2774–2780.

Hosoyamada, M., N. Yanai, K. Okumura, T. Uchihashi and N. Kimizuka. 2018. Translating MOF chemistry into supramolecular chemistry: soluble coordination nanofibers showing efficient photon upconversion. Chemical Communications 54: 6828–6831.

Hsu, S. -H., C. -T. Li, H. -T. Chien, R. R. Salunkhe, N. Suzuki, Y. Yamauchi et al. 2015. Platinum-free counter electrode comprised of metal-organic-framework (MOF)-derived cobalt sulfide nanoparticles for efficient dye-sensitized solar cells (DSSCs). Scientific Reports 4: 6983.

Huang, X., T. Shen, T. Zhang, H. Qiu, X. Gu, Z. Ali et al. 2020. Efficient oxygen reduction catalysts of porous carbon nanostructures decorated with transition metal species. Advanced Energy Materials 10: 1900375.

Jiang, H., X. -C. Liu, Y. Wu, Y. Shu, X. Gong, F. -S. Ke et al. 2018. Metal-organic frameworks for high charge-discharge rates in lithium-sulfur batteries. Angewandte Chemie International Edition 57: 3916–3921.

Jiang, J., Y. Sun, Y. Chen, Q. Zhou, H. Rong, X. Hu et al. 2020. Design and fabrication of metal-organic frameworks nanosheet arrays constructed by interconnected nanohoneycomb-like nickel-cobalt oxide for high energy density asymmetric supercapacitors. Electrochimica Acta 342: 136077.

Jin, S., H. -J. Son, O. K. Farha, G. P. Wiederrecht and J. T. Hupp. 2013. Energy transfer from quantum dots to metal–organic frameworks for enhanced light harvesting. Journal of the American Chemical Society 135: 955–958.

Kajiwara, T., M. Fujii, M. Tsujimoto, K. Kobayashi, M. Higuchi, K. Tanaka et al. 2016. Photochemical reduction of low concentrations of CO_2 in a porous coordination polymer with a ruthenium(II)-CO complex. Angewandte Chemie International Edition 55: 2697–2700.

Kalmutzki, M. J., N. Hanikel and O. M. Yaghi. 2018. Secondary building units as the turning point in the development of the reticular chemistry of MOFs. Science Advances 4: eaat9180.

Katayama, Y., M. Kalaj, K. S. Barcus and S. M. Cohen. 2019. Self-assembly of metal–organic framework (MOF) nanoparticle monolayers and free-standing multilayers. Journal of the American Chemical Society 141: 20000–20003.

Kaur, R., K. -H. Kim, A. K. Paul and A. Deep. 2016. Recent advances in the photovoltaic applications of coordination polymers and metal organic frameworks. Journal of Materials Chemistry A 4: 3991–4002.

Khalid, Mohd., A. Hassan, A. M. B. Honorato, F. N. Crespilho and H. Varela. 2018. Nano-flocks of a bimetallic organic framework for efficient hydrogen evolution electrocatalysis. Chemical Communications 54: 11048–11051.

Kim, D., D. R. Whang and S. Y. Park. 2016. Self-healing of molecular catalyst and photosensitizer on metal–organic framework: robust molecular system for photocatalytic H_2 evolution from water. Journal of the American Chemical Society 138: 8698–8701.

Kuyuldar, S., D. T. Genna and C. Burda. 2019. On the potential for nanoscale metal–organic frameworks for energy applications. Journal of Materials Chemistry A 7: 21545–21576.

Lee, Deok Yoon, D. V. Shinde, E. -K. Kim, W. Lee, I. -W. Oh, N. K. Shrestha et al. 2013. Supercapacitive property of metal–organic-frameworks with different pore dimensions and morphology. Microporous and Mesoporous Materials 171: 53–57.

Lee, Deok Yeon, D. V. Shinde, S. J. Yoon, K. N. Cho, W. Lee, N. K. Shrestha et al. 2014. Cu-based metal–organic frameworks for photovoltaic application. The Journal of Physical Chemistry C 118: 16328–16334.

Li, H., M. Eddaoudi, M. O'Keeffe and O. M. Yaghi. 1999. Design and synthesis of an exceptionally stable and highly porous metal-organic framework. Nature 402: 276–279.

Li, R., J. Hu, M. Deng, H. Wang, X. Wang, Y. Hu et al. 2014. Integration of an inorganic semiconductor with a metal-organic framework: a platform for enhanced gaseous photocatalytic reactions. Advanced Materials 26: 4783–4788.

Li, Y., Z. Di, J. Gao, P. Cheng, C. Di, G. Zhang et al. 2017. Heterodimers made of upconversion nanoparticles and metal–organic frameworks. Journal of the American Chemical Society 139: 13804–13810.

Li, Z., G. He, Y. Zhao, Y. Cao, H. Wu, Y. Li et al. 2014. Enhanced proton conductivity of proton exchange membranes by incorporating sulfonated metal-organic frameworks. Journal of Power Sources 262: 372–379.

Lim, Z. B., H. Li, S. Sun, J. Y. Lek, A. Trewin, Y. M. Lam et al. 2012. New 3D supramolecular Zn(II)-coordinated self-assembled organic networks. Journal of Materials Chemistry 22: 6218.

Linares, N., A. M. Silvestre-Albero, E. Serrano, J. Silvestre-Albero and J. García-Martínez. 2014. Mesoporous materials for clean energy technologies. Chemical Society Reviews 43: 7681–7717.

Liu, D., X. Zhang, Y. -J. Wang, S. Song, L. Cui, H. Fan et al. 2020. A new perspective of lanthanide metal–organic frameworks: tailoring Dy-BTC nanospheres for rechargeable Li–O_2 batteries. Nanoscale 12: 9524–9532.

Liu, J., W. Zhou, J. Liu, I. Howard, G. Kilibarda, S. Schlabach et al. 2015. Photoinduced charge-carrier generation in epitaxial MOF thin films: high efficiency as a result of an indirect electronic band gap. Angewandte Chemie International Edition 54: 7441–7445.

Liu, W., Y. Mi, Z. Weng, Y. Zhong, Z. Wu and H. Wang. 2017. Functional metal–organic framework boosting lithium metal anode performance via chemical interactions. Chemical Science 8: 4285–4291.

Liu, Y. -L., K. -F. Yue, B. -H. Shan, L. -L. Xu, C. -J. Wang and Y. -Y. Wang. 2012. A new 3-fold interpenetrated metal-organic framework (MOF) based on trinuclear zinc(II) clusters as secondary building unit (SBU). Inorganic Chemistry Communications 17: 30–33.

Lu, G., S. Li, Z. Guo, O. K. Farha, B. G. Hauser, X. Qi et al. 2012. Imparting functionality to a metal–organic framework material by controlled nanoparticle encapsulation. Nature Chemistry 4: 310–316.

Lu, Q., M. Zhao, J. Chen, B. Chen, C. Tan, X. Zhang et al. 2016. *In situ* synthesis of metal sulfide nanoparticles based on 2D metal-organic framework nanosheets. Small 12: 4669–4674.

Ma, X. and T. F. Scott. 2018. Approaches and challenges in the synthesis of three-dimensional covalent-organic frameworks. Communications Chemistry 1: 98.

Maza, W. A., A. J. Haring, S. R. Ahrenholtz, C. C. Epley, S. Y. Lin and A. J. Morris. 2016. Ruthenium(II)-polypyridyl zirconium(IV) metal–organic frameworks as a new class of sensitized solar cells. Chemical Science 7: 719–727.

Medishetty, R., J. K. Zaręba, D. Mayer, M. Samoć and R. A. Fischer. 2017. Nonlinear optical properties, upconversion and lasing in metal–organic frameworks. Chemical Society Reviews 46: 4976–5004.

Nasalevich, M. A., R. Becker, E. V. Ramos-Fernandez, S. Castellanos, S. L. Veber, M. V. Fedin et al. 2015. Co@NH_2-MIL-125(Ti): cobaloxime-derived metal–organic framework-based composite for light-driven H_2 production. Energy & Environmental Science 8: 364–375.

Nguyen, V. H., N. P. T. Nguyen, T. T. N. Nguyen, T. T. T. Le, V. N. Le, Q. C. Nguyen et al. 2011. Synthesis and characterization of zinc-organic frameworks with 1,4-benzenedicarboxylic acid and azobenzene-4,4′-dicarboxylic acid. Advances in Natural Sciences: Nanoscience and Nanotechnology 2: 025008.

Noor, T., N. Zaman, H. Nasir, N. Iqbal and Z. Hussain. 2019. Electro catalytic study of NiO-MOF/RGO composites for methanol oxidation reaction. Electrochimica Acta 307: 1–12.

Park, S. -K., J. K. Kim and Y. Chan Kang. 2017. Metal–organic framework-derived $CoSe_2$/(NiCo)Se_2 box-in-box hollow nanocubes with enhanced electrochemical properties for sodium-ion storage and hydrogen evolution. Journal of Materials Chemistry A 5: 18823–18830.

Qi, K., R. Hou, S. Zaman, Y. Qiu, B. Y. Xia and H. Duan. 2018. Construction of metal–organic framework/conductive polymer hybrid for all-solid-state fabric supercapacitor. ACS Applied Materials & Interfaces 10: 18021–18028.

Qu, G., X. Zhang, G. Xiang, Y. Wei, J. Yin, Z. Wang et al. 2020. ZIF-67 derived hollow Ni-Co-Se nano-polyhedrons for flexible hybrid supercapacitors with remarkable electrochemical performances. Chinese Chemical Letters 31: 2007–2012.

Rajwar, D., X. Liu, Z. B. Lim, S. J. Cho, S. Chen, J. M. H. Thomas et al. 2014. Novel self-assembled 2D networks based on zinc metal ion co-ordination: synthesis and comparative study with 3D networks. RSC Advances 4: 17680–17693.

Sasitharan, K., D. G. Bossanyi, N. Vaenas, A. J. Parnell, J. Clark, A. Iraqi et al. 2020. Metal–organic framework nanosheets for enhanced performance of organic photovoltaic cells. Journal of Materials Chemistry A 8: 6067–6075.

Schubert, U. S., C. Eschbaumer, O. Hien and P. R. Andres. 2001. 4′-Functionalized 2,2′:6′,2″-Terpyridines as building blocks for supramolecular chemistry and nanoscience. Tetrahedron Letters 42: 4705–4707.

Shen, L., H. B. Wu, F. Liu, C. Zhang, S. Ma, Z. Le et al. 2019. Anchoring anions with metal–organic framework-functionalized separators for advanced lithium batteries. Nanoscale Horizons 4: 705–711.

Shi, D., R. Zheng, C. -S. Liu, D. -M. Chen, J. Zhao and M. Du. 2019. Dual-functionalized mixed keggin- and lindqvist-type Cu_{24}-based POM@MOF for visible-light-driven H_2 and O_2 evolution. Inorganic Chemistry 58: 7229–7235.

Shi, Q., S. Fu, C. Zhu, J. Song, D. Du and Y. Lin. 2019. Metal–organic frameworks-based catalysts for electrochemical oxygen evolution. Materials Horizons 6: 684–702.

So, M. C., G. P. Wiederrecht, J. E. Mondloch, J. T. Hupp and O. K. Farha. 2015. Metal–organic framework materials for light-harvesting and energy transfer. Chemical Communications 51: 3501–3510.

Song, Z., N. Cheng, A. Lushington and X. Sun. 2016. Recent progress on MOF-derived nanomaterials as advanced electrocatalysts in fuel cells. Catalysts 6: 116.

Steele, B. C. H. and A. Heinzel. 2010. Materials for fuel-cell technologies. pp. 224–231. *In*: Materials for Sustainable Energy, by Vincent Dusastre. Co-Published with Macmillan Publishers Ltd, UK.

Sun, Y., C. Chen and P. J. Stang. 2019. Soft materials with diverse suprastructures via the self-assembly of metal–organic complexes. Accounts of Chemical Research 52: 802–817.

Tang, R., Z. Xie, S. Zhou, Y. Zhang, Z. Yuan, L. Zhang et al. 2016. Cu_2ZnSnS_4 nanoparticle sensitized metal–organic framework derived mesoporous TiO_2 as photoanodes for high-performance dye-sensitized solar cells. ACS Applied Materials & Interfaces 8: 22201–22212.

Tang, R., R. Yin, S. Zhou, T. Ge, Z. Yuan, L. Zhang et al. 2017. Layered MoS_2 coupled MOFs-derived dual-phase TiO_2 for enhanced photoelectrochemical performance. Journal of Materials Chemistry A 5: 4962–4971.

Tranchemontagne, D. J., J. L. Mendoza-Cortés, M. O'Keeffe and O. M. Yaghi. 2009. Secondary building units, nets and bonding in the chemistry of metal–organic frameworks. Chemical Society Reviews 38: 1257–1283.

Tripathy, R. K., A. K. Samantara and J. N. Behera. 2019. A cobalt metal–organic framework (Co-MOF): A Bi-functional electro active material for the oxygen evolution and reduction reaction. Dalton Transactions 48: 10557–10564.

Wang, C., K. E. deKrafft and W. Lin. 2012. Pt nanoparticles@photoactive metal–organic frameworks: efficient hydrogen evolution via synergistic photoexcitation and electron injection. Journal of the American Chemical Society 134: 7211–7214.

Wang, H., Q. -L. Zhu, R. Zou and Q. Xu. 2017. Metal-organic frameworks for energy applications. Chem 2: 52–80.

Wang, H. -N., M. Zhang, A. -M. Zhang, F. -C. Shen, X. -K. Wang, S. -N. Sun et al. 2018. Polyoxometalate-based metal–organic frameworks with conductive polypyrrole for supercapacitors. ACS Applied Materials & Interfaces 10: 32265–32270.

Wang, K. -B., Q. Xun and Q. Zhang. 2020. Recent progress in metal-organic frameworks as active materials for supercapacitors. EnergyChem 2: 100025.

Wang, Liyuan, N. Deng, Y. Liang, J. Ju, B. Cheng and W. Kang. 2020. Metal-organic framework anchored sulfonated poly(ether sulfone) nanofibers as highly conductive channels for hybrid proton exchange membranes. Journal of Power Sources 450: 227592.

Wang, Lu, X. Feng, L. Ren, Q. Piao, J. Zhong, Y. Wang et al. 2015. Flexible solid-state supercapacitor based on a metal–organic framework interwoven by electrochemically-deposited PANI. Journal of the American Chemical Society 137: 4920–4923.

Wang, Z., B. Wang, Y. Yang, Y. Cui, Z. Wang, B. Chen et al. 2015. Mixed-metal–organic framework with effective lewis acidic sites for sulfur confinement in high-performance lithium–sulfur batteries. ACS Applied Materials & Interfaces 7: 20999–21004.

Wee, L. H., C. Wiktor, S. Turner, W. Vanderlinden, N. Janssens, S. R. Bajpe et al. 2012. Copper benzene tricarboxylate metal–organic framework with wide permanent mesopores stabilized by keggin polyoxometallate ions. Journal of the American Chemical Society 134: 10911–10919.

Wei, T., M. Zhang, P. Wu, Y. -J. Tang, S. -L. Li, F. -C. Shen et al. 2017. POM-based metal-organic framework/reduced graphene oxide nanocomposites with hybrid behavior of battery-supercapacitor for superior lithium storage. Nano Energy 34: 205–214.

Wild, A., F. Schlütter, G. M. Pavlov, C. Friebe, G. Festag, A. Winter et al. 2010. π-conjugated donor and donor-acceptor metallo-polymers. Macromolecular Rapid Communications 31: 868–874.

Wild, A., A. Winter, F. Schlütter and U. S. Schubert. 2011. Advances in the field of π-conjugated 2,2′:6′,2″-terpyridines. Chemical Society Reviews 40: 1459.

Wu, J. and X. Guo. 2019. Nanostructured metal–organic framework (MOF)-derived solid electrolytes realizing fast lithium ion transportation kinetics in solid-state batteries. Small 15: 1804413.

Xin, N., Y. Liu, H. Niu, H. Bai and W. Shi. 2020. *In-situ* construction of metal organic frameworks derived Co/Zn–S sandwiched graphene film as free-standing electrodes for ultra-high energy density supercapacitors. Journal of Power Sources 451: 227772.

Xing, W., P. Ye, J. Lu, X. Wu, Y. Chen, T. Zhu et al. 2018. Tellurophene-based metal-organic framework nanosheets for high-performance organic solar cells. Journal of Power Sources 401: 13–19.

Xu, H. -Q., J. Hu, D. Wang, Z. Li, Q. Zhang, Y. Luo et al. 2015. Visible-light photoreduction of CO_2 in a metal–organic framework: boosting electron–hole separation via electron trap states. Journal of the American Chemical Society 137: 13440–13443.

Yan, W., Z. Guo, H. Xu, Y. Lou, J. Chen and Q. Li. 2017. Downsizing metal–organic frameworks with distinct morphologies as cathode materials for high-capacity Li–O_2 batteries. Materials Chemistry Frontiers 1: 1324–1330.

Yan, Y., I. Telepeni, S. Yang, X. Lin, W. Kockelmann, A. Dailly et al. 2010. Metal−organic polyhedral frameworks: high H_2 adsorption capacities and neutron powder diffraction studies. Journal of the American Chemical Society 132: 4092–4094.

Yan, Y., S. Yang, A. J. Blake, W. Lewis, E. Poirier, S. A. Barnett et al. 2011. A mesoporous metal–organic framework constructed from a nanosized C3-symmetric linker and [Cu24(Isophthalate)24] cuboctahedra. Chemical Communications 47: 9995–9997.

Yang, J., Z. Ma, W. Gao and M. Wei. 2017. Layered structural Co-based MOF with conductive network frames as a new supercapacitor electrode. Chemistry—A European Journal 23: 631–636.

Yang, L., B. Tang and P. Wu. 2015. Metal–organic framework–graphene oxide composites: a facile method to highly improve the proton conductivity of PEMs operated under low humidity. Journal of Materials Chemistry A 3: 15838–15842.

Yang, Q., Y. Liu, M. Yan, Y. Lei and W. Shi. 2019. MOF-derived hierarchical nanosheet arrays constructed by interconnected NiCo-alloy@NiCo-sulfide core-shell nanoparticles for high-performance asymmetric supercapacitors. Chemical Engineering Journal 370: 666–676.

Yuan, M., R. Wang, W. Fu, L. Lin, Z. Sun, X. Long et al. 2019. Ultrathin two-dimensional metal–organic framework nanosheets with the inherent open active sites as electrocatalysts in aprotic Li–O_2 batteries. ACS Applied Materials & Interfaces 11: 11403–11413.

Zhang, D., Y. Xu, Q. Liu and Z. Xia. 2018. Encapsulation of $CH_3NH_3PbBr_3$ perovskite quantum dots in MOF-5 microcrystals as a stable platform for temperature and aqueous heavy metal ion detection. Inorganic Chemistry 57: 4613–4619.

Zhang, H., J. Nai, L. Yu and X. W. (David) Lou. 2017. Metal-organic-framework-based materials as platforms for renewable energy and environmental applications. Joule 1: 77–107.

Zhang, L. -P., J. -F. Ma, J. Yang, Y. -Y. Liu and G. -H. Wei. 2009. 1D, 2D, and 3D Metal–organic frameworks based on bis(imidazole) ligands and polycarboxylates: syntheses, structures, and photoluminescent properties. Crystal Growth & Design 9: 4660–4673.

Zhang, Q., H. Wu, W. Lin, J. Wang and Y. Chi. 2019. Enhancing air-stability of $CH_3NH_3PbBr_3$ perovskite quantum dots by *in-situ* growth in metal-organic frameworks and their applications in light emitting diodes. Journal of Solid State Chemistry 272: 221–226.

Zhang, W., B. Wang, H. Luo, F. Jin, T. Ruan and D. Wang. 2019. MoO_2 nanobelts modified with an MOF-derived carbon layer for high performance lithium-ion battery anodes. Journal of Alloys and Compounds 803: 664–670.

Zhang, X., M. A. Ballem, Z. -J. Hu, P. Bergman and K. Uvdal. 2011. Nanoscale light-harvesting metal-organic frameworks. Angewandte Chemie International Edition 50: 5729–5733.

Zhang, Z., J. Ren, J. Xu, Z. Wang, W. He, S. Wang et al. 2020. Adjust the arrangement of imidazole on the metal-organic framework to obtain hybrid proton exchange membrane with long-term stable high proton conductivity. Journal of Membrane Science 607: 118194

Zhang, Z. -M., T. Zhang, C. Wang, Z. Lin, L. -S. Long and W. Lin. 2015. Photosensitizing metal–organic framework enabling visible-light-driven proton reduction by a wells–dawson-type polyoxometalate. Journal of the American Chemical Society 137: 3197–3200.

Zheng, Y. -R., H. -B. Yang, K. Ghosh, L. Zhao and P. J. Stang. 2009. Multicomponent supramolecular systems: self-organization in coordination-driven self-assembly. Chemistry—A European Journal 15: 7203–7214.

Zhou, J., X. Yu, X. Fan, X. Wang, H. Li and Y. Zhang. 2015. The impact of the particle size of a metal–organic framework for sulfur storage in Li–S batteries. Journal of Materials Chemistry A 3: 8272–8275.

Zhou, X., L. Qiu, R. Fan, J. Zhang, S. Hao and Y. Yang. 2020. Heterojunction incorporating perovskite and microporous metal–organic framework nanocrystals for efficient and stable solar cells. Nano-Micro Letters 12: 1–11

Zhu, B., R. Zou and Q. Xu. 2018. Metal-organic framework based catalysts for hydrogen evolution. Advanced Energy Materials 8: 1801193.

Zhu, K., Y. Liu and J. Liu. 2014. A fast charging/discharging all-solid-state lithium ion battery based on PEO-MIL-53(Al)-LiTFSI thin film electrolyte. RSC Adv. 4: 42278–42284.

Chapter 10

Piezoelectric Energy Harvesting and Piezocatalysis

Kaushik Parida[1,*] and *Ramaraju Bendi*[2,*]

Introduction

Since the 19th century, the world primarily depends on fossil fuels for its source of energy. However, over-consumption of these conventional forms of energy has led to its significant depletion and has played a major role in degrading the earth's atmosphere (Abbasi et al. 2011, Lu et al. 2015, Asumadu-Sarkodie and Owusu 2016). Renewable sources of energy are a viable alternative to overcome these challenges, however, at this current moment, it is not completely sufficient to sustain the energy crisis entirely by renewable sources. Thus, there has been a tremendous effort by researchers to develop novel power sources to harness energy from the environment as an alternative to non-conventional energy sources. Moreover, in the last few decades, there has been tremendous growth in the domain of portable smart electronic devices and wearable electronics. There is a tremendous need for an alternative to battery-based systems, with expanded durability and self-charging, self-maintenance capability (Bendi et al. 2016, Kumar et al. 2016, Parida et al. 2017, Kumar et al. 2017a and b, Park et al. 2017, Park et al. 2019). Due to the arrival of the 'Internet of Things', where the network of independent electronic units is interconnected, there is a huge requirement to develop self-powered systems. A self-powered system is a functionally integrated system comprising of an energy harvesting unit (to harness the ambient energy to useful electrical energy), an energy storage unit (to store the generated electrical energy) and an energy utilization unit (such as a sensor to consume the stored electrical energy). Figure 10.1 schematically depicts a self-powered system. A viable option is to tap energy from the ambient

[1] School of Materials Science and Engineering, Nanyang Technological University, 50 Nanyang Avenue, Singapore 639798.

[2] Department of Basic Science and Humanities (Chemistry), Aditya Institute of Technology and Management, Tekkali, Andhra Pradesh.

* Corresponding author: ramarajubendi@gmail.com; kparida@ntu.edu.sg

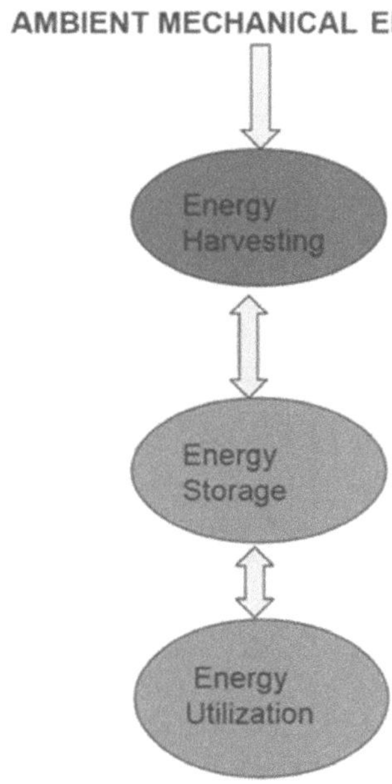

Figure 10.1: Schematic representation of the self-powered system.

environment to directly power these wearable sensors, wireless sensors, health-monitoring units (Parida et al. 2017, Xiong et al. 2019, Park et al. 2019, Gao et al. 2019, Lee et al. 2020). Various energy harvesting technologies such as mechanical, thermal, solar, water, wind and nuclear have evolved as an effective alternative for generating energy. Energy generation from a mechanical motion has evolved as a promising potential due to its ability to generate high power density, high output voltage and a better lifespan (Wang et al. 2013, Zhu et al. 2013, Yang et al. 2013, Wang 2014, Wang et al. 2014, Parida et al. 2017).

Among the various competing mechanical energy harvesting technologies, the piezoelectric nanogenerator has gained significant attention due to its superior performance and practical applicability. The concept of piezoelectric nanogenerators (NGs) was first demonstrated by (Wang and Song 2006). This pioneering demonstration opens a wide field for the exploration and development of various materials, functionalities and structural design for harnessing ambient mechanical energy. The NGs convert ambient mechanical vibrations into useful electricity energy. It is characterized by extremely high-power density; high efficiency and can operate in any place and at any time under all environmental conditions (Wang et al. 2015, Wu et al. 2019). Nanogenerators can generate energy from various random and arbitrary mechanical motions present in the ambient environment such as movements of the human body, vibrations of industrial equipment and vehicles, motions created by environmental sources such as wind, rainfall, ocean waves and water flows (Lin et al. 2016, Xiong et al. 2017, Chen et al. 2019). These devices can be effectively integrated on implanted and wearable devices required for sensing and remote monitoring (Dagdeviren et al. 2016).

The translation of ambient mechanical energy to useful electrical energy is associated with three different stages. The first stage is the translation of ambient mechanical to mechanical energy absorbed by the material. The mechanical properties of the material and the mechanical impedance match and govern the efficiency of this translation. The second stage is the conversion of absorbed mechanical energy to the generated electrical potential across the device. The efficiency of this stage depends on the piezoelectric coefficients (d_{33}), dielectric constant and mechanical modulus and electromechanical coupling (k_{33}). The third stage is the translation of generated electrical potential to the electrical energy transfer to the load, which depends on the electrical impedance matching and power management circuit used to the efficient transfer of charge (Uchino and Ishii 2010). Recently, the concept of the piezoelectric, pyroelectric and ferroelectric mechanism was explored in photocatalytic applications. The non-centrosymmetric crystal structure of the piezoelectric materials creates the built-in electric field owing to the change in the dipole moment when subjected to mechanical pressure, which provides a unique and excellent catalytic property (Kakekhani and Ismail-Beigi 2015, Wang et al. 2017). The internal potential difference created by these materials provides the built-in electric field to improve the separation of the charge carriers in the photocatalytic process (Damjanovic 1998). However, these concepts are still under investigation. One of the key applications in the utilization of the developed potential difference of piezoelectric materials (when subjected to mechanical activation) for the photocatalytic degradation of pollutants. The use of free energy present in the ambient environment for pollutant degradation is an exciting application.

This chapter highlights both the fundamental principles and the recent development in the domain of piezoelectric nanogenerator. Various materials (ceramics, polymers, nanocomposites), device structural designs, performances, strategies for the improvement of performance and the piezoelectric field and its influence on the catalytic and photocatalytic environmental remediation have been highlighted. Several recent publications as representative examples to indicate the various directions of research focus are explained. The key challenges encountered for further improving the energy harvesting performance and future perspectives for the development of next-generation mechanical energy harvesters are highlighted.

Fundamental theory and principles of nanogenerator

The piezoelectric effect was discovered by Pierre Curie and Jacques Curie. It is the property of the material based on which it generates a potential difference when mechanical pressure is exerted (Fig. 10.2). The direct piezoelectric effect is the conversion of the mechanical energy into electrical energy and the converse or inverse piezoelectric effect is the conversion of the electrical energy into mechanical energy. The energy harvester and the pressure-sensing devices are based on the principle of the direct piezoelectric effect and the actuator is based on the principle of converse piezoelectric effect.

All the ferroelectrics materials show piezoelectric behavior. Ferroelectrics materials are typically characterized by spontaneous polarization. Due to the multi-functionality of ferroelectric materials, it can be used for different applications such

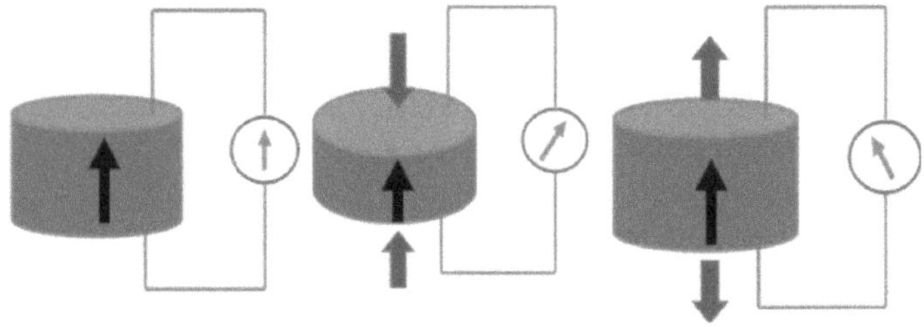

Figure 10.2: Schematic representation of the direct piezoelectric effect.

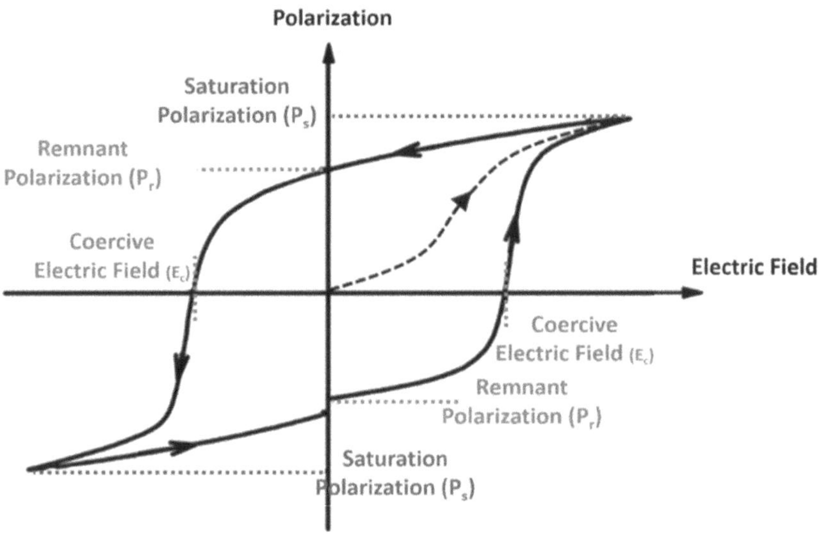

Figure 10.3: Typical P-E hysteresis loop of a ferroelectric material.

as memory devices, transistors, energy harvesters, and sensors. An external electric field can be used to reversibly switch the polarization states. The typical hysteresis loop (P-E loop) is the graph between the Electric field (E) and the Polarization (P). Figure 10.3 shows a typical P-E hysteresis loop of a ferroelectric material. Curie temperature (T_C), is the temperature below which the material exhibits ferroelectric behavior, above this temperature, the material acts like a normal dielectric. For ferroelectric materials such as lead zirconate titanate (PZT) and polyvinylidene fluoride (PVDF), the direction of polarization is governed by the direction of electrical poling. The minimum electric field required to pole the materials depends on the coercive field of the material. A higher coercive field requires a higher electric field to align the dipoles. For non-ferroelectric materials such as aluminum nitride (AlN) and zinc oxide (ZnO) the crystal orientation determines the direction of polarization.

The piezoelectric coefficient (d_{3i}) is defined as the ratio of the polarization (P) to the exerted applied mechanical pressure (σ). The unit of the piezoelectric coefficient is C/N. This quantifies the piezoelectric behavior of the material. Typically, the polarization direction is well-defined in piezoelectric materials and is referred to as '3' direction. The 'i' denotes the direction of the mechanical pressure. The direction of the mechanical stress relative to the direction of polarization affects the piezoelectric behavior. When the exerted mechanical pressure is perpendicular to the direction of polarization, the piezoelectric coefficient is referred to as 'd_{33}' and when the mechanical pressure is exerted parallel to the polarization the piezoelectric coefficient is referred to as 'd_{31}'.

Piezopotential is the potential generated due to the spatially separated electrical charges of opposite polarity across the end of the material. The piezoelectric effect can be attributed to the non-centrosymmetric crystal structure in various materials including polymers and ceramics. Due to the noncentrosymmetric crystal structure, these materials have an inherent dipole. When the mechanical stress (compression or tension) or vibrations are exerted, the molecular structure of the piezoelectric crystal gets disturbed thus causing a change in the dipole moment, and resulting in the development of the internal potential difference. This deformation of the internal crystal structure disrupts the internal charge balance. This results in the creation of charges of opposite polarity across the two ends of the piezoelectric material thus creating the surface charge density. When the metallic electrodes at the two ends of the material is connected, free charge carriers flow across the external connection to balance the internal piezoelectric potential. The number of electric dipoles in the piezoelectric materials governs the performance of the NGs.

To create the piezoelectric effect, the material is subjected to 'electrical poling'. It is the process of exerting an extremely high electric field (typically more than the coercive filed) at an elevated temperature (typically near the Curie temperature). Under the influence of the applied electric field the electric dipoles orient along the direction of the application of the field, causing a net polarization. Poling at elevated temperature allows the easy movement of molecules to orient in a single direction. On removing the exerted electric field, the electric dipoles remain permanently aligned thus exhibiting piezoelectric behavior. The piezoelectric property is affected by the crystal symmetry of the material. Single crystalline material with unidirectional orientation can generate piezoelectric potential without the application of electrical poling. However, polycrystalline materials, due to its different random orientation of dipoles, need to be aligned.

Materials for piezoelectricity nanogenerator

In terms of materials, there is a plethora of different piezoelectric materials including polymers, ceramics, nanocomposite and semiconducting materials (Brown et al. 1962, Groves 1986, Trolier-McKinstry et al. 2018, Liu et al. 2018). During the last 100 years of the discovery of the piezoelectric phenomenon, there have been more than 200 different piezoelectric materials reported (Priya et al. 2019). Apart from the piezoelectric performance, there are various factors such as mechanical modulus, flexibility, manufacturability, toxicity and cost are evaluated while considering the material for piezoelectric energy harvesters. Here various piezoelectric materials such as piezoelectric polymers, nanocomposites and ceramics are highlighted.

Piezoelectric ceramics

Inorganic piezoelectric materials are broadly categorized into piezoelectric crystals and piezoelectric ceramics (Anton and Sodano 2007, Kim et al. 2011). Piezoelectric crystals have a single crystal structure and show piezoelectric behavior even without electric poling (Zhou et al. 2014, Li et al. 2014, Mishra et al. 2019). The piezoelectric behavior of these single-crystal structure based materials originates from the orderly arranged opposite charges across the crystal structure. A few examples of such single-crystals are quartz crystal, lead-zinc niobate-lead titanate (PZN-PT), lead magnesium niobate lead titanate solid solution (PMN-PT) and lithium niobate ($LiNbO_3$) (Zhou et al. 2014, Li et al. 2014, Mishra et al. 2019). These materials are characterized by extremely high d_{33} (typically in the range of 2200 pC N^{-1}) and high k_{33} (typically in the range of 93%) (Chen and Panda 2005). These single crystals are used in various applications such as actuators, accelerometer, hydrophones, ultrasound generator and sonar transducers. On the other hand, piezoelectric ceramics, for example, barium titanate ($BaTiO_3$), lead-zirconate-titanate (PZT), aluminum nitride (AlN) and potassium niobate ($KNbO_3$), etc., consist of many small randomly oriented crystals, which shows the piezoelectric behavior after the polarization process (Li et al. 2014). Piezoelectric ceramic material with perovskite crystal structure such as PZT, have high d_{33} typically in the range of 2000 pC/N (Tang et al. 2014, Li et al. 2018). However, piezoelectric ceramics suffer from toxicity, rigidness, non-flexibility and brittleness (Ramadan et al. 2014). These factors significantly limit its applications. Recently several works have reported piezoelectric ceramic materials with improved mechanical properties and flexibility (Bellaiche and Vanderbilt 1999, Chen et al. 2010, Hwang et al. 2015, Jeong et al. 2017). Flexible PZT based NGs have been developed by depositing a flexible substrate utilizing Inorganic-based Laser Lift-Off (ILLO) method and Aerosol Deposition (AD) (Hwang et al. 2016). The device when subjected to mechanical bending motions generates a voltage output of 200 V and a current output of 35 μA, which can direct power 208 LEDs. However, due to the presence of lead (Pd), these materials are toxic and harmful, thus it is not suitable for epidermal or implanted energy harvesters. To overcome this, lead (Pb) free piezoelectric ceramics were developed (Jeong et al. 2014). Alkaline niobate-based particles (KNLN) were developed as lead-free piezoelectric materials that demonstrate a high d_{33} value close to 310 pC/N. When subjected to mechanical vibrations, these devices generate a voltage output of 140 V and current output of 8 μA (Misra et al. 2015). The development of nanomaterial-based piezoelectric ceramics due to the advance in nanofabrication techniques facilitates the incorporation of these high performing ceramic materials with mechanically flexible and robust polymeric substrate (Qi et al. 2011). The latest development in the 1D and 2D piezoelectric nanomaterials are highlighted next (Li et al. 2014).

One-dimensional (1D) piezoelectric materials

The piezoelectric behavior of the ZnO appears due to its wurtzite crystal structure (Wang and Song 2006, Yang et al. 2009, Cha et al. 2010, Zhu et al. 2012, Fan et al. 2016). Solution processing techniques such as the hydrothermal method (Greene et al. 2003, Ohshima et al. 2004, Polsongkram et al. 2008) have been used to fabricate

ZnO NWs for NGs. The wurtzite crystal structure is composed of stacked layers of tetrahedrally coordinated O^{2-} and Zn^{2+} (Ohshima et al. 2004). Before the application of mechanical force, the charge centers on the opposite charges overlap, thus there is no potential difference. When the mechanical force is exerted, the deviation of the charge centers generated a potential difference. Wang et al. demonstrated the first 1D nanogenerator using ZnO NW (Wang and Song 2006). They demonstrated that by mechanically deflecting aligned arrays of zinc oxide (ZnO) nanowires, a potential gradient can be developed owing to the inherent piezoelectric effect of the ZnO. This led to rapid and tremendous progress in the domain of the nanowire-based piezoelectric devices. However, the performance of the ZnO based NGs is not high due to its low d_{33} value which is around 5 to 10 pC/N (Qi and McAlpine 2010). Polymethyl methacrylate (PMMA) encapsulation over vertically aligned ZnO NWs fabricated on a conducting substrate was used to improve its mechanical strength. This enhances the robustness and strength of the nanogenerator to sustain mechanical deformations. The device generates a voltage output of 58 V and current output of 134 μA. To demonstrate the practical applicability of the generated voltage, the dynamic movement of a frog leg was stimulated (Zhu et al. 2012). Further improvement can be achieved by connecting many devices in various parallel and series configurations. The parallel connections improve the current density (Zhu et al. 2010) and the series connection improves the voltage output (Lee et al. 2013).

Two-dimensional (2D) piezoelectric materials

Two-dimensional (2D) piezoelectric materials are another promising material for high performing mechanical energy harvesters (Michel and Verberck 2009, Cao et al. 2012, Duerloo et al. 2012, Wu et al. 2014, Lee et al. 2017, Cui et al. 2017, Song et al. 2018, Xiong et al. 2018). The 2D structure of piezoelectric materials provides the mechanical robustness to realize flexible and mechanically robust NG. Mechanically exfoliated MoS_2 nanoflakes were also utilized as a 2D nanogenerator by (Wu et al. 2014) chromium/palladium/gold were used as electrical contacts with the interface of metal and MoS_2 is parallel to the y-axis of the NG. When the device is subjected to cyclic stretching and releasing, the piezoelectric charge was developed across the zigzag edges of the MoS_2 flake, thus generating outputs voltage across the external circuits. The device showed a very high energy conversion efficiency of 5.08% with voltage. The piezoelectric behavior of MoS_2 in d_{11} mode was also explored with a d_{11} value of 3 pm/V (Duerloo et al. 2012). Several other 2D materials were also used as a piezoelectric material such as lead iodide (PbI_2) (Song et al. 2018), tungsten diselenide (WSe_2) (Lee et al. 2017), and hexagonal boron nitride (h-BN) (Michel and Verberck 2009). However, the energy harvesting performance of most of the 2D based nanogenerator is still low and unsatisfactory. Therefore novel strategies need to be adopted to substantially enhance the performance of 2D NGs.

Piezoelectric polymers

Polymers are materials with a long carbon chain with excellent structural flexibility, easy manufacturability, simplicity of design and low manufacturing cost makes them a suitable candidate for NGs (Ramadan et al. 2014, Chen et al. 2017, Wan and Bowen

2017, Mishra et al. 2019). Room temperature and the solution-based process can be utilized to effectively tailor properties at the nanoscale and can be integrated with other nanomaterial and nanofabrication technologies. Moreover, these materials are lead-free unlike ceramic materials and they are biocompatible, thus they can be effectively integrated onto the human skin or organ for epidermal or implanted energy harvesters (Jung et al. 2014, Li et al. 2014). More importantly, these materials can sustain high mechanical stress and strain owing to its superior deformability and higher dielectric breakdown strength, thus making it possible to harness energy from different mechanical impacts such as bending, stretching, rolling and twisting (Mishra et al. 2019). Several piezoelectric polymers are used to fabricate NGs such as polyvinylidene fluoride (PVDF), polyvinylidene fluoride-trifluoro ethylene (PVDF-TrFE), polylactic acids (PLA), polyamides (PA) cellulose and their derivatives, etc. (Ramadan et al. 2014).

Polyvinylidene fluoride (PVDF)

Among the various piezoelectric polymers, polyvinylidene fluoride (PVDF) and its copolymer are extensively studied and used in various applications (Bhavanasi et al. 2016, Xin et al. 2018, Sappati and Bhadra 2018, Yan et al. 2019). PVDF and its copolymers show excellent piezoelectric properties such as high piezoelectric coefficient (d_{33}) typically in the range if 15–20 pm/V, high remnant polarization (usually in the range of 8–12 $\mu C/cm^2$) and a low coercive field (typically in the range of 30–50 MV/m) (Lovinger 1983, Horiuchi and Tokura 2008, Biomimetic materials: Fields of nanograss 2010, Kusuma et al. 2010). It also has superior thermal and chemical stability. Due to the excellent mechanical strength, superior flexibility, ease of manufacturing and processing and high chemical resistance to various solvents, acids and bases, PVDF and its copolymers are extensively studied and are an industrially applicable piezoelectric polymer (Xin et al. 2018, Yan et al. 2019). Compared to inorganic materials like barium titanate ($BaTiO_3$), lead zirconatetitanate (PZT) and zinc oxide (ZnO), piezoelectric polymers have excellent fracture strain (close to 2%) thus making them flexible, bendable, rollable, foldable and stretchable to be applied in epidermal and implanted energy harvesters. Its polymeric nature facilitates easy manufacturability, based on casting, molding, spinning and drawing technologies. In addition, recently it was also used in state-of-the-art 3D printing and inkjet printing (Lovinger 1983, Horiuchi and Tokura 2008, Zhou et al. 2020). Moreover, it is biocompatible, thus it is safe for biomedical applications to make implanted energy harvesters and sensors. Due to these excellent properties, apart from mechanical energy harvesters, piezoelectric polymers have been used for diverse applications such as actuators, sensors, non-volatile memories and transducers (Kang and Cao 2014, Naghizadeh et al. 2020). The piezoelectric effect in PVDF can be attributed to the molecular dipole (electropositive hydrogen atoms and electronegative fluorine attached to the carbon backbone). Energy is generated due to the change in the dipole moment when subjected to mechanical stress. However, one of the major drawbacks of these piezoelectric polymers compared to the single crystal piezoelectric ceramic is the additional electrical poling process required to effectively align the electric dipoles, which significantly limits its multidimensional design options and applicability (Zhou et al. 2020). Additionally, it requires further preparatory steps which increase the manufacturing cost.

The properties of the PVDF depend on the molecular conformation. PVDF exists in four different conformations. In the β-phase or form-1 structure, the hydrogen atoms are arranged on one side of the backbone and the fluorine atoms on the other, in all-trans (TTTT) conformations. This creates a net dipole moment, showing the piezoelectric effect. In the α-phase or form-2 structure, the hydrogen and fluorine atoms are alternatively arranged in trans–gauche (TGTG) conformation. Thus, creating net-zero dipole moment with no piezoelectric effect. In γ phase or form-3, the atoms are arranged in the (TTTG) conformation, thus causing a relatively weak piezoelectric effect, where the magnitude of the dipole moment is half of the magnitude of the dipole moment in β-phase. Van der Waal forces are the dominating forces in PVDF. As the space available for the fluorine atoms in β-phase is relatively small, it is thermodynamically unstable, thus PVDF favorably crystallizes into α-phase. Due to the similar energy of the various phases of PVDF, the non-piezoelectric phase can be converted to piezoelectric β-phase by thermal annealing, stretching and electrical poling. However, owing to the presence of an extra fluorine atom, poly (vinylidene fluoride-trifluoroethylene) P (VDF-TrFE) can easily form a piezoelectric β-phase. The copolymerization with PTrFE, replacing the bigger fluorine atoms by smaller hydrogen atoms, thus making the α-phase unstable and favoring the crystallization of β-phase. Thus, P (VDF-TrFE) is a more suitable candidate for piezoelectric devices compared to PVDF.

Piezoelectric polymer composites and nanocomposites

Nanocomposites of different piezoelectric materials have been shown to improve the energy harvesting performance (Chen et al. 2017). The schematic representation of the piezoelectrect device and the energy harvesting performance is depicted in Fig. 10.4. It has the advantages of both the phases, excellent piezoelectric properties of the ceramic's nanomaterials and the superior mechanical properties of the

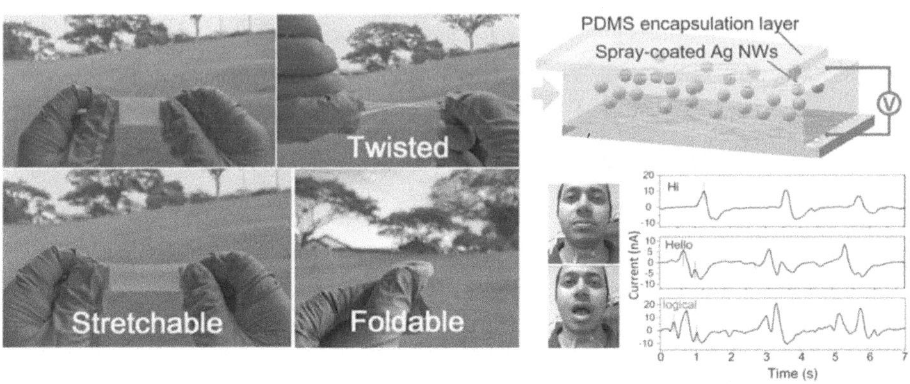

Figure 10.4: (a) Optical image of composite piezoelectric device. Scale bar 2 μm. (b) Schematic illustration of the composite piezoelectric device. (c) The device is attached to human skin to generate energy from different physiological conditions (Chen et al. 2017). Reproduced with permission.

polymers. The nanocomposites demonstrate superior energy harvesting performance compared to polymeric and ceramic piezoelectric materials (Li et al. 2014). Several nanomaterials including ceramics, carbon-based, metallic, semiconductor, hybrid and biomaterials have been used to enhance the energy harvesting performance. The geometry of the dispersion of the ceramic composite within the polymeric matrix affects the energy harvesting performance. The nanocomposite with the form of fiber (1–3 geometry) and laminates (2–3 geometry) (with the longitudinal direction of the composite parallel to poling direction) demonstrates better performance compared to that of the composite in the form of particles (0–3 geometry) (Narita and Fox 2018). Polymer nanocomposites can be further explored to develop piezoelectric nanogenerators with improved energy harvesting performance.

Piezoelectrets

Piezoelectrets are porous polymeric foams or sponges which shows piezoelectric behavior when subjected to corona poling (Savolainen and Kirjavainen 1989, Zhang et al. 2014, Wu et al. 2015, Zhong et al. 2015, Wang et al. 2017). These materials consist of a porous structure filled with air, thus making it spongy and flexible. The electrical dipoles are formed due to positive and negative charges trapped inside the porous structure. When these films are subjected to mechanical force, the compression of the porous structure, causes a change in the dipole moment thus generating voltage output (Wu et al. 2015). Kirjavainen et al. demonstrated the first work on piezoelectret devices with cellular propylene (PP) as the material (Savolainen and Kirjavainen 1989). The cellular PP was developed using the thermal expansion method displaying a piezoelectric coefficient close to ~ 200 to 600 pC/N (Zhong et al. 2015). These kinds of porous PP piezoelectret devices can be used both as energy harvesting and an actuation device. Wu et al. 2015 reported porous piezoelectret devices with extremely long cyclic stability and high output power density close to 52.8 mW/m^2. Recently, several novel kinds of piezoelectrets devices with extremely higher d_{33} have emerged. Zhong et al. 2017 reported a piezoelectret device with a sandwich-like structure with extremely high d_{33} of 6300 pC/N, generating a power output of 0.44 mW. The device can even operate in a humid environment with extreme cycle stability of 90,000 cycles. These devices are suitable for converting vibrational motion into useful electrical energy. Porous polydimethylsiloxane (PDMS) based piezoelectret was reported by Wang et al. (Liu et al. 2018) also demonstrated a high d_{33} upto 1500 pC/N. Thus, piezoelectrets device can be effectively utilized as a mechanical energy harvester.

Piezoelectric and catalysis

In the piezoelectric catalysis, mechanical pressure causes a generation of an in-built potential difference which drives the catalysis process. The first report on the piezoelectric catalysis was reported in 2010 (Hong et al. 2010). When the localized electrical field is modified on subjecting the material to mechanical pressure, the carrier generation, separation, transport and recombination can be controlled. This effect depends on the piezoelectric materials electronic states and the medium facilitating the reaction (Hamann et al. 2005, Hong et al. 2010). This piezocatalytic phenomenon is similar to that of the traditional electrocatalytic process, however in the former, the piezoelectric potential drives the catalytic process and in the latter an external power source is used to drive the catalytic process (Shi et al. 2012, Starr and Wang 2013, Starr and Wang 2015). When the piezoelectric potential is higher than 3 V (with respect to the Standard Hydrogen Electrode (SHE)), the potential will drive the redox reactions.

Hong et al. studied on piezoelectric $BaTiO_3$ microdendrites and reported the decomposition of an organic dye molecule of Acid Orange (AO7) by harvesting mechanical energy (Hong et al. 2012). When external mechanical energy is applied then strain-induced potential is generated which causes the deformation or bending of the $BaTiO_3$ dendrite. The reduction and oxidation reaction of AO7 molecules which adsorb onto the surface of $BaTiO_3$ dendrite occurs due to the strain-induced electric potential (Fig. 10.5).

To circumvent the limitation of the poor light absorption of piezoelectric ZnO nanowires, Hong et al. combined the piezoelectric material with the CuS (visible light active photocatalyst material), thus forming a heterostructure (Hong et al. 2016). As shown in Fig. 10.6, the obtained material removed Methelene Blue (MB) dye within 20 minutes. At the same time, due to the photocatalytic activity, there was a reduction in the concentration of MB by 37%. The superior catalytic activity of the obtained material can be attributed to the following factors: (i) large surface area, which is efficient for light-harvesting and (ii) effective coupling between the developed piezoelectric potential under stress and the in-built electric field of the heterostructure. As observed in Fig. 10.6b, when the material is subjected to mechanical stress, the potential gradient developed at the interface of CuS/ZnO propels the transfer of the photo-generated electrons from the conducting band of CuS into the conducting band of ZnO. At the same time, photogenerated holes transfer from the valance band of ZnO to that of CuS. The improvement in the photocatalytic activity can be attributed to the creation of the potential difference due to the exerted mechanical stress and separates the photo-generated electrons and holes and reduces the recombination rate. Hong et al. reported that when the bare

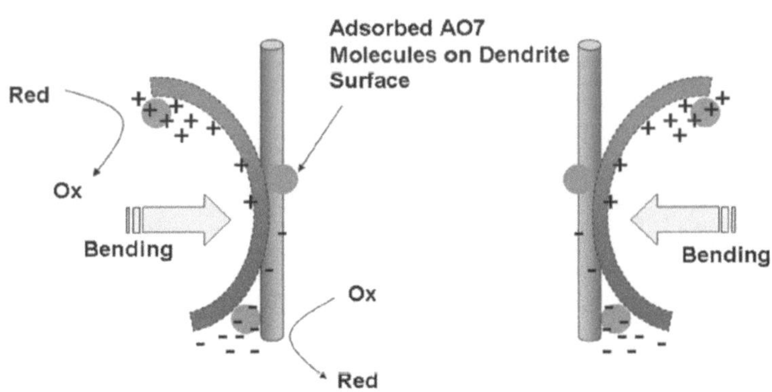

Figure 10.5: The schematic diagram demonstrating the development of charge on $BaTiO_3$ microcrystals owing to inherent built-in potential on bending. This leads to the decomposition of the AO7 due to the oxidation–reduction reactions when the $BaTiO_3$ is subjected to tensile strain (Hong et al. 2012).

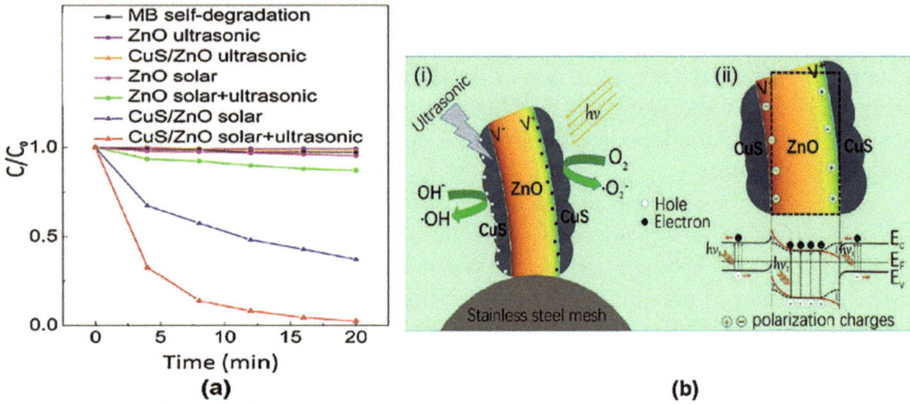

Figure 10.6: (a) Piezo-photocatalytic decomposition of Methylene Blue (MB) over a period of time for ZnO and CuS/ZnO nanowires, (b,i) Schematic illustration of the CuS/ZnO nanowires to demonstrate the piezo-photocatalytic process when simultaneously subjected to ultrasonic irradiation and solar energy, (b, ii) Schematic diagram to illustrate the energy band diagram of the CuS/ZnO heterostructure when simultaneously subjected to ultrasonic irradiation and solar energy. Reproduced with permission (Hong et al. 2016).

ZnO and CuS/ZnO composite nanowires were subjected to ultrasound, there was no decomposition of Methelene Blue dye. Xu et al. and Li et al. reported the possibility of the decomposition of Methelene Blue dye by utilizing ZnO and ZnO nanorods/MWCNT using the piezocatalytic effect (Vajda et al. 2011, Tan et al. 2015). Thus, a detailed and explicit investigation needs to be performed on the piezocatalytic process utilizing ZnO and ZnO composite nanostructures to clearly understand the actual contribution of ZnO in the piezocatalytic process.

Another alternative application of the piezoelectric materials is water splitting and the generation of hydrogen gas. By vibrating the ZnO micro-fibers and $BaTiO_3$ micro-dendrites Hong et al. demonstrated the generation of hydrogel gas (Hong et al. 2010). The length and aspect ratio of the ZnO fibers and $BaTiO_3$ micro dendrites determines the stoichiometric ratio of hydrogen and oxygen gas generation. The deflection caused due to the branch-like $BaTiO_3$ structures reduced the production rate of the gas. The piezocatalytic activity of the evolution of hydrogen gas by the gradient induced by the piezoelectric potential is a highly cost-effective procedure as it utilizes the ambient waste energy for its generation.

The concept of the generation of the piezoelectric potential can also be translated to the piezocatalytic degradation of organic pollutants (Starr and Wang 2013, Starr and Wang 2015). The fundamental mechanism of the piezocatalytic degradation of organic pollutants and the evolution of hydrogen gas is similar. On application of external mechanical perturbation, a piezoelectric potential is created, thus the reactive species present on the surface of the piezoelectric materials drives the water-splitting reactions. When the developed potential gradient is higher than 1.23 eV, then the water molecule dissociates to produces hydrogen and oxygen gas. To further enhance the charge separation Tan et al. combined semiconductor nanoparticles (Ag/Ag_2S)

with the piezoelectric material to be used in the water splitting reaction (Tan et al. 2015). The use of Ag/Ag_2S–ZnO/ZnS semiconductor-piezoelectric heterojunction resulted in improved photocurrent densities.

Conclusion

The invention of the piezoelectric nanogenerator is one of the significant achievements in the domain of energy harvesters. Owing to its high energy harvesting performance, high energy conversion efficiency, low material and manufacturing cost, ease of fabrication, lightweight and availability of a plethora of different piezoelectric materials. Additionally, the ability of the piezoelectric nanogenerator to hybridize with any other energy harvesting technologies such as solar cells (Zhang et al. 2019), electromagnetic (Arroyo et al. 2012), and triboelectric nanogenerators (Parida et al. 2019a and b) gives it an additional advantage to synergistically utilize the unique attributes of both the technologies. The piezoelectric material has been explored in several other applications such as pressure sensor (Parida et al. 2017a, Lin et al. 2018), memory, self-charging power pack (Parida et al. 2017b) water energy harvesting (Lin et al. 2016, Lin et al. 2017).

However, several issues need to be resolved before PNGs are practically used as a sustainable energy harvester. First, novel strategies need to be adopted to enhance the voltage output. Although the phenomenon of piezoelectric effect has been known for the last 100 years, the performance of energy harvesting is not enough to power usable portable electronic devices. Although there has been significant improvement in the voltage output still further work can increase the commercial viability of the PNGs as a power source. Second, is the improvement in the output current density of the PNGs. The current density is extremely low due to the high internal impedance. Several efforts have been made to improve the current density such as by incorporation of conducting fillers into the polymeric matrix. However, the performance is far from satisfactory. Third, further improvement is required in the long-term stability and device durability to different environmental conditions. The energy harvesting performance of the PNGs can be significantly degraded when exposed to high temperatures and other environmental conditions. Thus, proper encapsulation of the device is extremely crucial. The development of material with high durability to sustain multiple numbers of mechanical forces needs to be developed. Fourth, further development needs to be done in the area of Power Management Circuit (PMC). As the performance of the PNGs depends on the arbitrary mechanical motions present in the ambient environment, the generated voltage from the PNGs in actual practical conditions is unpredictable and pulsating. Thus, the PMC will regulate the generated energy according to the desired need. This will facilitate the effective transfer of generated power to energy storage and consuming units. It can be envisioned that with the remarkable effort to enhance the energy harvesting performance, nanogenerators will be soon translated into a commercially viable product for potential applications in various technologies.

References

Abbasi, T., M. Premalatha and S. A. Abbasi. 2011. The return to renewables: Will it help in global warming control? Renewable and Sustainable Energy Reviews [Internet] 15(1): 891–894. http://www.sciencedirect.com/science/article/pii/S1364032110003357.

Anton, S. R. and H. A. Sodano. 2007. A review of power harvesting using piezoelectric materials (2003–2006). Smart Materials and Structures [Internet] 16(3): R1–R21. http://dx.doi.org/10.1088/0964-1726/16/3/R01.

Arroyo, E., A. Badel, F. Formosa, Y. Wu and J. Qiu. 2012. Comparison of electromagnetic and piezoelectric vibration energy harvesters: Model and experiments. Sensors and Actuators A: Physical [Internet] 183: 148–156. http://www.sciencedirect.com/science/article/pii/S0924424712002725.

Asumadu-Sarkodie, S. and P. A. Owusu. 2016. Feasibility of biomass heating system in Middle East Technical University, Northern Cyprus Campus. Pham D, editor. Cogent Engineering [Internet] 3(1): 1134304. https://doi.org/10.1080/23311916.2015.1134304.

Bellaiche, L. and D. Vanderbilt. 1999. Intrinsic piezoelectric response in perovskite alloys: PMN-PT versus PZT. Physical Review Letters [Internet] 83(7): 1347–1350. https://link.aps.org/doi/10.1103/PhysRevLett.83.1347.

Bendi, R., V. Kumar, V. Bhavanasi, K. Parida and P. S. Lee. 2016. Metal organic framework-derived metal phosphates as electrode materials for supercapacitors. Advanced Energy Materials 6(3).

Bhavanasi, V., V. Kumar, K. Parida, J. Wang and P. S. Lee. 2016. Enhanced piezoelectric energy harvesting performance of flexible PVDF-TrFE bilayer films with graphene oxide. ACS Applied Materials & Interfaces [Internet] 8(1): 521–529. https://doi.org/10.1021/acsami.5b09502.

Biomimetic Materials: Fields of Nanograss. 2010. NPG Asia Materials [Internet]. https://doi.org/10.1038/asiamat.2010.198.

Brown, C., R. Kell, R. Taylor and L. Thomas. 1962. Piezoelectric materials, a review of progress. IRE Transactions on Component Parts 9(4): 193–211.

Cao, T., G. Wang, W. Han, H. Ye, C. Zhu, J. Shi et al. 2012. Valley-selective circular dichroism of monolayer molybdenum disulphide. Nature Communications [Internet] 3(1): 887. https://doi.org/10.1038/ncomms1882.

Cha, S. N., J. -S. Seo, S. M. Kim, H. J. Kim, Y. J. Park, S. -W. Kim et al. 2010. Sound-driven piezoelectric nanowire-based nanogenerators. Advanced Materials [Internet] 22(42): 4726–4730. https://doi.org/10.1002/adma.201001169.

Chen, J. and R. Panda. 2005. Review: commercialization of piezoelectric single crystals for medical imaging applications. pp. 235–240. *In*: IEEE Ultrasonics Symposium, 2005. Vol. 1. Rotterdam, The Netherlands.

Chen, Xiaoliang, K. Parida, J. Wang, J. Xiong, M. -F. Lin, J. Shao et al. 2017. A stretchable and transparent nanocomposite nanogenerator for self-powered physiological monitoring. ACS Applied Materials & Interfaces [Internet] 9(48): 42200–42209. https://doi.org/10.1021/acsami.7b13767.

Chen, Xiangfan, H. O. T. Ware, E. Baker, W. Chu, J. Hu and C. Sun. 2017. The development of an all-polymer-based piezoelectric photocurable resin for additive manufacturing. Procedia CIRP [Internet] 65: 157–162. http://www.sciencedirect.com/science/article/pii/S2212827117305346.

Chen, X., S. Xu, N. Yao and Y. Shi. 2010. 1.6 V nanogenerator for mechanical energy harvesting using PZT nanofibers. Nano Letters [Internet] 10(6): 2133–2137. https://doi.org/10.1021/nl100812k.

Chen, X., J. Xiong, K. Parida, M. Guo, C. Wang, X. Li et al. 2019. Transparent and stretchable bimodal triboelectric nanogenerators with hierarchical micro-nanostructures for mechanical and water energy harvesting. Nano Energy 64: 103904.

Cui, P., K. Parida, M. -F. Lin, J. Xiong, G. Cai and P. S. Lee. 2017. Transparent, flexible cellulose nanofibril–phosphorene hybrid paper as triboelectric nanogenerator. Advanced Materials Interfaces [Internet] 4(22): 1700651. https://doi.org/10.1002/admi.201700651.

Dagdeviren, C., P. Joe, O. L. Tuzman, K. -I. Park, K. J. Lee, Y. Shi et al. 2016. Recent progress in flexible and stretchable piezoelectric devices for mechanical energy harvesting, sensing and actuation. Extreme Mechanics Letters [Internet] 9: 269–281. http://www.sciencedirect.com/science/article/pii/S2352431616301092.

Damjanovic, D. 1998. Ferroelectric, dielectric and piezoelectric properties of ferroelectric thin films and ceramics. Reports on Progress in Physics [Internet] 61(9): 1267–1324. http://dx.doi.org/10.1088/0034-4885/61/9/002.

Duerloo, K. -A. N., M. T. Ong and E. J. Reed. 2012. Intrinsic piezoelectricity in two-dimensional materials. The Journal of Physical Chemistry Letters [Internet] 3(19): 2871–2876. https://doi.org/10.1021/jz3012436.

Fan, F. R., W. Tang and Z. L. Wang. 2016. Flexible nanogenerators for energy harvesting and self-powered electronics. Advanced Materials [Internet] 28(22): 4283–4305. https://doi.org/10.1002/adma.201504299.

Gao, D., K. Parida and P. S. Lee. 2019. Emerging soft conductors for bioelectronic interfaces. Advanced Functional Materials [Internet] n/a(n/a): 1907184. https://doi.org/10.1002/adfm.201907184.

Greene, L. E., M. Law, J. Goldberger, F. Kim, J. C. Johnson, Y. Zhang et al. 2003. Low-temperature wafer-scale production of ZnO nanowire arrays. Angewandte Chemie International Edition [Internet] 42(26): 3031–3034. https://doi.org/10.1002/anie.200351461.

Groves, P. 1986. A review of: "Piezoelectricity". Phase Transitions [Internet] 6(4): 329–330. https://doi.org/10.1080/01411598608220072.

Hamann, T. W., F. Gstrein, B. S. Brunschwig and N. S. Lewis. 2005. Measurement of the dependence of interfacial charge-transfer rate constants on the reorganization energy of redox species at n-ZnO/H_2O interfaces. Journal of the American Chemical Society [Internet] 127(40): 13949–13954. https://doi.org/10.1021/ja0515452.

Hong, D., W. Zang, X. Guo, Y. Fu, H. He, J. Sun et al. 2016. High piezo-photocatalytic efficiency of CuS/ZnO nanowires using both solar and mechanical energy for degrading organic dye. ACS Applied Materials & Interfaces [Internet] 8(33): 21302–21314. https://doi.org/10.1021/acsami.6b05252.

Hong, K. -S., H. Xu, H. Konishi and X. Li. 2010. Direct water splitting through vibrating piezoelectric microfibers in water. The Journal of Physical Chemistry Letters [Internet] 1(6): 997–1002. https://doi.org/10.1021/jz100027t.

Hong, K. -S., H. Xu, H. Konishi and X. Li. 2012. Piezoelectrochemical effect: a new mechanism for Azo dye decolorization in aqueous solution through vibrating piezoelectric microfibers. The Journal of Physical Chemistry C [Internet] 116(24): 13045–13051. https://doi.org/10.1021/jp211455z.

Horiuchi, S. and Y. Tokura. 2008. Organic ferroelectrics. Nature Materials 7(5): 357–366.

Hwang, G. -T., J. Yang, S. H. Yang, H. -Y. Lee, M. Lee, D. Y. Park et al. 2015. A reconfigurable rectified flexible energy harvester via solid-state single crystal grown PMN–PZT. Advanced Energy Materials [Internet] 5(10): 1500051. https://doi.org/10.1002/aenm.201500051.

Hwang, G. -T., V. Annapureddy, J. H. Han, D. J. Joe, C. Baek, D. Y. Park et al. 2016. Self-powered wireless sensor node enabled by an aerosol-deposited PZT flexible energy harvester. Advanced Energy Materials [Internet] 6(13): 1600237. https://doi.org/10.1002/aenm.201600237.

Jeong, C. K., K. -I. Park, J. Ryu, G. -T. Hwang and K. J. Lee. 2014. Large-area and flexible lead-free nanocomposite generator using alkaline niobate particles and metal nanorod filler. Advanced Functional Materials [Internet] 24(18): 2620–2629. https://doi.org/10.1002/adfm.201303484.

Jeong, C. K., S. B. Cho, J. H. Han, D. Y. Park, S. Yang, K. -I. Park et al. 2017. Flexible highly-effective energy harvester via crystallographic and computational control of nanointerfacial morphotropic piezoelectric thin film. Nano Research [Internet] 10(2): 437–455. https://doi.org/10.1007/s12274-016-1304-6.

Jung, S., J. Lee, T. Hyeon, M. Lee and D. -H. Kim. 2014. Fabric-based integrated energy devices for wearable activity monitors. Advanced Materials [Internet] 26(36): 6329–6334. https://doi.org/10.1002/adma.201402439.

Kakekhani, A. and S. Ismail-Beigi. 2015. Ferroelectric-based catalysis: switchable surface chemistry. ACS Catalysis [Internet] 5(8): 4537–4545. https://doi.org/10.1021/acscatal.5b00507.

Kang, G. and Y. Cao. 2014. Application and modification of poly(vinylidene fluoride) (PVDF) membranes—A review. Journal of Membrane Science [Internet] 463: 145–165. http://www.sciencedirect.com/science/article/pii/S0376738814002415.

Kim, H. S., J. -H. Kim and J. Kim. 2011. A review of piezoelectric energy harvesting based on vibration. International Journal of Precision Engineering and Manufacturing [Internet] 12(6): 1129–1141. https://doi.org/10.1007/s12541-011-0151-3.

Kumar, V., S. Matz, D. Hoogestraat, V. Bhavanasi, K. Parida, K. Al-Shamery et al. 2016. Design of mixed-metal silver decamolybdate nanostructures for high specific energies at high power density. Advanced Materials [Internet] 28(32): 6966–6975. https://doi.org/10.1002/adma.201601158.
Kumar, V., L. Liu, V. C. Nguyen, V. Bhavanasi, K. Parida, D. Mandler et al. 2017. Localized charge transfer in two-dimensional molybdenum trioxide. ACS Applied Materials & Interfaces [Internet] 9(32): 27045–27053. https://doi.org/10.1021/acsami.7b09641.
Kumar, V., S. Park, K. Parida, V. Bhavanasi and P. S. Lee. 2017. Multi-responsive supercapacitors: Smart solution to store electrical energy. Materials Today Energy 4.
Kusuma, D. Y., C. A. Nguyen and P. S. Lee. 2010. Enhanced ferroelectric switching characteristics of P(VDF-TrFE) for organic memory devices. The Journal of Physical Chemistry B [Internet] 114(42): 13289–13293. https://doi.org/10.1021/jp105249f.
Lee, J. -H., J. Y. Park, E. B. Cho, T. Y. Kim, S. A. Han, T. -H. Kim et al. 2017. Reliable piezoelectricity in bilayer WSe2 for piezoelectric nanogenerators. Advanced Materials [Internet] 29(29): 1606667. https://doi.org/10.1002/adma.201606667.
Lee, J., M. W. M. Tan, K. Parida, G. Thangavel, S. A. Park, T. Park et al. 2020. Water-processable, stretchable, self-healable, thermally stable, and transparent ionic conductors for actuators and sensors. Advanced Materials [Internet] 32(7): 1906679. https://doi.org/10.1002/adma.201906679.
Lee, S., S. -H. Bae, L. Lin, Y. Yang, C. Park, S. -W. Kim et al. 2013. Super-flexible nanogenerator for energy harvesting from gentle wind and as an active deformation sensor. Advanced Functional Materials [Internet] 23(19): 2445–2449. https://doi.org/10.1002/adfm.201202867.
Li, F., D. Lin, Z. Chen, Z. Cheng, J. Wang, C. Li et al. 2018. Ultrahigh piezoelectricity in ferroelectric ceramics by design. Nature Materials [Internet] 17(4): 349–354. https://doi.org/10.1038/s41563-018-0034-4.
Li, H., C. Tian and Z. D. Deng. 2014. Energy harvesting from low frequency applications using piezoelectric materials. Applied Physics Reviews [Internet] 1(4): 41301. https://doi.org/10.1063/1.4900845.
Lin, M. -F., K. Parida, X. Cheng and P. S. Lee. 2016. Flexible superamphiphobic film for water energy harvesting. Advanced Materials Technologies 2: 1600186.
Lin, M. -F., K. Parida, X. Cheng and P. S. Lee. 2017. Flexible superamphiphobic film for water energy harvesting. Advanced Materials Technologies [Internet] 2(1): 1600186. https://doi.org/10.1002/admt.201600186.
Lin, M. -F., J. Xiong, J. Wang, K. Parida and P. S. Lee. 2018. Core-shell nanofiber mats for tactile pressure sensor and nanogenerator applications. Nano Energy [Internet] 44: 248–255. http://www.sciencedirect.com/science/article/pii/S2211285517307711.
Liu, H., J. Zhong, C. Lee, S. -W. Lee and L. Lin. 2018. A comprehensive review on piezoelectric energy harvesting technology: Materials, mechanisms, and applications. Applied Physics Reviews [Internet] 5(4): 41306. https://doi.org/10.1063/1.5074184.
Lovinger, A. J. 1983. Ferroelectric polymers. Science (New York, NY) 220(4602): 1115–1121.
Lu, Y., N. Nakicenovic, M. Visbeck and A. -S. Stevance. 2015. Policy: Five priorities for the UN sustainable development goals. Nature 520(7548): 432–433.
Michel, K. H. and B. Verberck. 2009. Theory of elastic and piezoelectric effects in two-dimensional hexagonal boron nitride. Physical Review B [Internet] 80(22): 224301. https://link.aps.org/doi/10.1103/PhysRevB.80.224301.
Mishra, S., L. Unnikrishnan, S. K. Nayak and S. Mohanty. 2019. Advances in piezoelectric polymer composites for energy harvesting applications: a systematic review. Macromolecular Materials and Engineering [Internet] 304(1): 1800463. https://doi.org/10.1002/mame.201800463.
Misra, V., A. Bozkurt, B. Calhoun, T. Jackson, J. S. Jur, J. Lach et al. 2015. Flexible technologies for self-powered wearable health and environmental sensing. Proceedings of the IEEE 103(4): 665–681.
Nasrin, I., R. Hamid Reza and N. Rahim. 2020. Investigation on the effect of $BaCO_3$ on the physical and mechanical properties of 5Ni/10NiO.$NiFe_2O_4$ cermet. Functional Composites and Structures [Internet]. http://iopscience.iop.org/10.1088/2631-6331/ab9753.
Narita, F. and M. Fox. 2018. A review on piezoelectric, magnetostrictive, and magnetoelectric materials and device technologies for energy harvesting applications. Advanced Engineering Materials [Internet] 20(5): 1700743. https://doi.org/10.1002/adem.201700743.

Ohshima, E., H. Ogino, I. Niikura, K. Maeda, M. Sato, M. Ito et al. 2004. Growth of the 2-in-size bulk ZnO single crystals by the hydrothermal method. Journal of Crystal Growth [Internet] 260(1): 166–170. http://www.sciencedirect.com/science/article/pii/S0022024803016750.

Parida, K., V. Bhavanasi, V. Kumar, R. Bendi and P. S. Lee. 2017. Self-powered pressure sensor for ultra-wide range pressure detection. Nano Research 10(10).

Parida Kaushik, V. Bhavanasi, V. Kumar, J. Wang and P. S. Lee. 2017. Fast charging self-powered electric double layer capacitor. Journal of Power Sources [Internet] 342: 70–78. http://www.sciencedirect.com/science/article/pii/S0378775316316342.

Parida, K., V. Kumar, W. Jiangxin, V. Bhavanasi, R. Bendi and P. S. Lee. 2017. Highly transparent, stretchable, and self-healing ionic-skin triboelectric nanogenerators for energy harvesting and touch applications. Advanced Materials 29(37).

Parida, K., G. Thangavel, G. Cai, X. Zhou, S. Park, J. Xiong et al. 2019. Extremely stretchable and self-healing conductor based on thermoplastic elastomer for all-three-dimensional printed triboelectric nanogenerator. Nature Communications [Internet] 10(1): 2158. https://doi.org/10.1038/s41467-019-10061-y.

Parida, K., J. Xiong, X. Zhou and P. S. Lee. 2019. Progress on triboelectric nanogenerator with stretchability, self-healability and bio-compatibility. Nano Energy [Internet] 59: 237–257. http://www.sciencedirect.com/science/article/pii/S2211285519300977.

Park, S., K. Parida and P. S. Lee. 2017. Deformable and transparent ionic and electronic conductors for soft energy devices. Advanced Energy Materials [Internet] 7(22): 1701369. https://doi.org/10.1002/aenm.201701369.

Park, S., G. Thangavel, K. Parida, S. Li and P. S. Lee. 2019. A stretchable and self-healing energy storage device based on mechanically and electrically restorative liquid-metal particles and carboxylated polyurethane composites. Advanced Materials [Internet] 31(1): 1805536. https://doi.org/10.1002/adma.201805536.

Polsongkram, D., P. Chamninok, S. Pukird, L. Chow, O. Lupan, G. Chai et al. 2008. Effect of synthesis conditions on the growth of ZnO nanorods via hydrothermal method. Physica B: Condensed Matter [Internet] 403(19): 3713–3717. http://www.sciencedirect.com/science/article/pii/S0921452608002883.

Priya, S., H. -C. Song, Y. Zhou, R. Varghese, A. Chopra, S. -G. Kim et al. 2019. A review on piezoelectric energy harvesting: materials, methods, and circuits. Energy Harvesting and Systems [Internet] 4(1): 3–39. https://www.degruyter.com/view/journals/ehs/4/1/article-p3.xml.

Qi, Y. and M. C. McAlpine. 2010. Nanotechnology-enabled flexible and biocompatible energy harvesting. Energy & Environmental Science [Internet] 3(9): 1275–1285. http://dx.doi.org/10.1039/C0EE00137F.

Qi, Y., J. Kim, T. D. Nguyen, B. Lisko, P. K. Purohit and M. C. McAlpine. 2011. Enhanced piezoelectricity and stretchability in energy harvesting devices fabricated from buckled PZT ribbons. Nano Letters [Internet] 11(3): 1331–1336. https://doi.org/10.1021/nl104412b.

Ramadan, K. S., D. Sameoto and S. Evoy. 2014. A review of piezoelectric polymers as functional materials for electromechanical transducers. Smart Materials and Structures [Internet] 23(3): 33001. http://dx.doi.org/10.1088/0964-1726/23/3/033001.

Sappati, K. K. and S. Bhadra. 2018. Piezoelectric polymer and paper substrates: a review. Sensors (Basel, Switzerland) 18(11).

Savolainen, A. and K. Kirjavainen. 1989. Electrothermomechanical film. Part I. Design and characteristics. Journal of Macromolecular Science: Part A – Chemistry [Internet] 26(2–3): 583–591. https://doi.org/10.1080/00222338908051994.

Shi, J., M. B. Starr and X. Wang. 2012. Band structure engineering at heterojunction interfaces via the piezotronic effect. Advanced Materials [Internet] 24(34): 4683–4691. https://doi.org/10.1002/adma.201104386.

Song, H., I. Karakurt, M. Wei, N. Liu, Y. Chu, J. Zhong et al. 2018. Lead iodide nanosheets for piezoelectric energy conversion and strain sensing. Nano Energy [Internet] 49: 7–13. http://www.sciencedirect.com/science/article/pii/S2211285518302581.

Starr, M. B. and X. Wang. 2013. Fundamental analysis of piezocatalysis process on the surfaces of strained piezoelectric materials. Scientific Reports [Internet] 3: 2160. https://pubmed.ncbi.nlm.nih.gov/23831736.

Starr, M. B. and X. Wang. 2015. Coupling of piezoelectric effect with electrochemical processes. Nano Energy [Internet] 14: 296–311. http://www.sciencedirect.com/science/article/pii/S2211285515000452.

Tan, C. F., W. L. Ong and G. W. Ho. 2015. Self-biased hybrid piezoelectric-photoelectrochemical cell with photocatalytic functionalities. ACS Nano [Internet] 9(7): 7661–7670. https://doi.org/10.1021/acsnano.5b03075.

Tang, G., B. Yang, J. Liu, B. Xu, H. Zhu and C. Yang. 2014. Development of high performance piezoelectric d_{33} mode MEMS vibration energy harvester based on PMN-PT single crystal thick film. Sensors and Actuators A: Physical [Internet] 205: 150–155. http://www.sciencedirect.com/science/article/pii/S0924424713005529.

Trolier-McKinstry, S., S. Zhang, A. J. Bell and X. Tan. 2018. High-performance piezoelectric crystals, ceramics, and films. Annual Review of Materials Research [Internet] 48(1): 191–217. https://doi.org/10.1146/annurev-matsci-070616-124023.

Uchino, K. and T. Ishii. 2010. Energy flow analysis in piezoelectric energy harvesting systems. Ferroelectrics [Internet] 400(1): 305–320. https://doi.org/10.1080/00150193.2010.505852.

Vajda, K., K. Mogyorosi, Z. Nemeth, K. Hernadi, L. Forro, A. Magrez et al. 2011. Photocatalytic activity of TiO_2/SWCNT and TiO_2/MWCNT nanocomposites with different carbon nanotube content. Physica Status Solidi (b) [Internet] 248(11): 2496–2499. https://doi.org/10.1002/pssb.201100117.

Wan, C. and C. R. Bowen. 2017. Multiscale-structuring of polyvinylidene fluoride for energy harvesting: the impact of molecular-, micro- and macro-structure. Journal of Materials Chemistry A [Internet] 5(7): 3091–3128. http://dx.doi.org/10.1039/C6TA09590A.

Wang, B., C. Liu, Y. Xiao, J. Zhong, W. Li, Y. Cheng et al. 2017. Ultrasensitive cellular fluorocarbon piezoelectret pressure sensor for self-powered human physiological monitoring. Nano Energy [Internet] 32: 42–49. http://www.sciencedirect.com/science/article/pii/S2211285516305894.

Wang, S., L. Lin, Y. Xie, Q. Jing, S. Niu and Z. L. Wang. 2013. Sliding-triboelectric nanogenerators based on in-plane charge-separation mechanism. Nano Letters [Internet] 13(5): 2226–2233. https://doi.org/10.1021/nl400738p.

Wang, S., Y. Xie, S. Niu, L. Lin and Z. L. Wang. 2014. Freestanding triboelectric-layer-based nanogenerators for harvesting energy from a moving object or human motion in contact and non-contact modes. Advanced Materials [Internet] 26(18): 2818–2824. https://doi.org/10.1002/adma.201305303.

Wang, Z., J. Song, F. Gao, R. Su, D. Zhang, Y. Liu et al. 2017. Developing a ferroelectric nanohybrid for enhanced photocatalysis. Chemical Communications [Internet] 53(54): 7596–7599. http://dx.doi.org/10.1039/C7CC02548C.

Wang, Z. L. and J. Song. 2006. Piezoelectric nanogenerators based on zinc oxide nanowire arrays. Science [Internet] 312(5771): 242 LP–246. http://science.sciencemag.org/content/312/5771/242.abstract.

Wang, Z. L. 2014. Triboelectric nanogenerators as new energy technology and self-powered sensors—Principles, problems and perspectives. Faraday Discussions [Internet] 176(0): 447–458. http://dx.doi.org/10.1039/C4FD00159A.

Wang, Z. L., J. Chen and L. Lin. 2015. Progress in triboelectric nanogenerators as a new energy technology and self-powered sensors. Energy & Environmental Science [Internet] 8(8): 2250–2282. http://dx.doi.org/10.1039/C5EE01532D.

Wu, C., A. C. Wang, W. Ding, H. Guo and Z. L. Wang. 2019. Triboelectric nanogenerator: a foundation of the energy for the new era. Advanced Energy Materials [Internet] 9(1): 1802906. https://doi.org/10.1002/aenm.201802906.

Wu, N., X. Cheng, Q. Zhong, J. Zhong, W. Li, B. Wang et al. 2015. Cellular polypropylene piezoelectret for human body energy harvesting and health monitoring. Advanced Functional Materials [Internet] 25(30): 4788–4794. https://doi.org/10.1002/adfm.201501695.

Wu, W., L. Wang, Y. Li, F. Zhang, L. Lin, S. Niu et al. 2014. Piezoelectricity of single-atomic-layer MoS2 for energy conversion and piezotronics. Nature [Internet] 514(7523): 470–474. https://doi.org/10.1038/nature13792.

Xin, Y., J. Zhu, H. Sun, Y. Xu, T. Liu and C. Qian. 2018. A brief review on piezoelectric PVDF nanofibers prepared by electrospinning. Ferroelectrics [Internet] 526(1): 140–151. https://doi.org/10.1080/00150193.2018.1456304.

Xiong, J., M. -F. Lin, J. Wang, S. L. Gaw, K. Parida and P. S. Lee. 2017. Wearable all-fabric-based triboelectric generator for water energy harvesting. Advanced Energy Materials [Internet] 7(21): 1701243. https://doi.org/10.1002/aenm.201701243.

Xiong, J., P. Cui, X. Chen, J. Wang, K. Parida, M. -F. Lin et al. 2018. Skin-touch-actuated textile-based triboelectric nanogenerator with black phosphorus for durable biomechanical energy harvesting. Nature Communications [Internet] 9(1): 4280. https://doi.org/10.1038/s41467-018-06759-0.

Xiong, J., H. Luo, D. Gao, X. Zhou, P. Cui, G. Thangavel et al. 2019. Self-restoring, waterproof, tunable microstructural shape memory triboelectric nanogenerator for self-powered water temperature sensor. Nano Energy [Internet] 61: 584–593. http://www.sciencedirect.com/science/article/pii/S2211285519303908.

Yan, J., M. Liu, Y. G. Jeong, W. Kang, L. Li, Y. Zhao et al. 2019. Performance enhancements in poly(vinylidene fluoride)-based piezoelectric nanogenerators for efficient energy harvesting. Nano Energy [Internet] 56: 662–692. http://www.sciencedirect.com/science/article/pii/S2211285518309145.

Yang, R., Y. Qin, L. Dai and Z. L. Wang. 2009. Power generation with laterally packaged piezoelectric fine wires. Nature Nanotechnology [Internet] 4(1): 34–39. https://doi.org/10.1038/nnano.2008.314.

Yang, Y., H. Zhang, J. Chen, Q. Jing, Y. S. Zhou, X. Wen and Z. L. Wang. 2013. Single-electrode-based sliding triboelectric nanogenerator for self-powered displacement vector sensor system. ACS Nano [Internet] 7(8): 7342–7351. https://doi.org/10.1021/nn403021m.

Zhang, Q., A. Solanki, K. Parida, D. Giovanni, M. Li, T. L. C. Jansen et al. 2019. Tunable ferroelectricity in ruddlesden–popper halide perovskites. ACS Applied Materials & Interfaces [Internet] 11(14): 13523–13532. https://doi.org/10.1021/acsami.8b21579.

Zhang, X., G. M. Sessler and Y. Wang. 2014. Fluoroethylenepropylene ferroelectret films with cross-tunnel structure for piezoelectric transducers and micro energy harvesters. Journal of Applied Physics [Internet] 116(7): 74109. https://doi.org/10.1063/1.4893367.

Zhong, J., Q. Zhong, X. Zang, N. Wu, W. Li, Y. Chu et al. 2017. Flexible PET/EVA-based piezoelectret generator for energy harvesting in harsh environments. Nano Energy [Internet] 37: 268–274. http://www.sciencedirect.com/science/article/pii/S2211285517303063.

Zhong, Q., J. Zhong, X. Cheng, X. Yao, B. Wang, W. Li et al. 2015. Paper-based active tactile sensor array. Advanced Materials [Internet] 27(44): 7130–7136. https://doi.org/10.1002/adma.201502470.

Zhou, Q., K. H. Lam, H. Zheng, W. Qiu and K. K. Shung. 2014. Piezoelectric single crystal ultrasonic transducers for biomedical applications. Progress in Materials Science [Internet] 66: 87–111. http://www.sciencedirect.com/science/article/pii/S0079642514000541.

Zhou, X., K. Parida, O. Halevi, Y. Liu, J. Xiong, S. Magdassi et al. 2020. All 3D-printed stretchable piezoelectric nanogenerator with non-protruding kirigami structure. Nano Energy [Internet] 72: 104676. http://www.sciencedirect.com/science/article/pii/S2211285520302330.

Zhu, G., R. Yang, S. Wang and Z. L. Wang. 2010. Flexible high-output nanogenerator based on lateral ZnO nanowire array. Nano Letters [Internet] 10(8): 3151–3155. https://doi.org/10.1021/nl101973h.

Zhu, G., A. C. Wang, Y. Liu, Y. Zhou and Z. L. Wang. 2012. Functional electrical stimulation by nanogenerator with 58 V output voltage. Nano Letters [Internet] 12(6): 3086–3090. https://doi.org/10.1021/nl300972f.

Zhu, G., J. Chen, Y. Liu, P. Bai, Y. S. Zhou, Q. Jing et al. 2013. Linear-grating triboelectric generator based on sliding electrification. Nano Letters [Internet] 13(5): 2282–2289. https://doi.org/10.1021/nl4008985.

CHAPTER 11

Metal Hollow Spheres as Promising Electrocatalysts in Electrochemical Conversion of CO_2 to Fuels

Jayeeta Chattopadhyay[1,*] and *Rohit Srivastava*[2]

Introduction

The growing importance of renewable electricity with their falling prices, essentially deals with electricity storage in relation to the intermittent nature of renewable energy sources becomes urgent. Storing renewable electricity in chemical bonds ('electrofuels') is particularly attractive, as it allows for high-energy-density storage and potentially high flexibility (Birdja et al. 2019). It is well-known that, hydrogen is the most likely and realistic candidate for electricity storage in electro-fuels, research on the electrochemical conversion of carbon dioxide and water into carbon-based fuels has intrigued electrochemists for decades, and is currently undergoing a notable renaissance. In contrast to hydrogen production by water electrolysis, carbon dioxide electrolysis is still far from an emerging technology. Significant hurdles regarding energy efficiency, reaction selectivity and overall conversion rate need to be overcome if electrochemical carbon dioxide reduction is to become a viable option for storing renewable electricity (Whipple and Kenis 2010, Durst et al. 2015, Jones et al. 2014, Hori et al. 2008). Many electrocatalysts have been reported for the production of different compounds from the electrocatalytic carbon dioxide reduction reaction (CO_2RR). It is already reported that, CO and COOH are two-electron transfer products, which can be formed at comparatively lower overpotential and high Faradaic efficiency on suitable electrocatalysts, but substantially higher

[1] Department of Chemistry, Amity University Jharkhand, Ranchi 834002, India.
[2] Catalysis Research Lab, School of Petroleum Technology, Pandit Deendayal Petroleum University, Gandhinagar 382007, Gujarat, India.
* Corresponding author: jayeeta08@gmail.com

overpotentials and lower selectivities are observed for multi-electron transfer products such as methane, ethylene and alcohols (Bushuyev et al. 2018).

In a typical electroreduction reaction, CO_2RR is the activated product of CO_2 molecule in the first step. It has often been reported that, the activation and reduction steps of CO_2 is very difficult, as the first electron transfer to produce $CO_2^{\bullet-}$ radical intermediate possesses a very negative redox potential (–1.9 V vs. normal hydrogen electrode), or in other words one could say, the CO_2 molecule is very stable. This is the reason, why the role of electrocatalysts is so important in this process. In the electrocatalytic reaction, the first chemical bond will form between CO_2 and the electrocatalyst resulting into the stabilization of $CO_2^{\bullet-}$ radical, shifting to a less negative redox potential value. With the application of right electrocatalysts, the reduction process becomes more feasible from CO_2 to produce CO or COOH at low overpotential.

However, the existence of certain challenges hinders the practical application of carbon dioxide conversion: the electrochemical reduction process of carbon dioxide can be accomplished by obtaining $2e^-$, $4e^-$, $6e^-$, and $8e^-$ electrons from CO_2, and the products are diverse. The products of the CO_2 reduction reaction always depend on the applied electrode materials, which can be mainly divided into three categories: formate selective metals (Pb, Sn, In, etc.), CO selective metals (Au, Ag, Zn, etc.) and H_2 selective metals (Ni, Fe, Pt, etc.) (Qiao et al. 2014).

Hollow spherical structures have attracted great interest as a special class of materials in comparison to other solid counterparts, owing to their higher surface area, lower density, better permeation and physical and thermal stability (Xia et al. 2010, Xuan et al. 2009, Chattopadhyay et al. 2013). In a growing number of applications, such as catalysis, drug and gene delivery, cosmetics, hydrogen production and storage, chemical reactors, sensors, photonics, photovoltaics, rechargeable batteries and electrocatalysis, the chemical growth and distribution of matter within the particles play pivotal roles in function determinations (Chattopadhyay et al. 2013, Son et al. 2010, Ren et al. 2014, Wang et al. 2016, Zou et al. 2016).

The present chapter will emphasize on the single-shelled and multi-shelled hollow spherical nano-structures which are extensively utilized as electrocatalysts in electrochemical conversion of CO_2 useful fuels. The chapter will discuss various synthetic strategies of hollow spheres nano-materials and their structural relationship with performance will also be taken into account.

Synthetic strategies

i) **Hard Templating:** Hard templating methods were widely used in the earlier stage for the production of multi-shelled hollow structures due to its simple and straightforward concepts. In 1998, Caruso et al. introduced the colloidal templated electrostatic Layer-By-Layer (LBL) self-assembled structure of silica nanoparticles with polymer multi layers. These nanoparticles electrostatically self-assemble to the linear cationic polymer poly (diallyl dimethyl ammonium chloride) (PDADMAC) (Caruso et al. 1998). In this method, the core template structure was removed through the calcination process and/or exposure to solvent. The thickness of the wall and overall shape can be controlled with a number of layers of SiO_2-PDADMAC present in deposition cycles. In a typical multi-step process, the negatively charged PS latex particles were used as core material, on which three-layered smooth and positively charged polymer film was deposited. This multi-layer film in the actual sense induced subsequent adsorption of SiO_2. In the next step, the shell-by-shell assembly combined with colloidal templating to form multi-shelled structures was adopted. During this process, the sequential coating of precursor materials was developed on the surface of a hard template and another material as an inter-layer. By repeating the same coating process, multiple shells were formed. Subsequently, removing the template and inter layer resulted in the multi-layered hollow structure.

However, these traditional synthetic approaches suffer from serious limitations due to their complex and tedious methodology. Additionally, all these processes fail to establish a compatible relation between template and target material. At the same time, uniform coating could not be formed using these methods. By avoiding these constraints to synthesize multi-shelled hollow spheres, Dan Wang and co-workers established an outstanding approach using sequential templates. They demonstrated a remarkable new technique in which traditional tedious multi-step layer-by-layer methods can easily be avoided. Their synthetic technique had also proved that without creating a core-shell structure with multiple shells and solid templating material, a multi-layered hollow sphere can still be formed. In 2009, their group introduced a general method to form spinel ferrites hollow spheres (MFe_2O_4), where M = Zn, Co, Ni, Cd (Li et al. 2009).

ii) **Soft Templating:** The multi-shelled hollow structures were synthesized using soft-templating method with a relatively flexible multi-layer structure by using supramolecular micelles and polymer vesicles. Although the multi-shelled structure formed using a soft templating process was metastable thermodynamically and easily got raptured with the variation in different parameters resulting from shell growth, such as the pH value, temperature, solvent, ionic strength, concentration of organic templates and inorganic additives. In 2007, Xu et al. reported the formation of multi-shelled Cu_2O hollow spheres by self-assembling of surfactant molecules in an aqueous solution, which further led to the formation of micelles and closed aggregated vesicles (Xu et al. 2007). The method was processed with the assistance of CTAB vesicles and multi-lamellar vesicles. They had controlled the number of shells and structures by adjusting the CTAB concentrations. In recent times, template free methods have become very popular. In this method, the self-assembly of target materials are taken into account without using any template. One recent work also mentioned forming silica multi-layered spheres by Soltani et al. (Soltani et al. 2020).

iii) **Metal Organic Frameworks (MOFs):** Metal–Organic Framework (MOF) was recently used to prepare advanced transition metal oxide electrodes due to the capability of forming a well-organized nanostructure. By thermal annealing MOF materials, carbon-coated transition metal oxide with the desired structure can be achieved in a facile manner. In most of the studies, the MOFs were formed by the solvothermal reactions with a metal precursor solution, trimesic acid as organic ligand and polyvinyl pyrrolodine (PVP) (Zou et al. 2016). The concentration of PVP controls the size, shape, porosity and crystallinity of the end products.

iv) **Modified arc-discharge method:** The modified arc-discharge method with an air-annealing process has proved to be another promising synthesis method to synthesize metal nano-particle encapsulated onion-like carbon nano-capsules. In this typical process, metal powders were ball-milled with graphite powder, which was further compacted into a cylinder shaped under pressure and placed into one pit of a water-cooled carbon crucible. The carbon needle here was considered as a cathode, whereas metal-mixtures were used as an anode target. In this modified method, liquid ethanol was introduced the chamber (Chattopadhyay et al. 2020).

Hetero-structured nano-spheres in electrochemical reduction of CO_2

Electrochemical CO_2 reduction is a promising method in transformation of CO_2 to useful fuels, which can be processed in mild reaction conditions, e.g., room temperature, neutral solution and atmospheric pressure. It is well known that, CO_2 molecule is very stable and is very difficult to convert it electrochemically. Usually, Hydrogen Evolution Reaction (HER) is preferred over electrochemical reduction of CO_2 at similar potential ranges. Therefore, it becomes essential to develop effective electrocatalysts, which can reduce the overpotential value and increase the faradic efficiency. Au nano-wires, Au triangular, oxide-derived nanostructured Au and Ag in dendrites have been extensively reported in literature as effective and potential electrocatalysts in CO_2 reduction to CO (Mistry et al. 2014, Chen et al. 2012, Larrazábal et al. 2016). Recently, three dimensional metals coordinated with N-doped carbon catalysts became very attractive due to their lower toxicity, cost-effectiveness and remarkable stability, they have been widely used as electrocatalysts in CO_2 reduction (Ju et al. 2017, Varela et al. 2015, Jia et al. 2018). Experimental and theoretical models have been reported on the verification of catalytic activity of Ni-N/C (Fan et al. 2016, Yang et al. 2018).

i) N-doped carbon hollow spheres

In 2019, Ma et al. reported new atomic nickel coordinated N-doped carbon hollow spheres, which were fabricated using Melamine-Formaldehyde (MF) and Resorcinol-Formaldehyde (RF) as carbon precursors (Ma et al. 2019). These nanomaterials were further utilized as electrocatalysts in CO_2 reduction. These materials were synthesized by the carbonization of SiO_2@resorcinol-melamine co-polymer, and followed by washing with 10% HF to remove the core SiO_2 (Pan et al. 2018). In this typical method, F127, used as a surfactant, was dissolved in the solution to form micelle around the silica, which was yielded through TEOS hydrolysis. The micelle-silica composite was coated by a resorcinol-melamine co-polymer. The color of the resultant material transformed into reddish brown from white because to this reaction. Further the carbonization conducted at 700°C under inert atmosphere and 10% HF for etching of silica. In this method, the shell thickness was adjusted by

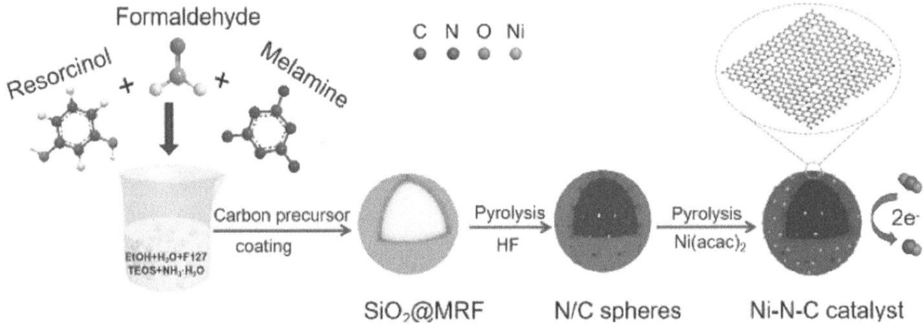

Figure 11.1: Schematic diagram of N-doped carbon hollow spheres Synthesis (Reproduced with the permission from Ma et al. 2019).

controlling the ratios of RF to MF. Figure 11.1 represents the synthetic approach of the materials.

They further investigated the reaction kinetics of CO_2 electroreduction with the presence of these materials. It was concluded from the Tafel study that, one electron reduction step of CO_2 to generate $CO_2^{-\cdot}$ had been determined as the rate determining step from Tafel plot studies. They concluded that, spherical structure of the electrocatalysts enhanced the overall performance of the CO_2 electro-reduction reaction.

ii) MgO anchored into carbon hollow spheres

In recent years, several efforts have been made to increase the local concentration of CO_2 at the catalyst surface in an aqueous solution. Li et al. reported an innovative material of MgO anchored hollow carbon spheres, which synergistically effect the adsorption and activation of CO_2 for electrochemical CO_2 reduction (Li et al. 2020). MgO is commonly used as a promising CO_2 adsorption reagent. However, HCS has found wide applications in electrocatalysis due to its unique structure, high electrical conductivity and good durability (Chattopadhyay et al. 2015).

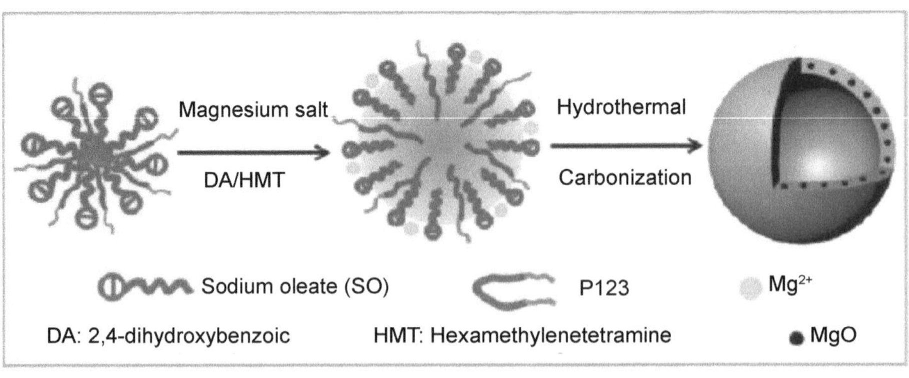

Figure 11.2: Schematic diagram of synthetic strategy of MgO/HCS (Reproduced from the permission from Li et al. 2020).

In this typical synthesis, MgO particles were embedded using the solvothermal method through carbonization with the application of Sodium Oleate (SO) and commercially available MgO. MgO had never been used as an electrocatalysts due to its insulating feature. But the mechanistic synergistic effect between HCS and MgO resulted into surface enhancement for adsorption/activation process of electrochemical reduction of CO_2. Tafel plots and Electrochemical Impedance Spectroscopy (EIS) with MgO/HCS spheres were analyzed. The Tafel slope value suggested that the initial CO_2 activation process served as a rate limiting step for HCS and MgO; on the other hand, it switched to surface reaction in the case of MgO/ HCS nano-materials. CO_2 adsorption is the pre-requisite step for triggering further CO_2 reduction process. Therefore, volumetric CO_2 adsorption measurement had been carried out to verify the potential adsorption characteristics of MgO. Finally, they performed the DFT calculations to reveal the mechanism by establishing a model to compare the performance of graphene with MgO/graphene nano-structure.

iii) SnO_2 coupled hollow carbon spheres

A large variety of metal or non-metal based CO_2RR electrocatalysts were reported with different potential ranges with applications of different catalyst composition, shape, particle size, porosity, defect, electronic structure, electrolyte, temperature and pH. Usually, Sn and Sn-based compounds are popular for their formate selectivity during electrochemical reduction. It is reported that, SnO_x layer with Sn^{2+} or Sn^{4+}, on the Sn film electrode works potentially as CO_2RR active sites, rather than Sn^0. In the case of SnO_2A strong interfacial Sn–O–C linkage between SnO_2 nanoparticles and carbon support was formed, which not only tuned the electronic structure of Sn^{4+} to promote its CO_2RR catalytic activity, but also benefited the catalyst durability. As a result, the CO_2RR catalytic performances of SnO_2/C hollow spheres was remarkably enhanced, with FE of formate and CO production as 54.2 and 21.8%, respectively, at –0.9 V vs. Reversible Hydrogen Electrode (RHE), and a much improved electrochemical stability of over 12 hours, which were significantly better than those of similarly structured SnO_x hollow spheres and carbon hollow spheres (Wang et al. 2018). In this work, SnO_2 couple carbon hollow spheres were synthesized using the solvothermal method from SiO_2 hollow spheres, with a continuous carbonization process.

Conclusions

Electrochemical CO_2 reduction is one of the promising pathways to transform CO_2 to valuable chemicals to mitigate the excess amount of CO_2 from the atmosphere, as the process can be processed at room temperature, pressure and at neutral solution. In the process of diminishing high overpotential value can be avoided with the application of electrocatalysts with bulk morphology with a wide range of geometry, morphology and size. In recent years, hollow spheres have received great attention because to their excellent properties due to the large surface area, tunable shell thickness and void volume. Very few research groups have reported on the hollow sphere electrocatalysts which are utilized in CO_2 electrochemical reduction. However, high surface area with heterogeneous interior structure can be promising materials and these materials can create an extensive application in this reaction.

References

Birdja, Y. Y., E. P. Gallent, M. C. Figueiredo, A. J. Göttle, F. C. Vallejo and M. T. M. Koper. 2019. Advances and challenges in understanding the electrocatalytic conversion of carbon dioxide to fuels. Nature Energy 4: 732–745.

Bushuyev, O. S., P. Luna, C. T. Dinh, L. Tao, G. Saur, J. Lagemaat et al. 2018. What should we make with CO_2 and how can we make it? Joule 2: 825–832.

Caruso, F., R. A. Caruso, H. Möhwald. 1998. Nanoengineering of inorganic and hybrid hollow spheres by colloidal templating. Science 282: 1111.

Chattopadhyay, J., R. Srivastava and P. K. Srivastava. 2013. Preparation of tin-doped carbon hollow spheres and their electrocatalytic activity in water electrolysis. International Journal of Electrochemistry 8: 3740.

Chattopadhyay, J., R. Srivastava and P. K. Srivastava. 2013. Ni-doped TiO_2 hollow spheres as electrocatalysts in water electrolysis for hydrogen and oxygen production. Journal of Applied Electrochemistry 43: 279.

Chattopadhyay, J., S. Singh and R. Srivastava. 2020. Highly efficient ternary hierarchical NiV_2S_4 nanosphere as hydrogen evolving electrocatalyst. Journal of Applied Electrochemistry 50: 207.

Chattopadhyay, J., T. S. Pathak, R. Srivastava, A. C. Singh. 2015. Chaos in chemical system. Electrochimica Acta 167: 429.

Chen, Y., C. W. Li and M. W. Kanan. 2012. Aqueous CO_2 reduction at very low overpotential on oxide-derived Au nanoparticles. Journal of American Chemical Society 134: 19969.

Durst, J., A. Rudnev, A. Dutta, Y. Fu, J. Herranz, V. Kaliginedi et al. 2015. Electrochemical CO_2 reduction—A critical view on fundamentals, materials and applications. Chimia International Journal of Chemistry 69: 769–776.

Fan, L., P. Liu, X. Yan, L. Gu, Z. Z. Yang, H. Yang et al. 2016. Atomically isolated nickel species anchored on graphitized carbon for efficient hydrogen evolution electrocatalysis. Nature Communications 7: 10667.

Hori, Y. 2008. pp. 89–189. *In*: Vayenas, C., R. White and M. Gamboa-Aldeco (eds.). Modern Aspects of Electrochemistry 42. Springer.

Jia, M., C. Choi, T. S. Wu, C. Ma, P. Kang, H. Tao et al. 2018. Carbon-supported Ni nanoparticles for efficient CO_2 electroreduction. Journal of Chemical Science 9: 8775.

Jones, J. P., G. K. S. Prakash and G. A. Olah. 2014. Electrochemical CO_2 reduction: Recent advances and current trends. Israel Journal of Chemistry 54: 1451–1466.

Ju, W., A. Bagger, G. Hao, A. S. Varela, I. Sinev, V. Bon et al. 2017. Understanding activity and selectivity of metal-nitrogen-doped carbon catalysts for electrochemical reduction of CO_2. Nature Communications 8: 944.

Larrazábal, G. O., A. J. Martín, S. Mitchell, R. Hauert and J. Pérez-Ramírez. 2016. Enhanced reduction of CO_2 to CO over Cu–In electrocatalysts: Catalyst evolution is the key. Journal of Catalysis 343: 266.

Li, C., W. Ni, X. Zang, H. Wang, Y. Zhou, Z. Yang et al. 2020. Magnesium oxide anchored into a hollow carbon sphere realizes synergistic adsorption and activation of CO_2 for electrochemical reduction. Chemical Communications.

Li, Z., X. Lai, H. Wang, D. Mao, C. Xing and D. Wang. 2009. General synthesis of homogeneous hollow core–shell ferrite microspheres. Journal of Physical Chemistry 113: 2792.

Ma, S., P. Su, W. Huang, S. P. Jiang, S. Bai and J. Liu. 2019. Atomic Ni species anchored N-Doped carbon hollow spheres as nanoreactors for efficient electrochemical CO_2 reduction. Chem. Cat. Chem. 10.1002/cctc.201901643.

Mistry, H., R. Reske, Z. Zeng, Z. Zhao, J. Greeley, P. Strasser et al. 2014. Exceptional size-dependent activity enhancement in the electroreduction of CO_2 over Au nanoparticles. Journal of American Chemical Society 136: 16473.

Pan, Y., R. Lin, Y. Chen, S. Liu, W. Zhu, X. Cao et al. 2018. Design of single-atom Co–N5 catalytic site: A robust electrocatalyst for CO_2 reduction with nearly 100% CO selectivity and remarkable stability. Journal of American Chemical Society 140: 4218.

Qiao, J., Y. Liu, F. Hong and J. Zhang. 2014. A review of catalysts for the electroreduction of carbon dioxide to produce low-carbon fuels. Chemical Society Reviews 43(2): 631–75.

Ren, H., R. Yu, J. Wang, Q. Jin, M. Yang, D. Mao et al. 2014. Multishelled TiO_2 hollow microspheres as anodes with superior reversible capacity for lithium ion batteries. Nano Letters 14: 6679.

Soltani, R., A. Marjani, R. Soltani and S. Shirazian. 2020. Hierarchical multi-shell hollow micro–meso–macroporous silica for Cr(VI) adsorption. Scientific Reports 10: 9788.

Son, J. E., J. Chattopadhyay and D. Pak. 2010. Electrocatalytic performance of Ba-doped TiO_2 hollow spheres in water electrolysis. International Journal of Hydrogen Energy 35: 420.

Varela, A. S., N. Ranjbar Sahraie, J. Steinberg, W. Ju, H. S. Oh and P. Strasser. 2015. Metal-doped nitrogenated carbon as an efficient catalyst for direct CO_2 electroreduction to CO and hydrocarbons. Angewandte Chemie 54: 10758.

Wang, Y., L. Yu and X. W. Lou. 2016. Formation of triple-shelled molybdenum–polydopamine hollow spheres and their conversion into MoO_2/Carbon composite hollow spheres for lithium-ion batteries. Angewandte Chemie 55: 14668.

Wang, Y. Z., C. Yang, A. Guan, L. Shang, A. M. Al-Enizi, L. Zhanga et al. 2018. Sub-5 nm SnO_2 chemically coupled hollow carbon spheres for efficient electrocatalytic CO_2 reduction. Journal of Material Chemistry A 6: 20121.

Whipple, D. T. and P. J. A. Kenis. 2010. Prospects of CO_2 utilization via direct heterogeneous electrochemical reduction. Journal of Physical Chemistry Letters 1: 3451–3458.

Xia, Y. and R. Mokaya. 2010. Templated nanoscale porous carbons. Journal of Materials Chemistry 153: 126.

Xu, H. and W. Wang. 2007. Template synthesis of multishelled Cu_2O hollow spheres with a single-crystalline shell wall. Angewandte Chemie International Edition 46: 1489–1492.

Xuan, S. H., W. Q. Jiang, X. L. Gong, Y. Hu and Z. Y. Chen. 2009. One-pot sequential synthesis of magnetically separable Fe_3O_4/AgCl photocatalysts with enhanced activity and stability. Journal of Physical Chemistry C 113: 553.

Yang, J., Z. Qiu, C. Zhao, W. Wei, W. Chen, Z. Li et al. 2018. *In situ* thermal atomization to convert supported nickel nanoparticles into surface-bound nickel single-atom catalysts. Angewandte Chemie 57: 14095.

Zou, F., Y. M. Chen, K. Liu, Z. Yu, W. Liang, M. S. Bhaway et al. 2016. Metal organic frameworks derived hierarchical hollow NiO/Ni/Graphene composites for lithium and sodium storage. ACS Nano 10: 377.

Index